Service Orientation

Winning Strategies and Best Practices

Companies face major challenges as they seek to flourish in competitive global markets, fueled by developments in technology, from the Internet to grid computing and Web services. In this environment, service orientation – aligning business processes to the changing demands of customers – is emerging as a highly effective approach to increasing efficiency. In this book, Paul Allen provides an accessible guide to service orientation, showing how it works and highlighting the benefits it can deliver. He provides roadmaps, definitions, templates, techniques, process patterns and checklists to help you realize service orientation. These resources are reinforced with detailed case studies Queensland Transport and Credit Suisse.

Packed with valuable insights, the book will be essential reading for CIOs, IT architects, and senior developers. IT facing business executives will also benefit from understanding how software services can enable their business strategies.

For the last 40 years, business has viewed IT as its principal tool to increase productivity, and rightly so. Industries as diverse as retailing and financial services have been transformed by technology, especially when used in a networked way. Yet very often, IT has constrained the evolution of those business processes by imposing rigid technology implementations on them. And even worse, as IT has grown and become more important, it has become more complex and difficult to manage, especially in larger enterprises. In some cases the sheer complexity of the IT environment has started to limit the ability of companies to innovate. However, as Paul Allen shows, recent advances in IT processes and technologies, especially service-oriented architectures are starting to free companies up from these constraints, and allow them to once again align their IT processes in support of their business processes. *Service Orientation: Winning Strategies and Best Practices* charts the course for handling what promises to be as profound a change to IT as any that has occurred in the last few decades.

John A Swainson
President and CEO
CA International

Paul Allen is the principal business–IT strategist at CA (www.ca.com), and is widely recognized for his innovative work in component-based development (CBD), business–IT alignment, and service-oriented architecture. Paul's detailed knowledge combines with a uniquely practical understanding of the problems companies face as they apply new technologies in search of business value. His pragmatism stems from over thirty years' experience in the management and development of large-scale business systems. His current focus of interest is in helping organizations' transition to service orienation.

Paul is a widely published author and has held the position of editor of Cutter Consortium's *Component Development Strategies*. He is also a popular speaker at industry conferences worldwide. His co-authored book *CBD for Enterprise Systems* was one of the first and most practical descriptions of what was involved in building component-based applications.

Sam Higgins is now with Forrester Research Inc. (www.forrester.com) as a senior analyst. Formerly he was a director of Encode Services, an Australian technology consulting firm, and managed the Innovation and Planning Unit of Queensland Transport's Information Services Branch. He has extensive knowledge of achieving winning strategies in service orienation.

Paul McRae is the application architect within the Innovation and Planning Unit of Queensland Transport's Information Services Branch. He has a wealth of experience in developing best practices in service oriented architecture.

Hermann Schlamann is a senior architect in the architecture group of Credit Suisse. Since starting his career in 1971, he has built a deep and extensive understanding in the practicalities of using methodologies to produce successful software.

Service Orientation

Winning Strategies and Best Practices

Paul Allen

with Sam Higgins, Paul McRae, and Hermann Schlamann

CAMBRIDGE UNIVERSITY PRESS

CAMBRIDGE UNIVERSITY PRESS
Cambridge, New York, Melbourne, Madrid, Cape Town, Singapore, São Paulo

Cambridge University Press
The Edinburgh Building, Cambridge CB2 2RU, UK

Published in the United States of America by Cambridge University Press, New York

www.cambridge.org
Information on this title: www.cambridge.org/9780521843362

First published 2006

Printed in the United Kingdom at the University Press, Cambridge

A catalog record for this publication is available from the British Library

ISBN-13 978-0-521-84336-2 hardback
ISBN-10 0-521-84336-7 hardback

Contents

PART 3 SERVICE-ORIENTED ARCHITECTURE

PART 4 SERVICE-ORIENTED MANAGEMENT

PART 5 CASE STUDIES

Foreword

In the broader scheme of things our use of computer systems is still in its infancy. In just fifty years we have made amazing progress, to the point that computer systems are essential to the smooth running of home and business. Yet any systematic or casual measurement of user satisfaction with enterprise systems will inevitably return unsatisfactory ratings. The problems are legion but, with monotonous regularity, surveys report that users believe that total costs of ownership are too high and systems are insufficiently responsive to constantly changing circumstances.

The root problem is that insufficient attention has been paid to the fundamental structure of software systems. In other engineering disciplines such as automotive, construction, or electronics, there is unambiguous understanding between the user requirements and the delivered product, such that the various stakeholders can engage with the architect in a highly meaningful manner. For example, the architect and designer of the automobile, bridge or silicon chip has well-understood responsibilities to optimize the design and economics by using standard patterns and components which deliver a product that has a well-understood life cycle from production engineering through to retirement.

In contrast, the typical software system has no coherent structural and economic model that governs its life cycle. Rather, the approach to acquiring software systems is usually a mixture of blind faith and optimism. Consequently the Chief Information Officer (CIO) role has become a high-risk career and the typically brief time in the position encourages short-term thinking, in contrast with the real requirements of enterprise systems which are always medium to long term. The result is that most enterprises buy their software systems from enterprise solution providers because it moves the responsibility to an external party. However until very recently the enterprise software providers have been unable to offer any better solutions than an enterprise might develop themselves, and the resulting chaos from ill-fitting systems being shoehorned into unsuitable situations has led to widespread dissatisfaction. This deep discomfort is often expressed as an urgent need for more adaptable systems.

Since the mid-1990s, many companies have worked to solve these problems by introducing *middleware* – technical software systems that allow disparate systems that were never intended to work together to cooperate at a fairly basic level. However the costs of integrating systems in this way has proved to be prohibitive, so that many information technology (IT) organizations report that 80% of their entire budget is consumed by this inherently unproductive activity.

Software engineering is in many ways very similar to automotive, construction and electronic engineering. While the resulting software product at an aggregate level is much less well defined and generally less stable, at what might be termed the "sub-assembly level" there is almost always a high level of stability. This suggests that the formal concepts of componentization, widely used in other engineering disciplines, is entirely appropriate.

Consider a personal computer (PC) system. The typical PC comprises many thousands of individual components that are clustered together into a manageable number of sub-assemblies. Each of these assemblies is linked to one of several *bus structures* that provide communications between the major functions such as the processor and disks. Each of the components and assemblies offer services – for example, a disk access service. These services are offered under an agreed service contract between the components. The underlying capability is offered purely on the basis of the service interface, and by design the consumer of the service should not need to know how the service is implemented. This encapsulation, together with stability of contract, allows many different parties to collaborate in the overall product, enabling the computer seller to create a highly optimized supply chain. Even more important, the seller can constantly evolve the overall specification of the end product, substituting parts and assemblies to respond to changes in the supply chain and to deliver improved capabilities to market.

In the long term, the existing monolithic structure and economic model underlying enterprise systems delivery was unsustainable. The software industry recognized that the inherent structural problems needed to be addressed through *industry standards* which would, over time, facilitate a fundamental reconfiguration of software products. The twin ideas of middleware and software components and services have been the basis of an unprecedented initiative by industry leaders to create standards for service interoperability. This technical backplane, referred to as Web services specifications, is today rapidly maturing and while it will take some years to become fully mature, is now in widespread use throughout industry.

The service backplane for enterprise systems exactly parallels the service bus in the personal computer. Business relevant services such as order taking and credit checking can be constructed as independent components and collaborate with other services such as master data, using formal contracts that govern their respective responsibilities. For the first time, the internal technical representation of systems functionality is exactly the same as the business perspective, providing enormous opportunity to establish common understanding between all stakeholders. Perhaps even more important, the services are implemented as an invokable resource that offer real-time, or near-real-time services, which enable software systems to mirror real-world activity. Like the personal computer the business components and services encapsulate the underlying implementation complexity and can be substituted, modified, and upgraded with a well-understood impact on other collaborating components and services, creating a constantly evolving information system that can more closely reflect business reality.

In the beginning, we talked about this approach as a service-oriented architecture or SOA. Not surprisingly, because the genesis of the concept is about fundamental restructuring of software systems, to break up monolithic designs into collections of moving parts that can be managed in a similar way to an automobile or PC. As our thinking matured we realized that the architecture was merely one discipline of a much broader approach that requires significant change right across the software supply chain and enterprise organizations impacting organization structures, business processes, and project management techniques as well as software architectures. We call this *service orientation*.

This book is important because it addresses the subject in depth. It provides a comprehensive and detailed treatment, probably for the first time, of the wider business and technology impacts of service orientation.

David Sprott
Founder and CEO, CBDI Forum

Preface

Why this book?

We live in a world of change. During my thirty-plus years in the software industry, I have witnessed some remarkable changes. One constant, however, seems to me to be that successful companies cope with change by nurturing their people, learning fast, building on what they have learned, and discarding things that no longer work. Surely, that's not rocket science! If that's the case then, "Why," I ask myself, "write a book about it?" It seems appropriate to begin by offering the reader some kind of answer!

Even moderately large companies can no longer afford to "just get by" as they seek to cope with the challenges of a world of unforgiving change. Business is increasingly moving toward a marketplace model. This model is in sharp contrast to the traditional view of an organization as a production line. In this world, organizations collaborate together, consuming and offering services to maximize efficiency, better serve customers, and achieve long-term advantage.

This is the world of service orientation, in which a hasty decision to outsource an activity may well be a decision that the firm repents at leisure. Equally, focusing the company's own energies on the wrong kinds of thing could be a disaster. The power of technology, from the Internet to grid computing and Web services, is the irresistible force that is fueling this change. Business is increasingly less supported by software and more enabled by it. This means that simply automating business activities in software is no longer enough. The software has to be agile enough to cope with change and foster innovation.

The really successful companies are companies that seem somehow to be acutely aware of these things. This is in some contrast to those that struggle! These organizations have, to a large degree, lost control over their software and the business processes that depend on that software. In domestic terms, it's like a mother that spends all her time cleaning her own windows while taking no interest in her children!

So what can we do about this?

A threefold response

- *First, improve business processes* to reflect the reality of service orientation, with an appropriate gearshift in approach to business architecture and process redesign.
- *Second, nurture software services* as the common threads between business and IT. Service-oriented architecture is the discipline that aims at reducing the risks posed by change by structuring an agile portfolio of software services.
- *Third*, introduce *effective control of the execution of software services* that extends their reach beyond the firewall to more and more external users and partners. "Service-oriented management" is the emerging discipline designed to monitor and control the live, executing software that transcends the old traditional boundaries.

The focus of the book

These three areas are often treated independently, resulting in a fragmented approach. This compartmentalizing of concerns is deeply ingrained in the IT psyche. One of my main aims in writing this book has been to tackle this issue by providing an holistic approach in which all three disciplines are integrated.

Most of the technologies that are enabling service orientation are developing out of previous well-established generations of technology. The long-term trend is toward greater and greater standardization of the interfaces through which software systems talk to each other.

The great paradox that we are faced with today is that developments in technology are rendering specific technology issues less and less important, but are hugely magnifying the potential scale and complexity of our business. At the same time, these developments promise greatly improved return on investment from utilization of technology in the form of on-demand computing – buying just enough software services as and when you need them – "by the drink" instead of "by the case," if you will.

The standards that are enabling these trends (specifically Web services) are both complex and constantly evolving. Rather than attempting the impossible task of reviewing and documenting all of these technologies and standards I have aimed at providing a badly needed technology agnostic approach to best practices, designed for maximum resilience to the pace of change.

The audience of the book

While this book is not a panacea, my hope is that it will at least provide you with some equipment (roadmaps, definitions, templates, techniques, process patterns, and checklists) to help you on your journey to service orientation. If you have already embarked on that journey, or are faced with making it, you are likely to be a CIO, IT architect or senior developer working in a large or medium-sized business. Equally, the book should be relevant to a second group: business executives with an interest in, and responsibility for, IT and managers who need to understand how to plan and develop cost-effective strategies to cater for service orientation. Thirdly, by clarifying concepts that have often been a source of past confusion, the book should be of value to teachers and students of both IT and business studies.

The structure of the book

The book is organized in five parts.

Part 1 sets the context for the remainder of the book by explaining the principles of service orientation, as a business phenomenon enabled by developments in technology.

Part 2 provides an approach to business architecture that is based on evolution of best practices and in tune with the need to provide useful business process improvements quickly.

Part 3 lays out the two major aspects of the service-oriented architecture: policy and structure. We explain how to achieve an integrated approach that is driven "top-down" by the business architecture, and equally that is supported "bottom-up" by service-oriented management.

Part 4 introduces a new discipline, service-oriented management (SOM), to tackle the run-time challenges of software services in tune with both business architecture and service-oriented architecture (SOA).

Part 5 provides two real-life case studies in service orientation that illustrate the *execution* of an adoption strategy.

Winning strategies and best practices

Although plans are essential, an organization's willingness and ability to execute its plans are the hallmarks of a *winning* strategy. The book is therefore slanted toward practical guidance above theory and recognizes that the approach that your organization chooses to take will depend on a range of factors, from its target business model to the state of its legacy systems. Pragmatic models that capture valuable organizational knowledge must be nurtured as a long-term investment, so that this knowledge does not walk out the door every time that the CIO or any of her software architects leaves the company for another job!

The case studies are particularly important for understanding how to execute the winning strategies and best practices contained in the book.

In chapter 14, Sam Higgins and Paul McRae describe how Queensland Transport developed a service-oriented approach that is used to allow third-party software developers to integrate regulatory processes seamlessly within commercial motor vehicle dealer business processes. In chapter 15, Hermann Schlamann describes how Credit Suisse uses service orientation to enable money transfers for customers over different channels, and to provide various foreign exchange and accounting services to the smaller banks.

Both case studies are the result of ongoing improvement programs that have successfully applied the core techniques explored within this book. They both show how architectures and models can be nurtured, with a focus on getting the most out of your existing software assets, so that service orientation becomes integral to the organization's business strategy.

"You can't measure what you don't specify"

In my visits to many organizations and in my discussions with hundreds of colleagues and professionals a constant theme emerges: there is a resounding need for clear guidance on service-oriented concepts, unfettered by detailed technical jargon and unwieldy procedures. There seem to be no end of processes and procedural guidance available, but relatively little in the way of plain common sense approaches to specification, particularly specification of services. Over twenty years ago Tom DeMarco (1982) told software project managers with his usual elegance that: "You can't control what you don't measure."

As a software project manager at the time, those words have made a lasting impression on me. So much so that one of the main messages that you will find repeated again and again in this book is that: "You can't measure what you don't specify."

If there is one thing above all else that I would like the you to take away from this book it is that simple, yet critical, message.

Acknowledgments

Many of the ideas in this book have been shaped by previous work done with many client organizations and individuals too numerous to name individually. While I extend my thanks to them all there are some individuals that I would also like to thank personally.

Much of the thinking behind the book emerged through the ebb and flow of discussions at CA with John Dodd (now an independent methods architect). These were sometimes passionate, never dull! I am particularly thankful to John for being at once my strongest critic yet greatest ally, for doing his best to keep my feet on the ground, and for his insightful reviews. Bruno Lefever and Danny Saro also deserve thanks for the part that they played in these discussions.

I also thank Bruno Lefever, Michael Stephenson and Janique Vandekerkhove of Computer Associates for their review comments. Champion reviewer honors however go to Danny Saro for his extremely thorough and often downright annoying comments. It is perhaps ironic that an Englishman is thanking a Belgian for keeping him true to his own language!

I extend my thanks to the case study contributors: Sam Higgins and Paul McRae of Queensland Transport, and Hermann Schlamann of Credit Suisse. Their influence on me stretches much further than the case studies, in that all have worked closely with me, during the progress of this book, in both discussion and through reviews.

Thanks are also extended to Art Sedighi, solutions architect, of Platform Computing for helping to guide my thinking and for his review comments.

Credit is also due to the following individuals who have stimulated or educated my thinking in various ways that have played a part in shaping this book: Jason Bloomberg, John Daniels, Paul Harmon, Richard Solely, David Sprott, Borys Stokalski, Richard Veryard, Tom Welsh, and Lawrence Wilkes.

I also thank the management team at CA for their support in enabling me to write this book. In particular, I would like to thank Colin Bannister for his help and encouragement.

Thanks are also extended to my editor at Cambridge University Press, Emily Yossarian.

Last but not least, I thank my wife and children for their patience with my distractions while writing this book, and above all for helping to keep me sane.

Acronyms and abbreviations

AHS	American Hospital Supply
APS	Analytic Systems Automated Purchasing
API	application program interface
APPo	Application Portfolio
APQC	American Productivity and Quality Centre
ASAP	aligning services and priorities
ASP	application service provider
AST	agreed service time
ATM	automatic teller machine
B2B	business-to-business
BA	business architecture
BAM	business activity manager
BCM	business capacity management
BIAT	business–IT alignment table
BLA	business-level agreement
BPEL4WS	Business Process Execution Language for Web Services
BPM	business process management
BPMI	Business Process Management Initiative
BPMN	Business Process Modeling Notation
BPR	business process re-engineering
BRM	business rules manager
BSB	business service bus
CBD	component-based development
CEO	chief executive officer
CEP	complex event processing
CIO	chief information officer
CMDB	configuration management database
CMMI	Capability Maturity Model® Integration
CRC	class responsibility collaborator
COBOL	Common Business Oriented Language
COM	Component Object Model

COM+	an extension of COM
CORBA	Common Object Request Broker Architecture
CRC	class responsibility collaborator
CRM	customer relationship management
CRODL	create, read, update, delete, and list
CS	Credit Suisse
CIS 3.1	Component Standard 3.1
DAIS	Dealer Agency Interface System
DIS	Dealer Interface System
DCOM	Distributed Component Object Model
DSDM	Dynamic Systems Development Methodology
DMTF	Distributed Management Task Force, Inc.
EAI	enterprise application integration
EDI	electronic data interchange
EFQM	European Foundation for Quality Management
EJB	Enterprise JavaBeans
ERP	enterprise resource planning
ESB	enterprise service bus
eTOM	enhanced Telecom Operations Map
EU	European Community
FIOM	Financial Instruments Object Model
FTP	File Transfer Protocol
GGF	Global Grid Forum
HR	human resources
HTTP	Hypertext Transfer Protocol
IDL	Interface Definition Language
ISB	infrastructure service bus
ISO 9000	International Organization for Standardization 9000
IT	information technology
ITIL	Information Technology Infrastructure Library
it SMF	IT Service Management Forum
J2EE	Java 2 Enterprise Edition
JAP	Java Application Platform
JIT	just-in-time
JMS	Java Message Service
KPI	key performance indicator
M & A	mergers and acquisition
MDA	Model Driven Architecture
MOM	message-oriented middleware
MOWS	management of Web services
MSMQ	Microsoft Server for Message Queuing

MTBF	mean time between failures
MTTR	mean time to repair
MUWS	management using Web services
OASIS	Organization for the Advancement of Structured Information Standards
OGC	Office of Government Commerce
OGSA	Open Grid Services Architecture
OLA	operational-level agreement
OMG™	Object Management Group
OO	object-oriented
PC	personal computer
QC	quality check
QoS	quality of service
QT	Queensland Transport
R & D	research and development
RAEW	responsibility, authority, expertise, work
RCM	resource capacity management
RMI	Remote Method Invocation
ROI	return on investment
SAML	Security Assertion Markup Language
SBS	Services Booking System
SCOR	Supply Chain Operations Reference
SCM	service capacity management
SDB	Service Database
SDLC	systems development lifecycle
SEI	Software Engineering Institute
SDM	software development methodology
SEM	service execution management
SFTP	Secured File Transfer Protocol
SLA	service-level agreement
SLM	service-level management
SOA	service-oriented architecture
SoS	specification of service
SOT	security outage time
SOV7	seven service-oriented viewpoints
SOAP	Simple Object Access Protocol
SOM	service-oriented management
SSL	Secure Sockets Layer
SSO	single sign-on
TAC	Technical Advisory Committee
TC	technical committee
TCO	total cost of ownership

TIA	technical infrastructure architecture
TRAILS	Transport Registration and Integrated Licensing System
UDDI	Universal Description, Discovery, and Integration
UML	Unified Modeling Language
W3C	Worldwide Web Consortium
WS-CDL	Web Services Choreography Description Language
WS-I	Web Services Interoperability Association
WAP	Wireless Application Protocol
WS80	Wertschriften System Asset Management
WSBPEL	Web Services Business Process Execution Language
WSDL	Web Services Description Language
WSDM	Web Services Distribution Management
XML	extended mark-up language
xP	eXtreme programming

Part 1 Overview

This first part of the book sets the context for the remainder of the book by explaining the principles of service orientation, as a business phenomenon enabled by developments in technology.

Chapter 1 provides a conceptual overview, defines the core terminology and introduces the pivotal idea of service-oriented architecture (SOA). Service orientation presents some massive cultural and technical challenges that cross three areas that have traditionally worked largely in isolation from one another: business process improvement, application development, and software operations. We provide a simple example of how the idea of a service can provide a unifying thread for drawing together these areas, along with some guidance for selling this approach to business management.

The remainder of part 1 maps out the foundation technologies required for practical application. Most of these technologies are developing out of previous well-established generations of technology. At the same time, they are based on standards that are complex as well as changing. Rather than attempting the impossible task of reviewing and documenting all of these technologies and standards we provide an essentially abstract primer designed for maximum resilience to the pace of change; a list of useful evolving Internet information sources is provided at the back of the book. For example, our emphasis is on the general idea of a *software service*, rather than specifically on a Web service, unless of course the context demands it.

Chapter 2 surveys the overall technology trends toward on-demand computing and sets Web services in context. We consider the execution management technologies and the gear shifts in service-level management (SLM) that are required in order to realize the promises of service orientation.

Chapter 3 parallels chapter 2, in that it considers the business process management (BPM) technologies and the associated gear shifts in business modeling that are required in order to effectively tackle the challenges of service orientation. In particular, we introduce the key concept of *service-oriented viewpoints*.

1 Basics of service orientation

1.1 The idea of service orientation

The phrase, "Easy to do business with" has become a major driver for most companies. The idea of service orientation is to provide customer value by contracting others to do what a company has to do just to get by, and by focusing the company's own resources on what it does best. By subscribing to the *commodity functionality* provided by service providers who can perform it better, faster, and cheaper an organization can minimize its cost of market participation. This strategy releases energy for an organization to concentrate on its core competencies, thus bringing value to market through what the organization does best, thus making it easy to do business with. By taking a service-oriented approach, in an increasingly complex and competitive market, products can then be taken to market quickly and business processes conducted in agile response to change through multiple channels.

Such is the compelling attraction of service orientation. But "Wait a minute," we hear top management say, "We might be serving the same customer, but we are a complex organization with multiple business units each with different targets, and supported by different technologies with different capabilities. We'll let the faster-moving, more profitable business units continue to do their own thing and just outsource the rest without worrying too much about the provider. That way we'll be service-oriented but we can't afford an integrated approach."

However there is no free lunch – current practices are simply not going to get them there. The main business driver of service orientation is *agility*: the speed, cost-effectiveness, accuracy, and flexibility required for organizations to prosper. Achieving agility requires a service-oriented architecture (SOA) that aligns technology with business goals, allowing the company to move in new and exciting ways that open up new markets. Very briefly,[1] an SOA refers to the software structures and policies that are required to enable the business phenomenon of service orientation. Managing agility requires appropriate processes that connect consumers and providers of services in a

[1] A great deal more detail is provided as the book unfolds. SOA is introduced further in this chapter and is the central topic covered in part 3.

cohesive fashion. Above all, service orientation requires effective execution strategies for dealing with inefficient ingrained business processes and the legacy software that holds them together. In this chapter, we introduce the concepts that are necessary to understand the change in mind-set mandated by service orientation.

1.1.1 The economic imperative

Increasingly, many of our business leaders see IT as a cost that must be trimmed to the bone. As a result of the ongoing quest to cut costs, more and more organizations are outsourcing large slices of both software development and business operation. This "economic imperative" is most strongly embodied in an article in the May 2003 edition of the *Harvard Business Review* by Nicholas G. Carr (Carr, 2003), who contends that IT has ceased to provide competitive advantage. He compares the growth of IT through the late twentieth century to the global growth of rail track in the mid-to-late nineteenth century and with the expansion of electric utility generating capacity in the U S in the early part of the twentieth century. Periods of massive growth in these industries followed huge investment but subsequently resulted in falling prices and in commoditization. Carr argues that IT has similarly become a commodity today, and that companies should therefore look to spend much less on it.

You may well, like me, see this argument as based on a flawed view of IT as nothing more than a set of bytes that is part of market participation cost – for example, you simply have to use human resources (HR) software to pay your employees; Smith and Fingar (2003a) provide a particularly good counter argument. At the same time, it is important to realize that the overgeneralization implicit in this kind of argument is alive and flourishing today in the minds of many of our business leaders. It is important therefore that CIOs and senior architects equip themselves with clear concepts and good business arguments as they seek to convince their business leaders of the need for better software architectures.

One of the most important tasks of CIOs and senior architects is to articulate to the business leaders why Carr's argument is a danger to the organization's health. In particular, it is important to explain that while it is true that some aspects of IT are indeed commoditized that does not mean that companies no longer have any IT applications that produce value for them that their competitors cannot easily copy.

At the same time, there is an important measure of truth in Carr's argument that we would be foolish to ignore. There is an economic imperative that is forcing many large organizations to become smaller and more specialized as they focus on what they do best and on what differentiates them from the competition. In turn, this involves outsourcing many of the support functions that have traditionally been managed in-house to third parties who can provide these services more cost-effectively and at higher quality through their own focused competencies. For example, a whole business process may be outsourced to others who are better equipped to deal with it efficiently. Payroll is the classic example. Another is the insurance company that chooses to outsource its claims function. In essence, these "treadmill" services simply represent the cost of market participation. Utility work such as call centers, billing, and claims processing is – courtesy of the Internet – following manufacturing jobs by transferring to places such as China, India, and the Philippines and other low-wage countries.

These business trends in service orientation are paralleled at the IT level. For example, packaged software, from small components to large-grained enterprise resource planning (ERP) solutions, provides a long-standing example of automated treadmill services. There is also a groundswell of legacy software, from COBOL systems to back-end databases that perform much of the treadmill work. In addition, the running of this software may be outsourced to an application service provider (ASP). The Internet and now the emergence of utility computing (Hughes, Bader, and Corrigan, 2004), which we discuss shortly, is accelerating this trend of focusing on one's core competencies and letting providers do the rest.

The problem that many organizations face today is that because their current software is not organized in a service-oriented way they have lost control of their software. And, in some cases, where economic pressures reach boiling point, it is all too easy to succumb to the "IT Doesn't Matter" line of argument and outsource all of IT: "Act in haste, repent at leisure" is a maxim that comes to mind. These organizations do not have a clear picture that allows them to make the right decisions on service provision. The SOA is designed to provide that picture.

1.1.2 The competitive imperative

If becoming "leaner and meaner" is one side of the service-oriented coin, the need to compete and to attract and keep customers is the other. Customers are growing more service-oriented in that they are demanding much smoother and better experiences from the companies that they choose to do business with. Forward-thinking companies are applying service orientation in using technology innovatively to support the customer experience.

For example, it is one thing to integrate existing legacy systems so that their services are exposed consistently across different channels. However, what we are now seeing is the emergence of new *multi-channel business processes*, in which a business process is supported by a variety of channels along its route from inception to closure. For example, like many others, I now purchase my air tickets by credit card online via my laptop and then use my credit card to complete the transaction at a self-service airport kiosk, which dispenses my tickets and itinerary without the need for me to suffer a long wait at the check-in queue.

Amazon.com has evolved into a multi-channel retailer, interacting with customers using Web portals, call centers, email contact centers, and catalogs. Through kiosks placed strategically in its various partners' stores, Amazon.com even has a significant presence in the bricks-and-mortar channel.

These kinds of approaches require significant investment in the SOA and the associated technologies required to manage and run it. As a business process changes, so must the services supporting it. Moreover, customers from any channel value *visibility*. For example, they want to know the status of their order. The Internet has further shaped customer expectations by enabling visibility into the process. However it is the quality of the underlying SOA that determines the success of the process.

In addition, those companies that become the most successful in their channels open up further opportunities for themselves. For example, a growing number of organizations including Toys 'R' Us, Circuit City, Target, Office Depot, and Nordstrom, are outsourcing their online channel interaction to Amazon.com. In a relatively short period, Amazon.com has built a lucrative new outsourcing business selling goods on behalf of other merchants. Amazon.com collects commissions on those third-party transactions with neither the risks nor the costs of owning inventory. Amazon's example presents compelling evidence that a service-oriented approach can deliver the strategic advantage that simply breaks apart conventional business models.

While Amazon.com is perhaps an extreme example, it is indicative of the overall change that is taking place in the way businesses are organized. In the 1980s, businesses were organized along strict departmental lines. In the 1990s, the emphasis shifted toward end-to-end processes, or value chains, in an effort to streamline and optimize, but still very much within the traditional boundaries of the business. The "noughties" are witnessing a further evolution in the form of federated processes[2] fueled by the emergence of the economic and competitive imperatives. This is the world of service orientation in which traditional boundaries quickly disappear and in which new "virtual" boundaries can emerge in response to new opportunities and changes such as mergers and acquisitions (M&As).

The economic factors that underlie the movement toward service orientation are extremely compelling in the long run. In most industries, specialization occurs as the industry matures. Those who focus on providing a particular kind of service and learn to do it best end up doing it on behalf of others. By acquiring commodity services at the lowest cost, an organization can focus much more strongly on its core competencies that add most business value. Equally there are competitive factors that underlie the movement toward service orientation in the form of multi-channel processes and the agility to compete in different channels. A company needs the ability to take its core competencies to market in new and innovative ways that beat the competition.

The move toward a federated business model involves dealing with the often conflicting drivers of economic stability and competition. On the one hand, companies

[2] As we examine in more detail in 1.3.1, a federated process allows largely independent parts to act with the unity of a whole, toward a common purpose, such that the whole is more than the sum of the parts. In contrast, a tightly integrated process, involves highly dependent parts.

require constructive strategies for dealing with the trend toward commoditization of software services that is fueled by the economic imperative. On the other hand, they need techniques for applying software services to improve the experience of customers and to extend business in competitive fashion.

Organizations are improving business processes wherever possible to drive productivity higher and compete more effectively. However, even if isolated parts of business processes are performing well, too often the business process as a whole is not. That is because many processes – with customers, suppliers, providers, partners, and employees – remain largely disjointed, papered together with a myriad of information flows: telephone calls, faxes, spreadsheets, and FedEx packages. Connecting and improving these fragmented processes has taken on a new urgency due to the speed and change of business today toward service orientation.

1.1.3 The legacy jungle

Of course, these changes in business process do not happen in a vacuum! Many organizations are faced with hybrid software and hardware environments that have evolved over a number of years. Most of these organizations are not early adopters: there is a need to minimize risk and maximize business return from existing software assets, both internally developed and externally acquired. In many situations, these software systems are acquired from and outsourced to a wide variety of vendors. These external systems need to be interfaced with many internally developed systems. The result is usually a complex patchwork of applications, a "legacy jungle" if you will, that is costly to maintain, inflexible to change, and causes unacceptable development times.

At the same time, these systems have stood the test of time: they "do a job." They may not do it in the most adaptable and efficient manner. They may be a maintenance nightmare. Nevertheless they are tried and tested. There is a comfort factor embodied in the slogan "Better the devil you know." The fact is that "legacy" is not a bad term, despite the fact that it is often read that way in the context of software. A legacy represents something bequeathed by a predecessor that has a value not commonly apparent. There is an onus on the inheritor to do some work to realize the value. As well as facilitating the trend toward federated processes a service-oriented approach must also support low-risk migration and integration of the legacy software portfolio to achieve ease of maintenance, flexibility, and responsive solution delivery.

1.1.4 The need for balance

Cutting a path out of the legacy jungle is a pressing challenge for most IT departments especially in large organizations. Enterprise application integration (EAI) technology can provide some help, but this is limited by the proprietary nature of the products. The introduction of *portals* that provide customers or users with a consolidated view

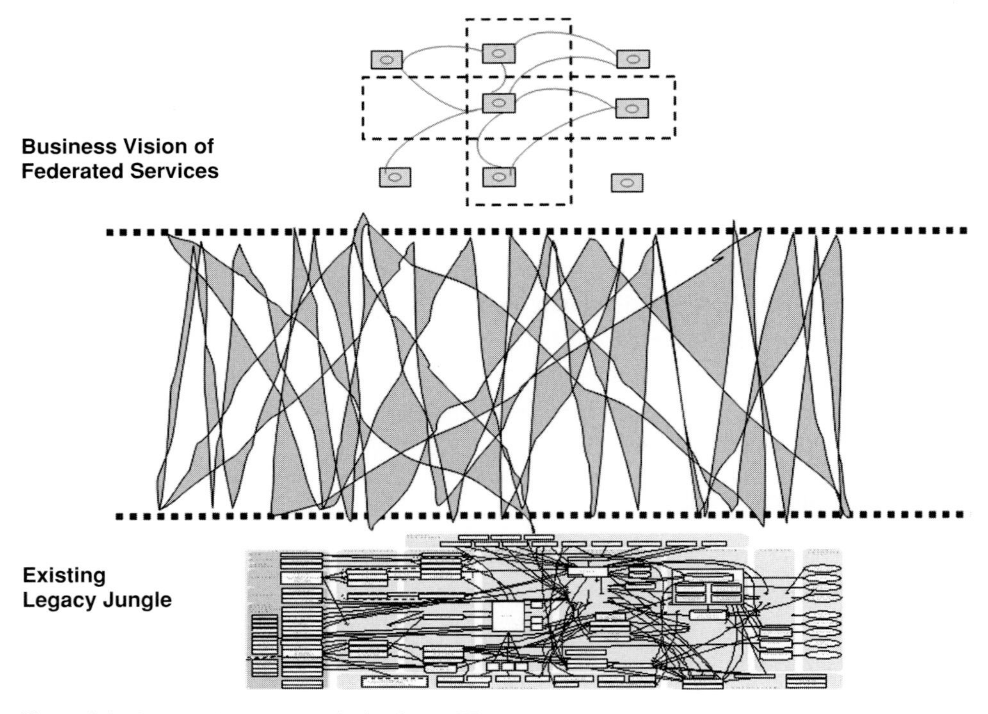

Figure 1.1 Severe churn across the business–IT gap

of business functionality can provide some relief. At the same time, such "integration on the glass" must be seen for what it is: a short-term measure. The danger with such approaches is that a whole new generation of legacy systems is born as portals spring up on a local departmental basis with little heed to adaptability, consistency, or reuse.

The vision of federated services (discussed in 1.3.1) is one in which a firm can respond with agility to changes in business scope, as illustrated by the dashed boundaries in figure 1.1. However, the gap between the vision of where the business wants to go and the current state of IT only widens and grows more difficult to address with each layer of technology that is applied to the problem. "Getting anything done round here is just such a hassle," complain the business users. A single software change request can have multiple knock-on effects resulting in severe "churn" across the business–IT gap, as illustrated in figure 1.1.

Only by aligning its supporting software with its business processes in such a way that software can be reused across those processes and then replaced when things change is it possible to reduce churn and provide long-term balance. Design of the SOA is the discipline that organizations must master if they are to achieve this balance, and if they are to win and maintain control of their software portfolio and ensure that it aligns with business needs.

1.2 Some definitions

At this point, we need to take a brief step back and firm up on some definitions.

1.2.1 Services

A *service* is offered by a provider to a consumer through its *interface*. An interface describes the contract between the provider and the consumer. In other words, it should specify what the provider is obliged to do on behalf of the consumer and what responsibilities the consumer agrees to in using the interface. Everyday life is full of examples of service interfaces. Sometimes these are written down in detail, as in the case of buying a TV and receiving a product booklet and written guarantee. At other times the interface is assumed, as in the case of a car wash (although it is usual to see a brief description of what types of wash are on offer as well as instructions describing what the driver must do to use the car wash).

> A **service** is functionality that must be specified in the business context and in terms of the contracts[3] between the provider of that functionality and its consumers. Implementation details should not be revealed. The implementation of the service does not have to be automated – it could consist of purely human activity.

The concept of a service is akin to a reusable chunk of a business process that can be mixed and matched with other services.

1.2.2 Business processes

The traditional view of a business process is as follows.

> A **business process** is a set of activities that is initiated by an event, transforms information or materials, and produces an output. These sets of activities are either value chains that produce outputs valued by customers or infrastructure processes that produce outputs that are valued by other processes.

[3] Whereas functionality is traditionally expressed in terms of procedural steps (if A then do B, then do C until $X = Y$), the contracts of a service are expressed nonprocedurally (given input A with $K = L$ then output B with $M = N$ is produced).

Perhaps the most significant aspect of the role of services is that they start to reshape this view of business processes, taking us toward a more *federated business model.* A business process is usefully pictured as being composed of re-configurable services.

1.2.3 Software services

A **software service** is a type of service that is implemented by software and that offers one or more operations (or software functions).

In this sense, a software service becomes a commodity bought, sold, and delivered in a similar manner to any other kind of service – for example, electricity or telecommunications. The consumer of a service is not concerned with implementation detail. The implementation of the service can vary from one supplier to another while still delivering the same service through its interface. At the same time the consuming software, invoking the service, can be implemented using any technology we choose providing it calls the services using the right interface. This situation is illustrated in figure 1.2. The same service, Transfer Funds, is provided by two alternative suppliers, each of which uses different implementations (COBOL on a mainframe, and C++ on a Unix platform). This service may be invoked via three different channels, cell phone, ATM, and PC. The implementation is transparent to the users of the service.

The potential for reuse of the same software service in different contexts is readily apparent here. This is especially important when a business wants to move a software service into a new sales channel. Equally, one implementation technology is replaceable with another. The possibility of upgrading to better technologies without disruption is hugely attractive. And from a utility computing perspective, discussed in chapter 2, the possibility of being able to switch implementations according to various criteria (for example cost, level of quality, or user demand), while retaining the same functionality is again hugely attractive.

1.2.4 Web services technology

Web services technology is a set of XML-based industry standards and specifications that specify a communication protocol,[4] a definition language,[5] and a publish–subscribe registry.[6]

[4] Simple Object Access Protocol (SOAP). [5] Web Services Description Language (WSDL).
[6] Universal Description, Discovery, and Integration (UDDI).

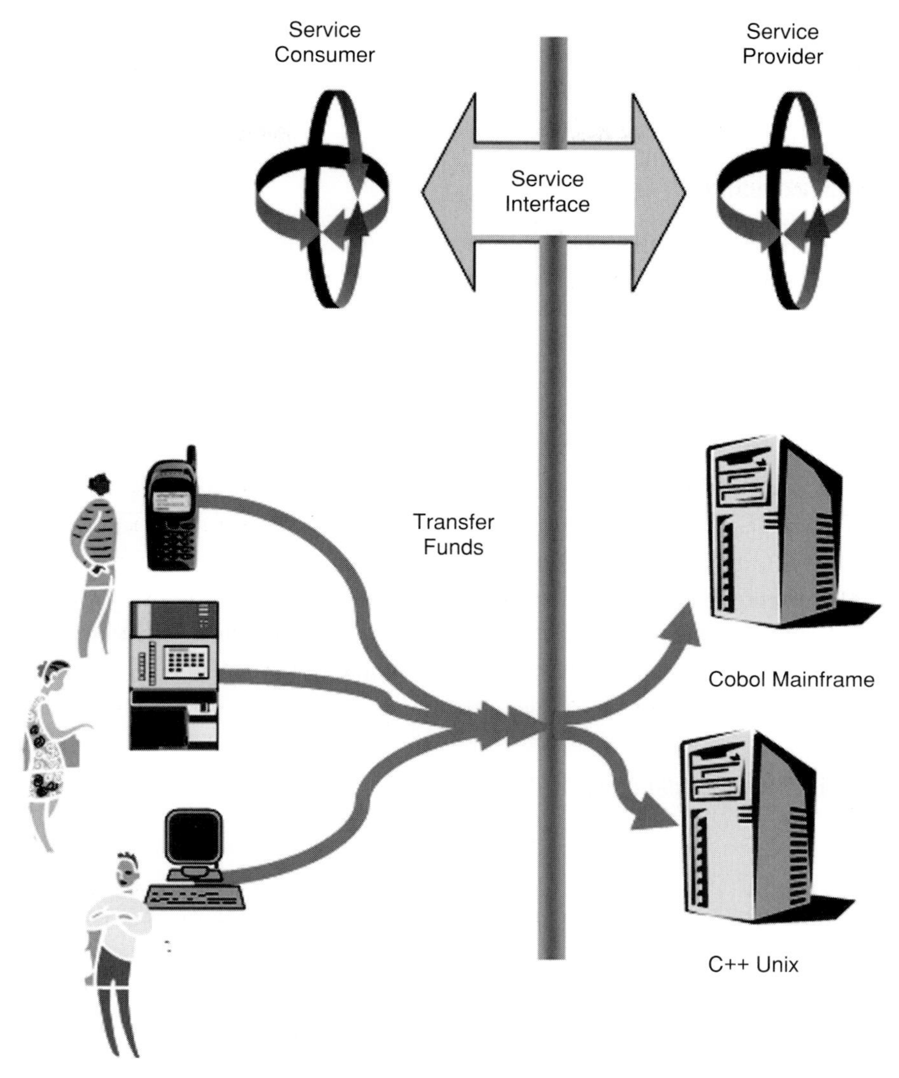

Figure 1.2 Multiple consumers and suppliers of the same service

These are the core features of Web services, though there is a proliferation of associated emerging standards required for industrial-strength application. Web services provide a particularly good means of offering software services, in that they provide the enabling technology standards for languages, protocols, and registries. However, it is perfectly conceivable to use other (or alternative) enabling technology standards – who knows what is around the corner in the world of emerging technology standards? There is nothing sacrosanct about Web services. Therefore in this book we use the general term "software service" unless the context demands otherwise. At the same time, because

Web services are such a core technology for utility computing, we do provide more details and references in chapter 2.

1.2.5 Loose coupling

One of the biggest attractions of software services is their potential for loose coupling. The following definition adapted from Doug Kaye (Kaye, 2003, p. 31) provides a good succinct explanation.

> **Loose coupling** is a feature of software systems that allows those systems to be linked without having knowledge of the technologies used by one another.

Loosely coupled software can be plugged together in different ways, depending on requirements, at run-time. In contrast, tightly coupled software is soldered together at design time, rendering its constituents resistant to reuse.

1.2.6 Service-level agreements (SLAs)

> A **service-level agreement** (SLA) is an agreement between a customer and provider about what services are to be offered by the provider and to be used by the customer. It must specify the measurable levels of those services that the provider must achieve, and the terms of use that the customer must comply with.

Notice that the SLA is a back-to-back agreement that covers obligations on *both* the provider and the customer.

1.2.7 Suppliers and providers

One of the most important distinctions of service orientation is the distinction between suppliers and providers. This distinction is commonly overlooked, and therefore we spell it out in brief now, though we return to this topic as part of the discussion of roles in chapter 13.

A **supplier** implements services. This may involve the implementation design and/or the actual execution of services.

A **provider** agrees SLAs with customers and makes services available to users in compliance with the SLAs.

1.2.8 Customers and users

The term "consumer" may refer to either a customer or a user of a service. The distinction between "customers" and "users" is an important one, which is commonly overlooked. Therefore we spell it out in brief now, though we return to this topic as part of the discussion of roles in chapter 13.

A **customer** agrees SLAs with providers, ensures the services address the correct business requirements, and provides funds for using services.

A **user** invokes services[7] in compliance with SLA terms of usage.

1.3 The overall approach

The integrated, tightly coupled traditional world should be contrasted with the federated, loosely coupled world of service orientation. The future direction of the latter needs to be in realistic harness with the current capabilities of the former. In this section, we sketch out an approach that balances these two forces of *integration* and *federation*

As well as providing balance between the push toward future ways of doing business and using software, and the need to pull the best out of what we have achieved so far, our approach will need to unite three areas that have traditionally been kept very separate:

- Business process improvement
- Application design and development
- Software operations.

[7] In the case of a software service it is a software application that invokes the Web service. Therefore we adopt the convention "User via software application" in describing the user.

It is the concept of *software service* that provides the new thread that can unite these disciplines. In this section we first provide a quick taster of how services can provide this unifying thread in terms of integration and federation. We then preview what has to be done to both improve and bring together the three areas above. In doing so, we outline the agenda for the rest of the book.

1.3.1 Integration and federation

Integration allows inter-dependent parts to be made to look independent. The integrated parts are tightly coupled and tuned toward a highly focused objective. Integration is a useful approach to help exploit existing investments in legacy systems by exposing these legacy systems as software services. It can also be a useful strategy at the business process level where the need is for a highly unique and efficient process.

Federation allows largely independent parts to act with the unity of a whole, toward a common purpose, such that the whole is more than the sum of the parts. The federated parts are loosely coupled and tuned toward reuse in different contexts. Traditional software modules are coupled at design time. One of the advantages of software services is that they may be coupled at run-time. Software services therefore have a crucial role to play in federated business processes that distribute their functionality across different partner organizations according to their capability to execute.

A service-oriented approach needs to achieve a balance between the forces of integration and federation. Let us take a quick preview of what such an approach involves. Figure 1.3 illustrates how we might take a service-oriented approach to improve the business process Make Travel Arrangements. The Buy Ticket activity is shown with an embedded circle to highlight that it is the first activity within the process to receive the service-oriented treatment. A two-pronged treatment is applied.

The Find Customer and Locate Flight parts of the Buy Ticket activity are integrated "bottom-up" by preparing and then exposing the legacy systems in the form of software services – Customer Management and Flight Locator – for consistent once-only use. Integration can be done in various ways. Customer Management acts as a clean pathway if you will between the business process and the various legacy systems that are required to provide the customer information. Flight Locator replaces parts of existing systems so that flight details are provided through a clean pathway to the business process.

The Make Booking piece of Buy Ticket is designed "top-down" so that it can satisfy requirements not only for this particular type of booking, but also other types of booking. It is supported by a new software service, Bookings, which is a federated and loosely coupled service that can be used by other business processes.

Of course, the illustration is just meant to provide a flavor of what's involved in moving to service orientation. While this may appear simplistic, it is necessary to give a "broad brush" view, if only to counter the tendency to think in terms of localized

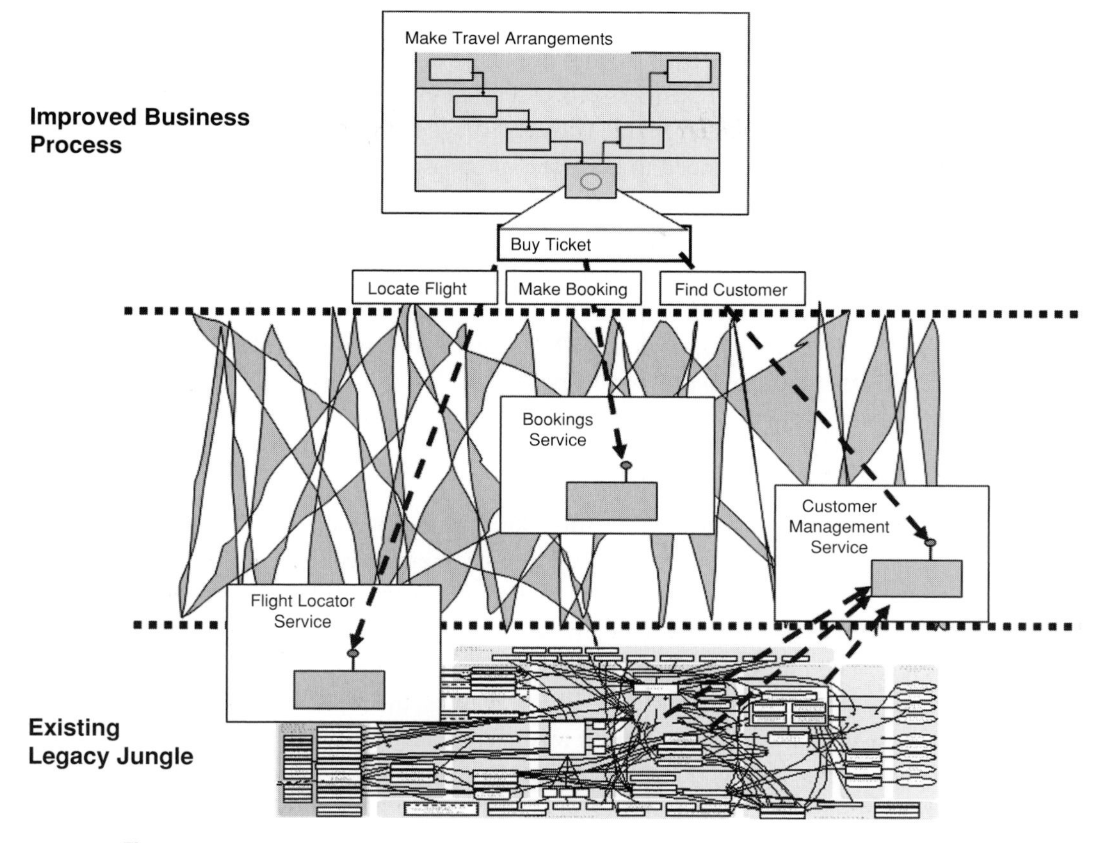

Figure 1.3 Integration and federation: a two-pronged approach

solutions to very specific user needs. Personally, I find that useful as the world I was originally bought up in was the world of highly integrated software tuned to very specific tasks and it is easy to slip back into old ways!

We can now move on to briefly address our three main areas of focus:

- Business process improvement
- Application design and development
- Software operations.

1.3.2 Business process improvement

First, we must ensure that we are addressing the right business problem. For example, in seeking to improve the Make Travel Arrangements process, are we concerned only with airline bookings or bookings in a more general sense? Do we need to think about how to communicate with the business processes of partners?

Our approach to the design of the Booking service must cover the different contexts in which the service is to be used. Examining these different contexts involves appraising and redesigning different business processes that use the service. One of these (the one shown in figure 1.3) is the Make Travel Arrangements process. However there may be many others that we need to consider and possibly invent as the business seeks to use ever-more innovative means of gaining customers, reducing costs, and increasing profits. Services take on a central significance as the business increasingly seeks to "plug and play" services together in a flexible fashion to meet these demands (Veryard, 2001).

In other words *a service-oriented approach to business process improvement* is required. (These considerations actually take us into the world of *business architecture* (BA), the chief topic of part 2 of this book.)

1.3.3 Application design and development

Second, we need to consider application design and development. The Booking service has dependencies on information such as customers and flights that will be provided using the "integration" software services (Flight Locator and Customer Management). We will need to take care with the overall design of these parts and their dependencies as well as with the specification of the Booking service.

In many everyday situations, such as car washes, the "specification" is implicit – we know very well if the car hasn't been washed properly. In other situations, such as a TV guarantee, terms are assiduously recorded but we seldom bother to read the details – just having the piece of paper is enough. Unfortunately with software the situation is often very different in that we cannot afford to assume the contractual details governing usage and provision of a piece of software are "OK." This problem becomes acute with the development of mission critical business software, especially if that software is to be offered over the Internet to potentially thousands of users!

So while the prizes are great, the risks must be managed. Software services and the dependencies that obtain between them must be clearly identified and defined in the first place. The service specification must, like all contracts between providers and customers, be precise and unambiguous. This applies to the functionality and quality of the service from a customer viewpoint and to the implementation constraints on the provider.

In other words *an SOA* is required. (This is the chief topic of part 3 of this book.)

1.3.4 Software operations

Third, we must consider the operational side of things. An SLA must specify the commercial terms that govern the execution of the service. The Bookings SLA must be agreed by both customers and providers of this service. Service-level management (SLM) is the discipline that oversees this process of agreement. However:

You cannot manage what you don't measure.

And more than that:

You cannot measure what you don't specify.

It is therefore critical for SLM to raise its game and achieve a very close tie-in between the service specifications of SOA and the SLAs.

We must also have confidence that our solution can be "scaled up" to deal with larger and larger volumes of transactions at speeds that are acceptable to customers. The execution of software services must be carefully monitored and controlled, in line with the specification, across increasingly complex distributed networks. Software service execution management (SEM) becomes increasingly important as your organization moves upward through the technology adoption curve to process increasingly larger numbers of transactions, and as these transactions increase in possible size and complexity. Security becomes an ever-increasing concern, too, as the company extends its reach beyond the firewall to more and more external users and partners.

In a nutshell *service-oriented management (SOM)* – a combination of both SLM and SEM – is required. (This, is the chief topic of part 4 of this book.)

1.3.5 Putting it all together

Gear shifts will be needed in coordinating all three of our key elements – BA, SOA, and SOM – in a unified way, as indicated in figure 1.4.

A clearly defined industrial-strength asset inventory will be required to ensure that our approach is underpinned with consistent specifications. The right execution strategies will need to be applied in the right contexts. "No small challenge," I am sure you will be thinking. However this is what service orientation is all about, and it is a long journey. A large part of the agenda for this book is to make some progress along the first step of this journey. The view here is that organizations can afford to delay no longer and must at least make a start.

1.3.6 Old wine in new bottles?

"Is this really new," you may also be thinking. In one sense there's nothing new about service orientation. Indeed, I was advocating a service-based component architecture in the late 1990s in relation to component-based development (CBD). A service-based

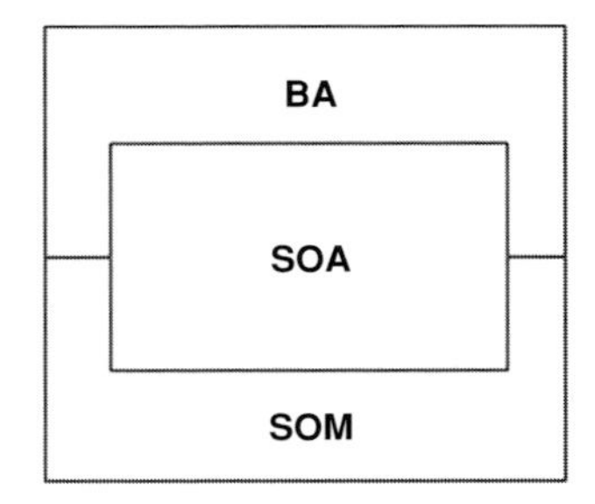

BA	Business Architecture
SOA	Service-Oriented Architecture
SOM	Software-Oriented Management

Figure 1.4 The building blocks of service orientation

component architecture abstracts a logical layer of components that provide functionality via their interfaces. The functionality should be provided as services. That means that implementation details must not be revealed. Services are specified in terms of contracts. Components are layered to provide increasing use of services downward through the layers. Component-based design techniques are used to structure components within components, in building-block fashion. Significantly, this idea is applied to re-engineer legacy systems into more resilient and easily maintainable pieces: these design techniques are an important valuable lesson that must not be lost.

1.3.7 The pivotal role of SOA

At the same time, certain things have changed significantly. Services take us up a level of abstraction from components, moving software abstraction closer to the business and opening up the possibility of a more federated business model. While the challenges, particularly in the areas of standards, are significant, so are the opportunities. Figure 1.5 builds on our earlier discussion of the business–IT gap. It depicts the pivotal role of SOA as a vehicle for greatly reducing that gap and achieving balance between the business vision, on the one hand, and the need to manage the legacy jungle, on the other. A good SOA should in fact help to greatly diminish the legacy jungle (as illustrated in figure 1.5) while at the same time making effective use of available commodity services, such as those offered by software service providers.

Clearly, there is a level of complexity here that will need careful management. One of the major differences between a service-oriented approach and previous generation approaches such as CBD is the greatly increased profile of *specification*, not just in the software process, but also in the business process. Later in this book we shall explore specification of services and the relationship to SLAs and business-level agreements (BLAs) in some detail.

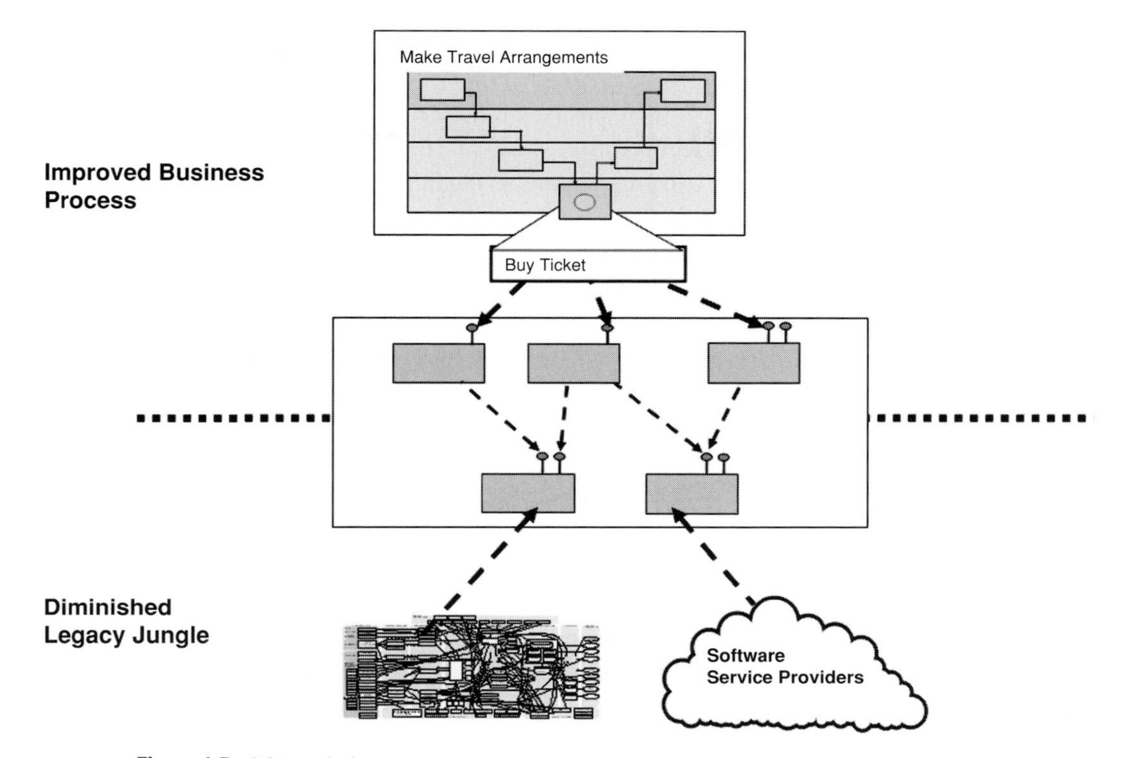

Figure 1.5 SOA as balance

1.4 The business question

I'm sure all readers will know the situation. You walk into the elevator only to be joined by the CEO. "Ah you're John Smith, the guy that's working on this service-oriented approach. I can't see it so just why should I pay for it?" You may feel like stressing the importance of technologies such as extended mark-up language (XML) and Web services, but my advice is not to go there on this occasion. In this chapter we have deliberately avoided the technology because service orientation is essentially a business phenomenon. The best way to address the question is to focus on the business benefits of a service-oriented approach. In this section, we offer some advice first for dealing with the elevator question and then for the slightly longer scenario represented by the water cooler.

1.4.1 The elevator question

Our company is too often a victim of change. For example, changes in regulation caused us to spend huge amounts of time and money on fixing our systems last year. Our competitors have been able to reach customers more quickly through the Internet and streamline their operations. Our business processes are hard to change because they are hardwired into inflexible systems. A service-oriented approach will allow us to make these changes much more easily, cheaply, and faster.

This can't happen overnight because of the scale of the task. But we're making a start by reworking some of our existing systems as flexible services. Right now we have six separate product systems that have to be changed every time product details change. The product streamline project will rationalize these systems. Details are changed through a service just once. This will result in saving of time and money in several business processes.

1.4.2 The water cooler question

If your encounter with the CEO takes place at the water cooler you might have more time to frame your answer in terms of one or more of the following themes:

Living with change

Change, once viewed as a short period of transition between two (much longer) periods of relative stability, is nowadays a continuous process. Unlike the closed systems of the mainframe, client-server and early-distributed eras, software is now increasingly never "complete." A service-oriented approach builds on what has worked in the past while providing new and better techniques, architectures and execution strategies for coping with business change. Here are three examples:

- **Channel neutrality:** A service-oriented approach allows the same software functionality (or software service) to be offered via different, alternative technologies – for example, desktop, kiosk, mobile, or digital TV. The organization can move more easily into a new channel to counter competitive threat or to take advantage of new markets.
- **Reusability:** A service-oriented approach promotes reuse of the same software functionality by different business processes, enabling cost savings. This helps promote business agility through reconfiguration of business processes from the same software building blocks in different combinations.
- **Replaceability:** A service-oriented approach allows the same software functionality to be realized in different, alternative ways. This means that redundant technologies can be replaced with faster and better ones without affecting the business functionality of a software service. This can lead to increased efficiency of business process with minimum disruption and is in sharp contrast to the excessive pain that is often caused to the business when technology needs to be "upgraded," with systems down for perhaps days while the "techies" do their work.

Managing your providers

A service-oriented approach helps in comparing and evaluating alternative implementations of the same service and choosing the best one. The organization is decoupled from vendor lock-in. This leads to much better control of the provider relationship, with consequent business benefits. These benefits often center on cost reduction. For example, I am reminded of the company who paid several times for the same claims component from the same supplier (I have actually changed the example so as not to embarrass the organization!), thus overspending several million dollars. They had lost control over the provider relationship. A good service-oriented approach would have prevented that!

Ensuring governance

A service-oriented approach helps an organization gain control over its assets by clearly marking out lines of accountability. For example, if services are not clearly defined and if organizational responsibility and authority for a service are fragmented, there's usually a governance problem. A service-oriented approach helps by demarcating services and ensuring that responsibility and authority cohere as much as possible, and by helping to identify missing or inadequate control points. Having mapped out, documented, and then implemented software services, manually prepared spreadsheet summarizations can be eliminated in favor of direct system-to-system communication, yielding tightly controlled audit trails. See (Veryard, 2004a) for further details.

1.5 Where to next?

This chapter provides a starting point for moving on to address technology management factors and to discuss the emergence of services (as opposed to business processes) in more detail in chapters 2 and 3. We then move on to present a roadmap, with guidelines, for redesigning business processes along service-oriented lines. This will take us naturally into the area of how to plan and execute a strategy for the SOA that balances long- and short-term needs, and that is integrated with SLM.

Assuming that your early attempts at service orientation are successful, there will soon be pressure to scale up those early solutions to cater for mission critical functionality over larger groups of users. Sooner or later the early prototype will not be enough. As transactions grow larger and more frequent their execution must be carefully monitored and controlled, in line with the specification of the service. *Execution management* becomes increasingly important as your organization moves upward through the technology adoption curve, across increasingly complex distributed networks. Security becomes an ever-increasing concern, too, as the company extends its reach beyond the firewall to more and more external users and partners. The phrase "Lack of planning on your part does not justify an emergency on mine" comes strongly to mind. In chapter 2 we therefore examine these issues.

2 Execution management

2.1 Web services in context

Your early attempts at service orientation are likely to be within the organization firewall, aimed at low-risk demonstrations of business capability, and involve modest numbers of users. These early solutions may well expose legacy systems in the form of software services, but without great heed to the scalability of these services. If successful, there will soon be pressure to scale up those early solutions to cater for mission critical functionality over larger groups of users. It is Web services that provide the group of emerging technology standards necessary to industrialize your services. There is a core group of established Web services standards that are required to climb the first rung of the adoption ladder. Other emerging Web services standards are required to progress further.

At the same time, as transactions grow larger and more frequent, their execution must be carefully monitored and controlled, in line with the specification of the service. Security becomes an ever-increasing concern, too, as we saw in chapter 1, as the company extends its reach beyond the firewall to more and more external users and partners. While Web services standards are necessary to deal with these issues, good execution management becomes increasingly important as your organization moves upward through the technology adoption curve, across increasingly complex distributed networks.

"Yes," we hear top management say, "but if this stuff needs to be ramped up we'll just outsource it to the experts. Or maybe we can simply use some of this on-demand computing that the big vendors keep telling us about and just tap into that. That way we'll be service-oriented but we can't afford to worry too much ourselves about technical subjects like Web services and execution management."

Again, however, current technologies and associated management practices are not going to get them there! As we explained in chapter 1, the right architectures and processes are part of what is needed.

In this chapter, we turn our attention to surveying the technology side of the service-oriented equation: the necessary enabling technologies, standards, and execution management practices. In particular, we look at the place of Web services in relation to the broader picture of service orientation that we introduced in chapter 1, before surveying the need for SOM: the harnessing together of SEM with enhanced approaches to SLM.

Ultimately software services should enable your business objectives through clearly defined BLAs that govern all types of service. It will therefore be important to clarify what is currently a poorly understood area: the relationship between SLM and BLAs.

2.1.1 Outsourcing

It's no secret that senior executives have grown increasingly frustrated with the shortcomings of IT departments to respond to business challenges. Outsourcing is a common result of these frustrations and is a strategy that can take many forms, ranging from outsourcing of development work to software operations management and, as we shall examine later in this book, to outsourcing of whole business processes.

However outsourcing is far from the panacea that it might seem. With half of all outsourcing deals not living up to expectations, according to Gartner (Best, 2003), organizations need to tread very carefully: for example, companies can enter into outsourcing deals that they are forced to renegotiate substantially a year or eighteen months later when they find out that the agreement isn't in line with their expectations (Best, 2003).

This situation has led some organizations to re-evaluate their outsourcing arrangements and bring their IT functions back in-house. The term "backsourcing" (Kaplan, 2005) is sometimes used to refer to this process.

One of the major reasons for failure of outsourcing arrangements is a failure to clearly specify the contracts that obtain between customers and providers of services. If responsibilities are not agreed unambiguously when contracts are agreed then it will not be surprising if disputes arise when things inevitably go wrong downstream.

The large-grain nature of many outsourcing arrangements is also at the root of many of these problems. For example, outsourcing your customer resource management (CRM) or human resources (HR) software represents a much larger risk than outsourcing one element of these software suites such as lead generation or benefit payments. The relevance of Web services is that they now provide an increasingly attractive option for breaking this functionality down into more manageable and reusable chunks. Moreover, it is also interesting to note that application package vendors are now modularizing their offerings as Web services. And yet more significantly, as we shall discuss shortly, we are seeing the emergence of on-demand computing with the promise of being able to take just enough software services when you need them – "by the drink," instead of "by the case" if you will.

Outsourcing is therefore changing toward a service-oriented model in which an organization can start to unburden itself of running and maintaining its commodity software by mixing and matching a range of provisioning options – from subscribing to a Web service to outsourcing the whole business process. The firm can then concentrate its internal resources on value-add services that are perceived as critical for retention and attraction of customers through activity that differentiates the company from its competitors, and on core services that are subject to change.

2.1.2 Introducing Web services

Some readers may be surprised that we have only just got round to the topic of Web services. There are a couple of good reasons for this. First, this is primarily a book about service orientation, not Web services. Web services are an important enabling technology but they are not the main subject of interest. Second, there are already many excellent sources of information around on Web services. Many of these are on the Internet and through regular journals that provide an excellent means of keeping pace with what is a fast-emerging technology. At the same time, there are also some excellent books that articulate the conceptual underpinnings very well. It is not my intention to retread any of this ground, so I hope readers will not be disappointed at the brevity of this section. Suffice it to say that a set of recommended information sources, many of which are updated on a regular basis, is provided at the back of this book.

So let's cut to the chase! Simply put, as we stated in chapter 1, Web Services technology is a set of XML-based industry standards and specifications that specify a communication protocol[1] and definition language,[2] plus a publish-subscribe registry.[3] These are the core features of Web services, though there is a proliferation of associated emerging standards required for industrial-strength application. At a technology level the promise is that Web Services will provide the universal glue for finally solving the well-documented problems of enterprise application integration (EAI) technologies. This promise extends beyond the organization firewall to easily accessible, highly reusable and replaceable software on a global level.

The underlying idea of Web Services is very much akin to that of component middleware technology, such as Common Object Request Broker Architecture (CORBA), Enterprise JavaBeans (EJB) or Distributed Component Object Model (DCOM). Contrary to what is often said by vendor marketers, Web services have actually evolved from these technologies, the idea of which was also to link together different software modules located on different platforms to create seamless distributed applications. However, the component technologies suffered from close coupling of modules to platforms – successes were therefore largely confined inside the organization firewall. And the promised component marketplace that would facilitate the vision of "plug and play" software failed to materialize except in specialized niches.

Web services promise to break this mold by providing the potential for loose coupling and standards and protocols for connectivity, not just across any platforms but

[1] Simple Object Access Protocol (SOAP). [2] Web Services Description Language (WSDL).
[3] Universal Description, Discovery, and Integration (UDDI).

also across the Internet itself. Moreover, they promise software modules connected in real time via a universal registry for publishing and subscription. In short, the interoperability of Web services make them an ideal mechanism for assembling applications from loosely coupled components distributed widely over the Internet. It is interesting to note that this loosely coupled feature of Web services is strongly associated with asynchronous messaging. However, Web services are simply the standards and there is nothing in those standards that mandates that only asynchronous processing must be used. Indeed, all of the early implementations of Web services that I can think of have been based on synchronous processing, mainly because it is much easier to manage within the confines of established execution environments.

Again, it is not our purpose here to provide a detailed up-to-date account of all of the many Web services standards. The reader should consult the set of recommended information sources at the back of this book.

2.1.3 Toward utility computing

Utility computing is commonly positioned as on-demand computing, in one form or another, by many of the leading software vendors. Associated products are rapidly becoming prevalent in the marketplace. Computer Associates, HP, IBM, Microsoft, and Sun Microsystems are among the more prominent on-demand vendors. These companies refer to their on-demand products and services by a variety of names, but the basic idea is the same.

The idea of on-demand computing is actually very simple: computing resources are made available to the user as needed. However, for our current purposes, it is the reasons behind this idea that are more interesting. The concept was developed in order to meet fluctuating demand for computing resources efficiently. Because an organization's demand for computing resources can vary drastically from one time to another, maintaining sufficient resources to meet peak requirements can be costly. Conversely, if the organization cuts costs by maintaining only minimal computing resources, there will not be sufficient resources to meet peak requirements.

On-demand computing depends on certain other concepts, the two most important of which are autonomic computing and grid computing:

- *Autonomic computing* is a self-managing computing model named after, and patterned on, the human body's autonomic nervous system. An autonomic computing system controls computer applications and systems without input from the user, in the same way that the autonomic nervous system regulates body systems without conscious input from the individual. The goal of autonomic computing is to create systems that run themselves, capable of high-level functioning while keeping the system's complexity invisible to the user.
- *Grid computing* refers to the use of the resources of many computers in a network that are applied to problems that require typically huge-scale computer processing cycles

or access to large amounts of data.[4] Grid computing requires the use of software that can divide and farm out software to as many as several thousand computers. Overall, the idea is that the resources of many computers can be managed toward a common objective.

Utility computing broadens out the idea of on-demand computing by including outsourced computing resources and infrastructure management with a usage-based payment structure. A service provider makes computing resources and infrastructure management available to the customer as needed, and charges them for specific usage rather than a flat rate. Again the idea is to maximize the efficient use of resources and/or minimize associated costs. The word "utility" is used to make an analogy to other services, such as electrical power, that seek to meet fluctuating customer needs, and charge for the resources based on usage rather than on a flat-rate basis.

To understand the relevance of Web services to this picture, it is important to realize that the first versions of grid computing platforms used proprietary resource management protocols and communication mechanisms. As a consequence, two organizations that wanted to share resources had to share a grid computing platform. This severely limited the scope of grid computing initiatives and often made sharing of resources across organizational boundaries impossible. The advent of the Internet coupled with Web services are what now make such sharing increasingly possible. In fact, the latest version of the Open Grid Services Architecture (OGSA), developed by the Global Grid Forum (GGF), maps IT resources to grid services and defines grid services as a special kind of Web service.

So it looks very much as if utility computing will play an important role in the move toward service orientation. It also is interesting to note that application of utility computing fits very well with a pragmatic and incremental route to business improvement by introducing services a step at a time, as and when needed. In contrast, traditionally computing services come in large chunks where costs are locked in place over several years. ERP packaged software is the classic case in point.

2.1.4 Virtual service networks

The technology vision of service orientation is of one global virtual service network in which business processes comprise Web services that collaborate on an as needed basis. In this world, organizations do not own their own platforms. They use provider platforms that are supplied on a "pay as you go basis." The idea is that a provider can supply a hugely more reliable and cheaper platform than an individual enterprise. Now,

[4] A number of corporations, professional groups, university consortiums, and other groups have developed or are developing frameworks and software for managing grid computing projects. The European Community (EU) is sponsoring a project for a grid for high-energy physics, earth observation, and biology applications. In the US, the National Technology Grid is prototyping a computational grid for infrastructure and an access grid for people.

there is nothing particularly new about that. What is new, however, is the offering of business services, via these platforms, to provide virtual service networks.

You may think that sounds fanciful and unrealistic. However the market is already taking shape. Grand Central (www.grandcentral.com) is an organization that has moved from simply providing a platform to focusing on populating their service network with value-adding services that deliver business services designed to work in collaboration with other business functionality. For example Grand Central hosts business-focused services from Salesforce.com. For more details of this emerging market the reader is referred to Sprott (2004).

2.1.5 The standards jungle

So the vision of service orientation promises much. However, the hype and noise that surrounds Web Services must be tempered by the fact that the associated standards are still evolving. We have come a long way from the starting point of XML as a simple mark-up language that would provide a means of universal data communication. The number of standards extends well beyond the basic WSDL, SOAP, and UDDI into a complex and growing jungle of acronyms. We would do well to maintain a healthy skepticism in this regard (Welsh, 2004).

No fewer than four organizations – Liberty Alliance, OASIS (Organization for the Advancement of Structured Information Standards), W3C, and WS-I – are involved in the various dimensions of Web services standards and guidelines. While each has different goals and spheres of interest, there is inevitably some jockeying for industry position and influence. Uneasy and fragile alliances quake under the pressure of enormous revenue potential, with a consequent proliferation of sometimes confusing and overlapping Web services standards initiatives. The patent and licensing implications of the resulting specifications present a critical issue for CIOs of user organizations.

The danger is that the user community is forced to choose from among multiple Web services standards that may not interoperate, may have limited life spans, and may come with expensive patent licensing requirements. It is therefore important for user organizations to increase their involvement in the standards process, if only to counteract vendor domination of the standards initiatives. At the same time, it must be recognized that such involvement demands lots of time and money. So what are CIOs to do?

While it is important for user organizations to increase their involvement in the standards process, it is perhaps more important to be aware of the economic trends and find a practical strategy for progressing. CIOs must introduce the principles of service orientation into their organizations in such a way that an investment for the future is provided in terms of software architectures that are agile in the face of inevitable change. At the same time, they must do this quickly and achieve results. The average tenure for a CIO is eighteen–twenty four months! There is therefore a need to balance strategy and delivery in a way that provides continuity of execution of service orientation.

This is to a large extent a "soft," cultural, and educational exercise. The bottom line is that business leaders must understand that while Web services are an important emerging set of standards, it is the gear shift away from application development to service orientation that is the really significant factor.

The "hard" substantial part of the gear shift is twofold. On the one hand, it involves maintaining and running legacy systems while at the same time improving, replacing, or reusing those systems to cater for service orientation. On the other hand, it involves structuring, specifying, and managing the services. This requires pragmatic models that capture valuable organizational knowledge as a long-term investment, so that this knowledge does not walk out the door every time that the CIO or any of her software architects leaves the company for another job!

2.1.6 Adoption strategy in a nutshell

Transitioning to service orientation is a complex and rapidly changing field, and there are no easy formulas for success. The developments in standards and utility computing have an important bearing on the pace of the transition from an industry viewpoint. And from a company viewpoint the adoption strategy will depend on a range of factors from the target business model to the state of the legacy systems.

At the same time, this book aims to provide you with some equipment (roadmaps, definitions, templates, techniques, process patterns, and checklists) to at least make a start on your journey. And having done all that – in chapters 3–13 – we shall also provide two real-life case studies to illustrate adoption strategy. Before progressing, however, it is perhaps useful to provide a broad picture of the possible stages for Web services adoption, as part of the overall trend toward service orientation. This will also serve to provide an overall context for our discussion of SOM later in this chapter (2.3).

Web services, like any other new technology, are likely to have several possible stages of evolution and adoption, as illustrated in figure 2.1.

A typical early approach is to expose legacy data as a Web service for presentation in a *portal* or the creation of a *point-to-point bridge* between two different applications. Portals and Web services share a common goal: enabling a company's previous software investments to be combined and exploited easily in unanticipated, value-added ways. These early integration projects are relatively low-risk and represent "low hanging fruit." As these first-phase deployments establish the benefits, the complexity of Web services integrations will increase.

Having established viability within the firewall, Web services can then "reach out" to become available externally to partners and customers. As security standards mature in this second phase, enterprises will begin to deploy Web services more broadly, addressing high-value integrations with their trading partners. Even today a few innovative companies such as Amazon, Dell, Dollar, and Orbitz have aggressively deployed Web

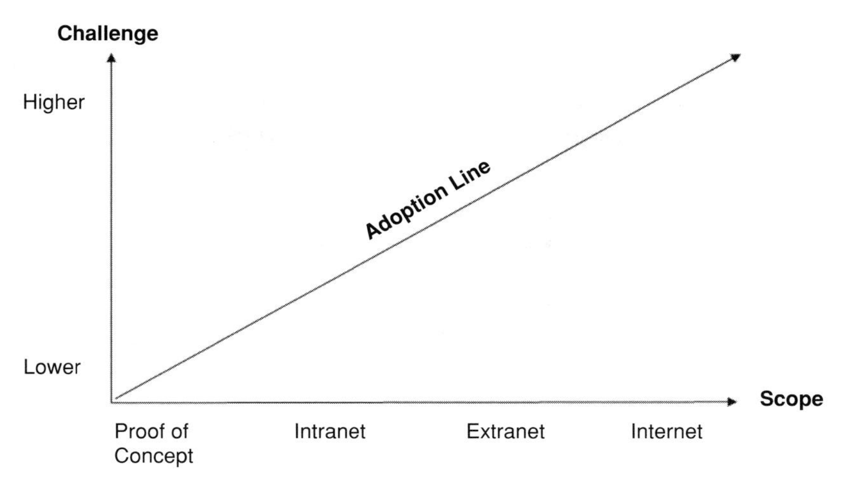

Figure 2.1 The web services adoption line

services to provide clean interfaces to back-end systems. This approach has enabled the "easy to do business with" philosophy through integrated supply chains and innovative customer-facing Web applications.

Finally, Web services will become ubiquitous. All new systems and all packaged applications will expose their functionality via Web services. Business processes will be dynamically configured out of services. Web services will form the enabling technology for services implemented as software, distributed across network and administrative domains, and discovered and consumed in a dynamic fashion. Unlike previous attempts to deliver on this vision, such as CORBA and DCOM, this time the "vision" is already showing some reality in the form of enterprises such as Grand Central, discussed above.

2.2 Execution management

Moving through the stages of adoption presents a number of management challenges. These challenges apply to the concept of software services in general, not only Web services, and therefore we now resume our convention of specifically referring to Web services only where contexts demands it. Many of these challenges relate to the need to evolve the SOA and use this as a roadmap to control development and reuse of services in relation to business needs. In fact, Credit Suisse (see the case study in chapter 15) refers to this approach as one of "managed evolution."

For example, it is vital to know when, *and when not*, software services are appropriate to help solve a business problem. As software services "scale up" in this way so projects can start to proliferate, again further increasing the need for careful planning and management. Equipped with easy access to freely available software services, hasty and thoughtless developers now have the ability to "really foul up" thousands of computer systems all around the world. Unless we are very careful, Web Services could become a Pandora's box of troubles. This raises the stakes for sound SOA planning and adoption management if such software anarchy is to be avoided through sound programs of corporate governance.

In addition there are run-time management challenges that concern the execution of software services. The need for high quality[5], in terms of factors such as capacity, availability, and security, only increases with adoption[6]. And it is a major critical success factor for utility computing. On the one hand, as emphasized in chapter 1, it is important to specify such quality of service (QoS) requirements clearly and accurately. On the other hand, it is important to be able to monitor and control these factors in the execution environment. Just as there is a shift in mind-set toward SOA, so also there must be an accompanying shift in management of execution environments.

2.2.1 Limits of execution management environments

Previous generations[7] of monitoring and management technology were designed to solve a different set of problems to those posed by software services. These previous-generation solutions were designed around a static IT infrastructure often taking several months to initially deploy and several more months to customize to specific customer needs. Because the managed systems were static and the interfaces between systems were proprietary, correct functioning was defined as "the server is up and the expected process is running." Software services, on the other hand, are dynamic with services being added and removed from applications in an adaptive way.

The notion that all resources are controlled and owned by a central organization was presumed in the design of old-generation solutions. In the world of software services, resources can be spread across different organizational domains. While Web services can be simple synchronous calls, the real spirit of software services rests in the notion of asynchronous messaging across heterogeneous technologies.

2.2.2 The need for SEM

Let us just consider a couple of examples of how software services raise the bar for execution management. First, consider the customer viewpoint. Increasingly software

[5] For example, some Web service hosts are more reliable than others. How can this reliability be specified and communicated? Most business-related services use encrypted communications with authentication. How does a Web service authenticate users? When a system comprises many Web services whose location and qualities are potentially dynamic, testing and debugging takes on a whole new dimension. How do you achieve predictable response times? How do you debug Web services that can come from different vendors, hosted in different environments, and on different operating systems?

[6] Capacity, availability, and security of software services are discussed in some detail in chapter 10.

[7] Network management solutions focus on the *throughput* and *latency* characteristics of networks. Systems management solutions focus primarily on the *hardware* and the health of the *operating system*. Application management solutions focus on the specific *server processes* involved in serving an application and characterizing the availability and performance of applications.

services are crossing traditional organizational boundaries. In these situations, the ability to understand and trace the cause of failures of a software service across multiple nodes in a supply chain becomes very important. Cross-enterprise usage of software services requires a level of visibility into resources that are outside of your control.

Second, consider the provider viewpoint. The increased sharing of software services results in a need to track resource consumption by the various service requestors. For example, there is the potential for one service requestor to overload the system, causing unacceptable performance for other users of the same service.

A great deal of activity is taking place to define standards that will help address these challenges, under the auspices of the Organization for the Advancement of Structured Information Standards (OASIS) Web Services Distributed Management (WSDM) technical committee (TC).[8] These developments are ongoing and outside our current scope; the reader is referred to the sources of information section at the end of this book for up-to-date information. Right now, we focus on the essentials of SEM and how this needs to join up effectively with both business process redesign and SOA. To do that, we first take a step back and draw an analogy.

2.2.3 The asset inventory

The above considerations also illustrate the need for an asset inventory to provide the required information to manage these complex situations. Without an asset inventory, it is impossible to manage software services in any meaningful fashion, especially in an organization of any size.

Surveying and cataloging of assets is part of a much wider topic. For example, it is an absolute necessity for effective corporate governance. The asset inventory should summarize both existing assets and possible future assets. These assets include all resources, not only services, including processes, business units, and groups of individuals, as well as software such as databases, systems, and packages.

The asset inventory is discussed in more detail in 6.5.

2.2.4 An analogy

We have emphasized that although an SOA operates at a logical level there are important connections and relationships between the SOA and the run-time management of software services.

[8] The WSDM TC was formed at OASIS by a number of vendors to define Web services management, including using Web services architecture and technology to manage distributed resources. They have identified two distinct areas that they will work on. Management of Web services (MOWS) concerns the run-time monitoring of SOAP messages. Management using Web services (MUWS) concerns the management of any resource (not just Web services), using Web services as the standard interface to expose management information regarding that resource.

One way of thinking about this is in terms of metro maps, as I was reminded in a recent visit to Prague. At a logical level the metro map simply shows the different metro lines, in different colors, the stations and intersection points. Everything else is abstracted out. Similarly an SOA simply shows services and their dependencies. Services are grouped into service buses[9] for use by different groups of business users, partners, or customers. If we liken services to metro stations then service buses are akin to the metro lines. Ideally the logical metro map will have a certain "shape" that is readily recalled. The Prague map had a particularly good shape: a simple variation on a cross; see figure 2.2.

In addition, just as it is important to provide for customers to plan their metro journeys, itineraries for metro journeys from one station to another, so we model how services are used to support different business processes. Business process modeling techniques traditionally belong to a separate domain, that of business strategists and analysts, but it is a major mistake to divorce this domain from software architecture. One of the key characteristics of SOA is that these kinds of technique are included in harmony with those used to build the SOA.

Just as with a logical metro map, both customers and providers are interested in the SOA. Customers need to know what services are offered and how they can be put to work for the business. Providers need to know what services they are contracted to offer and also the dependencies that obtain between services.

On the other hand, a physical metro map shows the actual paths taken by the lines in relation to city features such as roads, parks, and rivers. Again Prague had a good one, shown in figure 2.2, which proved useful for determining the full detailed picture.

Similarly it is important to have the right components in place in relation to the SOA; the problems in implementing and deploying the underlying software components do not magically disappear, as has often been erroneously assumed in the past. Software components, including components that provide technical infrastructure services such as security and database connectivity, are an important part of the overall enterprise architecture of which the SOA is part. These component architectures set design guidelines for providers of services.

To extend the analogy a little further, customer demands on the metro vary throughout the day. During the rush hour the metro network must run at peak capacity whereas in the middle of the night the train service is down to a bare minimum. Just as the metro management need to plan for optimum utilization of the metro and to ensure that the service runs smoothly without hiccups or security problems, so software planners and architects should do the same with the software services that are specified by the SOA. They must ensure that software services are employed in the optimum way so that valuable resources are not wasted. And they must do this in such a way that factors such as reliability and security are taken into full consideration.

[9] A service bus is a grouping of deployed services that must conform to defined corporate policies and standards; see chapter 9 for a detailed discussion.

Logical Metro Map

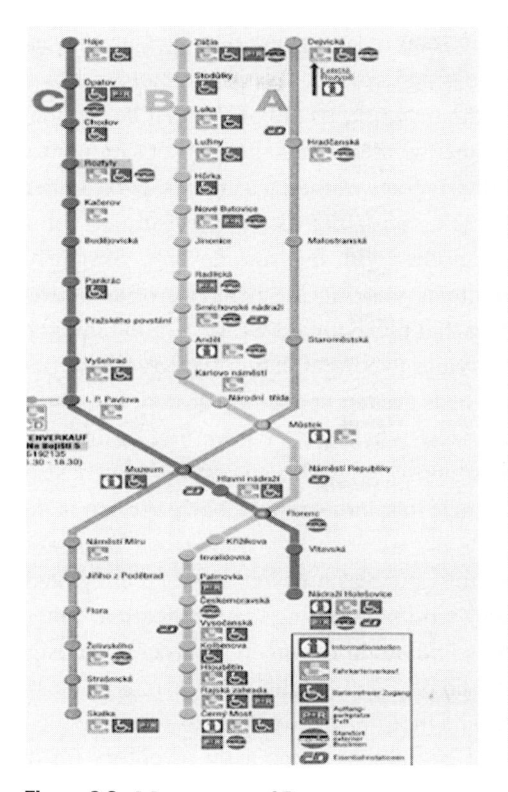

Physical Metro Map

Figure 2.2 Metro maps of Prague

As the metro becomes more and more popular with increased numbers of visitors and commuters, efficiency of service must be maintained. Similarly, software services are becoming increasingly popular as the standards and technologies upon which they run become increasingly stable. The performance of software services must be maintained.

Another consequence of increased usage is the need to keep information about equipment, trains, tracks, and so on up to date in a registry: the metro needs to manage its assets! Similarly, information about software services must be managed within an asset inventory.

In short, just as QoS is a huge issue in the running of the metro, so the same is true in the running of sets of services that are being used to support an organization's business: hence the need for SLM.

As far as *software* services are concerned SLM becomes a bigger and bigger challenge as the software services become increasingly distributed beyond the traditional comfort boundary of the organization firewall. SEM of these distributed services is an order of magnitude more complex than with traditional operations management. The application now has, perhaps surprisingly, many points of failure.

Effective SEM software is therefore required to ensure that QoS levels can be guaranteed while scaling up to meet increasing demand. Just as the metro maps and plans must be in tune with the running of the network, so the SOA must be in tune with our approach to SEM. We can also imagine the metro managers observing metro network traffic on their consoles and analyzing reports of metro usage. A good system should enable as much automation of control as possible: for example, trains automatically run a controlled distance from one another with fail-safe devices. Similarly, we can imagine software planners and architects viewing the behavior of software services in relation to the SOA, and analyzing that behavior via SLM. A good integrated SEM system should enable the software services to correct themselves on failure or to invoke more powerful software programs where greater speed is required. Finally, we might also picture our metro planners assessing whether they have met business objectives by viewing alerts and reports on their consoles. Again a good integrated SEM system should provide similar alerts and reports that indicate whether a service is meeting its business objectives.

In summary, just as SOA and business process redesign must be integrated so SEM must ensure that the software actually works according to specifications laid out in the SOA and delivers the business results laid out in the business process redesigns.

Now, think of the maps as logical and physical architecture designs of the metro network. Think of running, controlling, and monitoring the trains as run-time management of the metro. Think of the registry of equipment as the asset inventory for the metro. Clearly, the greater the scale and complexity of an undertaking, the higher the price for getting it wrong, and the greater the benefits from getting it right, then the greater the need for both good design and good execution management.

With a city metro the case for good design and good execution management is pretty self-evident. In the case of software services it is much harder to communicate. Let us begin to address this issue by considering the major dimensions of SEM:

- Quality
- Consume–provide
- Temporal.

2.2.5 The quality dimension

Monitoring and control of QoS involves analyzing QoS types, such as capacity, availability, and security[10] and alerting users where the measured QoS falls below acceptable threshold levels. As software services usage scales up, there is a problem of sheer scale

[10] Clearly agreed standards for Web services security in the form of languages, models, and definitions are a necessity. These concepts are continuously evolving (Welsh, 2004, pp. 5–7).

here that is difficult to envisage. For example, monitoring and controlling extends from isolated software services to whole portfolios of applications, each making numerous calls to services on behalf of perhaps hundreds of users. Each call should be logged against a user account and the response time recorded. As the system grows, the automatic alerting of key managers when failures occur, or when performance falls below a predetermined threshold, becomes essential.

The measurement of QoS also demands some subtler shifts. For example, the measurement of software service performance requires a finer level of granularity than the determination of application, system, or network performance. In particular, the overall performance of a Web service is a relatively useless metric – you need visibility into the specific operations defined in a Web services WSDL interface description to understand whether observed delays might have a business impact. Operation-level information is also necessary in order to isolate and optimize performance bottlenecks. Characterization of operation performance may sometimes require visibility into the parameters associated with the operation calls. These requirements mean the XML data stream must be observed; this itself poses a significant operational challenge.

Similarly, the measurement of software services reliability requires a finer level of granularity than the determination of application, system, or network reliability. All software services are not created equal, and some are more critical to the operation of a business than others. For instance, key software services that perform such functions as a credit check or execution of a market sell order are vital components of a business process whose failure brings the entire business process to a halt. Others, such as those that provide supplemental information like product information or weather details, may be merely "nice to have."

For simple scenarios such as synchronous look-up of nonsensitive information, security is seldom an issue. However as we move through the stages of Web services adoption elements of security such as authentication and access must be seriously considered and an appropriate security strategy put in place. And again this needs to be at the level of operations.

Last but not least older-generation solutions are not focused on the business information contained in the XML streams. It is becoming increasingly important to monitor this business content so that business goals and rules can be verified at execution. For example, a fund transfer between bank accounts might trigger an alert that is immediately sent to user as soon as the transfer hits an amount of say over $1,200. In this way SEM moves up to the business level, offering many possibilities for real-time monitoring and control of business goals and rules.

2.2.6 The consume–provide dimension

The execution management side of service orientation also takes on consumer (both customer and user) and provider aspects. A provider needs to manage users' access, to control which users are allowed to access which services, and to determine the customer who will be billed. Service levels must be monitored so that quality can be assured against SLAs. A provider will also need to manage which suppliers implement which services in order to provide flexibility by separating the actual implementation from the delivery of the service.

A customer of a software service can only monitor the software service upon which it may have a critical dependence, not the implementation behind it. A customer must be able to control access and identify users of services so that departments can be billed accordingly. It will also be important to monitor service usage so that levels of service can be measured and assessed against SLAs. A customer should also be aware of possible alternative providers of these services.

A common mistake is to think of SEM purely in terms of ensuring that the customer gets the service promised by the provider. For example, a provider agrees to make a service available 99.9% of each day. Equally, however, it should be employed to ensure that users do not abuse the service by breaching terms of use.

For example, Queensland Transport (QT) provides a vehicle registration software service that is used by motor dealers. However one of these customers wrote calling software motivated by trying to capture new insurance business, and not to register vehicles. This motivation meant that they "skimmed" on testing. The result was that the customer's small user community nearly bought the provider's services to their knees due to the high rate of "business errors" that resulted from their software. The QT case study is discussed further in chapter 14.

Note that the customer is always responsible for how the software service is used within the context of his or her organization's business processes. This kind of responsibility is separate from the responsibility of the provider for ensuring the availability of the service; see 13.4.2 for details.

2.2.7 The temporal dimension

If these challenges were not enough, they are multiplied still further by the utility computing dimension of time, especially with respect to monitoring, control, and routing of software services. For example, different performance thresholds may be applied to a software service at different times, requiring different configurations of supporting software. The routing function focuses on dynamic, context-sensitive switching based on different variables; Sedighi (2004) provides a useful discussion.

It is interesting to note that these kinds of challenges are a direct consequence of the principle of separation of interface from implementation that service orientation brings. The good thing about this is that redundant, or poor-performing, implementations can be smoothly replaced by newer and better ones. However, this also puts an onus on routing mechanisms to select implementations dynamically in response to QoS values reaching various thresholds.

2.3 The need for SOM

While the challenges of SEM must not be underestimated, software services bring new potential opportunities. More generally, there is the opportunity to fine-tune software support for business processes so that the software is utilized in the optimum way. This has been euphemistically referred to in some quarters as the "Real-Time Enterprise." What is commonly overlooked, however, is that the business requirements must be identified and the software specified in such a way that it meets those requirements and can be accurately measured against them. In fact, software services should be the subject of SLAs and must be monitored for compliance. It will be important to ensure that such SLAs are clearly specified in such a way that business metrics can be applied to software services where appropriate.

SLM is a term which has been around for a while now: the setting up, agreeing, monitoring, and controlling of SLAs. It is a discipline that involves comparing actual performance of services with pre-defined expectations, analyzing any gaps between the two, and determining appropriate actions.

SOM is a term introduced in this book to refer to the combination of SEM with approaches to SLM enhanced to handle the challenges of service orientation.

2.3.1 The need for SLAs

Contracts between providers and customers must be agreed if a service is to work as a commercial arrangement to the satisfaction of all parties. This is especially the case with regard to utility computing. The promise to customers that they can take just what they need at the right time means that services must be monitored and controlled so as to ensure optimum utilization over time. In particular, the organization will need to consider the distribution of service usage over time. For example, the number of financial transactions to be processed by a bank might fluctuate throughout the day and the week. It makes sense to apply high-capacity processing power when transaction volumes are at their height – say, in the middle of the morning and afternoon throughout the working week. In contrast, night time and weekends might require much lower capacity.

In the traditional world of application development we can be reasonably sure that the software evaluated before purchase will continue to operate as it did when it was tested because it will be used in an environment that is under the control of the organization that purchased it. The world of service orientation takes that comfort blanket away because the software used by an organization is no longer products but services. The quality of those services can change day-to-day, hour-to-hour, and even second-to-second. While pre-purchase evaluations and due diligence remain on the agenda, the requirements for execution management far outstrip anything we have become accustomed to until now. In particular, just as testing and warranties are critical to managing packaged software,

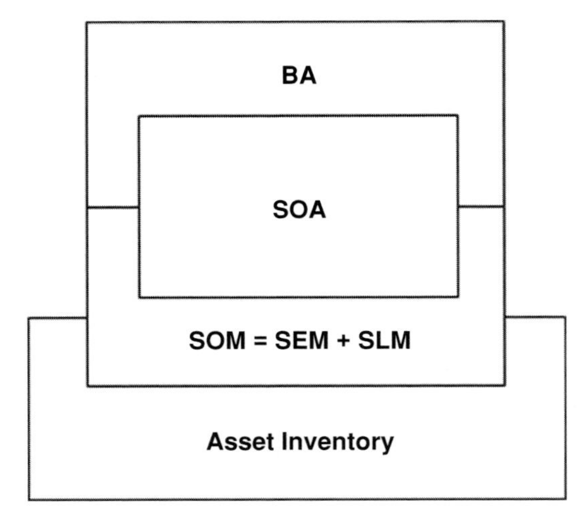

Figure 2.3 SOM in context

so SLAs are vital in managing software services. Services, particularly those provided on a utility basis, must be specified clearly.

2.3.2 The process of SOM

The process of SOM depends, then, on a clear definition of the SLA. The SLAs must be recorded and maintained in an asset inventory that underpins the process. SOM couples together the service management disciplines, as indicated in figure 2.3, and ensures that our architectures are not just "theory": they are reflected in real production running software services. We take a close look at this area in chapter 11.

There are three primary ways of initiating this process, according to market context. In a *provider-driven* approach, a provider of services publishes a service specification in the manner of marketing and selling a product. In a *customer-driven* approach, a customer publishes a service specification in the manner of an invitation to tender to providers. Finally, in a *collaborative* approach, a customer and provider work together in partnership. Market contexts are covered in 13.5.

It is important to note that there are operational aspects of SOM, such as problem and incident management, that are absolutely vital. There is a fairly comprehensive literature that covers these aspects; for example, the Information Technology Infrastructure Library (ITIL) standards, which we discuss in 11.4. However, these operational aspects must be clearly linked with the processes that govern SOA and service specification. Unfortunately these have been lacking. SLM is commonly understood as very much a specialized operational management discipline in isolation from the world of service orientation.

We therefore focus on the process of connecting these worlds, on the process of managing the customer–provider relationship. That, in turn, depends upon clear and workable SLAs that do not require parties to take a diploma in linguistics and do not suffer from the complexity and inconsistency that seems to dog this subject. A connected process based upon clear SLAs can only help foster the trust of customers in providers that is so often lacking. SLAs are covered in detail in chapter 12.

2.3.3 Business alignment

Business processes that smoothly cross organizational boundaries have been an elusive goal for many years that has been inhibited by immature software protocols and standards. In particular, security issues, and the difficulty of managing long-running asynchronous transactions, are significant barriers to the vision of enterprises collaborating through virtual value chains: the vision of the *networked enterprise*. This situation is changing with the standards for choreography of Web services (see 3.3.1) across multiple partners in a rapidly evolving business process. We shall look at this situation some more in chapter 3 in our discussion of business process management (BPM). We mention it now to draw attention to the business alignment aspect of service management.

Software services bring new opportunities to align business and IT in a fashion that until now has not been possible. Business processes are hard to change when they are codified into inflexible monolithic applications. An SOA offers the promise of business agility by enabling the rapid and dynamic configuration of software services in coordination with choreography software. The prize is the ability to respond easily to changes in business and technology and to initiate such changes for competitive advantage. However, even if we assume the ideal in terms of sufficiently mature standards, excellent service management software and a smooth running top-of-the-range choreography engine, there is another gate that must be opened. All of this great technology is for naught unless and until parties can agree on the *structure and meaning of the business documents* that will be exchanged via a software service. Yet more fundamentally, it is also for naught if the software service is not aligned with the business process through clearly agreed definitions of the functionality and information provided by it. It is this lack of alignment, far more than the immaturity of the choreography technology, that will hamper the vision of the networked enterprise.

Achieving alignment requires *traceability* between the artifacts developed. This is a major reason for taking the kind of model-based approach to service orientation that we describe in this book. Traceability of business requirements to real executing services is achieved through the models which are underpinned by the asset inventory.

2.3.4 Business semantics

It is perhaps instructive at this point to restate the obvious: XML and WSDL are *languages*. Languages require vocabularies that enable their users to communicate clearly. Software services can be interpreted in many different ways, depending on their context of use. Many potentially different business domains, cultures, and partnerships might define the context of use. XML vocabularies, for different business domains, go some way to alleviating this issue, but commonly do not define business semantics in any detail. There is also a proliferation of different vocabularies, making standardization difficult. Therefore the problem of actually agreeing business semantics among the participants in Web services exchanges remains. For example, take the tag "shipping," which is used in an XML document by a supplier to mean the date that its product leaves its factory. Perhaps the carrier interprets the same tag to mean the date that the product leaves the loading bay. And unfortunately the buyer perhaps interprets the tag "shipping" as the date that the product leaves the port. The various Web services standards help with lots of issues and at a technology level they go along way to establishing common definitions – as in the case of topics such as run-time management, security, and reliability. However someone still has to do the job of defining business semantics. A large part of the purpose of this book is to help tackle this issue by providing suitable techniques, so often lacking, to help do just that!

2.3.5 Toward BLAs

A BLA specifies the business goals and rules that govern the functionality and information described by an SLA, and the commercial criteria (such as financial penalties) attaching to these business goals and rules. Let us look briefly at this idea; the reader is referred to 12.3.3 and 12.4.4 for the details.

The underlying idea here is that a BLA should be monitored by SEM software so that customers know when (and *when not*) a service is meeting its business goals and so that users know when (and *when not*) the service is performing according to business rules.

Traditionally systems management software has focused on the monitoring of non-functional requirements. Now, however, in the world of service orientation the bar is raised for management software, in that business features must also being monitored.

This may seem somewhat unrealistic, given what we have already said about the state of standards and business semantics. However, as Winston Churchill once famously said, "Perfection is spelt P-A-R-A-L-Y-S-I-S." Even if the Web services standards and technologies are not yet fully mature and the business semantics problems have not been solved, this should not keep us from deploying Web services for simpler applications at the earlier stages of the adoption cycle. As we move along the adoption line it is important to always keep in mind that Web services are about enabling business processes. Through WSDL and SOAP, monitoring tools for the first time have a practical way of gaining visibility into the business context of messages. Older-generation solutions are not focused on the business information contained in the XML streams. Instead, they remain focused on system metrics and so cannot provide a business perspective on the systems they manage and monitor.

2.4 Where to next?

The early adoption of software services has often been in the area of *internal integration* where the benefits are usually quite straightforward. Grand visions of everything being connected in dynamic real-time scenarios are one thing, but most organizations have more specific tasks and mundane problems to solve. As a result, we have emphasized that a certain balance is needed in taking a middle road of one form or another.

In order to achieve this balance it will be important to establish the *business drivers* in terms of required services, of which software services are but one (albeit important) example. This is a lot more than re-engineering of business processes in terms of streamlining a production line or value chain, as is often understood today. In chapter 3, we examine the background to these factors, as the first step in getting to grips with service orientation.

In particular, we shall examine the requirements for BPM. In this chapter, we have stressed the importance of good SOM as a prerequisite for the serious implementation of BPM tools based on SOA. While it is technically feasible to use BPM tools and approaches to assemble business processes once the service portfolio is in place, one can achieve the benefits of BPM only if the portfolio is rich and reliable. Otherwise, business operations may be in jeopardy.

3 Business process management

3.1 Cultural shifts

In this chapter we outline the BPM technologies and the associated gear shifts in business modeling that are required for effective service orientation. We first take a brief detour into the history of BPM technology before outlining the key elements of an effective BPM solution and previewing the main differences between a service-oriented approach and traditional approaches to business modeling. In this section, by way of prelude, we take a look at some of the cultural shifts involved.

It is important to restate some ground rules up front. We are focusing in this chapter on services as a way of enabling federated business processes (note for brevity we will refer to "business process" as "process" throughout, unless the context requires otherwise). Software services are services offered in software. Web services are but one manifestation – albeit a very compelling one – of software services. We shall therefore (as before) refer specifically to Web services, rather than software services, only where context demands it – as, for example, in the technology background to BPM.

3.1.1 Cultural shifts in IT

IT organizations continue to find themselves under increased pressure to find ways of doing more with less. In many cases, the development and integration efforts of the 1980s were designed to increase organizational efficiency by automating departmental functions. In the 1990s, the agenda moved toward using automation to help streamline internal processes and to removing the "white space" (Rummler and Brache, 1995) on the organization chart. Nowadays, the Internet allows companies to collaborate and to share information far more easily than they could at the beginning of the 1990s. These developments have been paralleled by an increased role for business people in IT capabilities that were previously only for the technically initiated. The use of email has become ubiquitous. Business people use the Internet as part of their daily lives to increase knowledge, watch what competitors are doing, and check for stock information. These developments in technology allow them to work across organizational boundaries in ever more adaptive fashion. The result is hugely increased expectations of business people regarding IT.

A consequence of failure to meet these expectations is that many of the traditional development tasks of IT departments are outsourced to specialist application service providers (ASPs). This puts the onus on internal IT development to focus on value-added activities that confer competitive advantage. Similarly, it creates an environment in which IT management must switch toward overseeing the broad strategy, away from projects and toward programs. Equally, it puts the onus on IT managers to be proactive in business process improvement. "Technospeak" is unlikely to receive much sympathy in the boardroom: the language of IT increasingly needs to be the language of business.

This is a huge challenge for most IT shops in that a "blue sky" strategy is not an option given the legacy software application baggage which has been inherited through the generations and upon which organizations depend. At the same time, despite its sheer scariness, it is not a challenge from which organizations can afford to walk away. The bottom line is that once we have some idea of the business drivers we can start to work toward SOA, while at the same time coping with the legacy portfolio and keeping the business happy. That will be much less about planning and more about execution as we shall see later in the book, but we have to have the right approach to business modeling.

3.1.2 Cultural shifts in modeling

Service orientation is a business model that centers on the buying and selling of services to achieve increased profits and long-term commercial advantage. This is in some contrast to the traditional view of an organization as a production line that converts raw materials into finished goods. Of course, most organizations will naturally combine elements of both views and there will be organizations that continue to operate more in the manner of a production line than a service. However, the undeniable trend is increasingly toward service orientation.

The dominant approach to business process modeling that has persisted since the early 1990s is predicated on the view of an organization as a production line that converts raw materials (or data) into finished goods (or knowledge). This approach must change in order to cater to the world of service orientation, so that we can focus on how services combine and interact to make other services. In this chapter, we provide some tactics for doing just that.

Another consequence of the production line approach to business process modeling is that software is relegated to the status of a production support function. This fosters a separation of processes and software programs, and more generally of business and IT organizational roles. In particular, it helps explain why SOA is commonly treated as a largely technical subject that is regarded solely as an advanced form of component architecture. This view has severe shortcomings in relation to the world of service orientation in which most organizations are, or will be, forced to compete.

While SOA builds on many of the ideas of component architectures its underlying relevance rests in the fact that business itself is increasingly service-oriented. Organizations have struggled for many years with the more general issue of business–IT alignment. The root cause of this struggle is that traditional approaches treat processes and the applications that support them at two different abstraction levels. An SOA has the potential to greatly diminish this abstraction gap and open up the possibility of business–IT convergence (Taylor, 1995).

Again, realizing this potential requires the right approaches to business process improvement and business modeling. That means emancipation from the concept of an organization as a production line and a move toward understanding businesses as *collaborations of services* that must be defined to contract.

3.2 A little history

There is a paradox at work in that developments in technology have increased the expectations of business people, but at the same time the limitations of currently used technologies work against these expectations.

However, enabling technologies are fast maturing out of what we have already learned. Contrary to what we sometimes read in the popular press this body of knowledge is not a "Web services revolution," it is growing naturally out of distributed computing, CBD, and SOA, as well as Web service standards and business process modeling languages. The IT and business trends fuel each other: business pressure to realize the full potential of the Internet means greater prizes for the IT vendors. Conversely as the technology "opens up," so business awakens to the cost savings promised by utility computing as well as to increasingly imaginative ventures involving hitherto undreamed-of partnerships.

One of the most important enabling technologies to realize service orientation is BPM. The concept of BPM is not new but has grown naturally out of previous-generation technologies. In this section, we provide a brief historical overview.

3.2.1 Workflow and EAI

In its earliest form, a workflow system was designed to automate the flow of documents through an organization from one employee to another. The original document is received and scanned into a computer. Then, an electronic copy of the document is sent to the desk of any employees who need to see or approve the document. Copies are routed according to a combination of automated conditions and employee input. Eventually the flow of documents results in either a further document or set of documents sent in response to the original request and sometimes prompts an action on the part of an employee to perhaps call a customer to say that further documents are needed. The workflow is thus concluded in analogous fashion to a production line in a factory.

Workflow application tools are more or less directly aimed at supporting workflow system developers or managers in controlling the order that electronic documents show up on employees' computers by modifying a diagram depicting the workflow. These tools normally include mechanisms for routing, forks, and joins to allow for tasks in a workflow to be executed concurrently and then synchronized at a later point in time, message transformation facilities, and some level of support for maintaining consistency of transactions.

The principle is that, by using these tools, the business process designer can focus on defining processes and the way information flows between applications, and let the tools take care of the technical details.

Hot on the heels of workflow applications, the large application package vendors began to organize the software modules within their products so that they could be represented as *processes*. They applied the idea of document-based workflow to the linking of software modules used by a process. Vendors such as J. D. Edwards, Oracle, PeopleSoft, and SAP all offered systems of this kind, which were usually called enterprise resource planning (ERP) systems, that offered "templates" that depicted the way that several modules could be linked together to form a process. A process designer was able to use the template to effectively form a process by excluding some modules and changing some of the rules controlling the actions of some of the modules. Many of the modules included customer interface screens and therefore controlled employee behaviors relative to particular modules. The flexibility was quite limited, however – the process designer could choose and sequence only pre-existing software modules. In many cases, it was actually easier to constrain the process itself by stitching together software modules in pre-defined sequences, rather than having to go through the pain and expense of introducing new software modules, interfacing mechanisms, and routing criteria.

In short, ERP systems have received a lot of criticism for overconstraining processes. At the same time, there are processes that are quite rigid and constrained, such as payroll, sales ledger and predictive production line automation. ERP systems thus tended to be most successful in these kinds of areas.

The salient point is that an ERP system is controlled by another kind of workflow system. Instead of moving documents from one employee workstation to another, the ERP systems offered by SAP and others, allowed managers to design processes that moved control from one software module to another. ERP systems allowed companies to replace older software applications with new applications, and to organize the new applications into an organized process. This worked best for routine processes that were well understood and common between companies.

The first-generation workflow solutions were very much constrained by point integration between enterprise applications and the workflow solution. Integration between individual applications across the workflow often required developer intervention and hard coding, which meant that the changes were often out of date the minute they were deployed, and often resulted in broken connections.

To combat the inefficiencies of patched-up integrations, enterprise application integration (EAI) products emerged. These provided an infrastructure for application integration and were scalable for enterprise use. These first-generation EAI products provided only the most rudimentary workflow. They did not typically include much if anything in the way of modeling capability and employed proprietary architectures for messaging, application adapters, and scripting languages. These products were also internally focused on integrating back-office applications.

In the late 1990s a new class of Internet business-to-business (B2B) "gateway" product emerged to exploit the growth of the Internet using low-level technologies such as Hypertext Transfer Protocol (HTTP) and extended mark-up language (XML) to automate cross-enterprise application integration. Nevertheless the limitations of previous generations of product remained. Standards development was (as indeed it still is to a large extent) in its infancy, and these products still implemented application interfaces and response-request messaging protocols in a proprietary way. Many of the EAI solutions facilitated integration through message-oriented middleware (MOM), which often proved to be expensive to deploy and maintain. The associated consultancy services also added significantly to the costs. And, like its predecessors, MOM lacked the higher-level capabilities such as business process modeling.

3.2.2 The emergence of BPM

Business process modeling tools emerged in parallel with the emergence of workflow tools, quite often in terms of families of business modeling techniques offered alongside analysis and design techniques as part visual modeling tools. In fact, the two classes of tool overlapped in some areas with the workflow tools, usually including limited diagramming capabilities to help specify how the document moves from one employee to the next. At the same time, business process modeling tools sometimes offered limited process execution facilities, an extreme example of which would be a business process simulation tool.

It is also interesting to note that the business process modeling tools grew out of the business process re-engineering movement of the early 1990s (Hammer and Champy, 1993) which I shall come back to shortly. These tools typically include some means of analyzing a process into sub-processes (or activities) to varying levels, and a diagram of some kind to model the flow of events and/or data between the activities that together constitute a process. These diagrams, effectively a kind of workflow diagram, are sometimes known as "swim-lane diagrams," as the roles that are responsible for carrying out the activities are depicted as bands that stretch either horizontally across or vertically

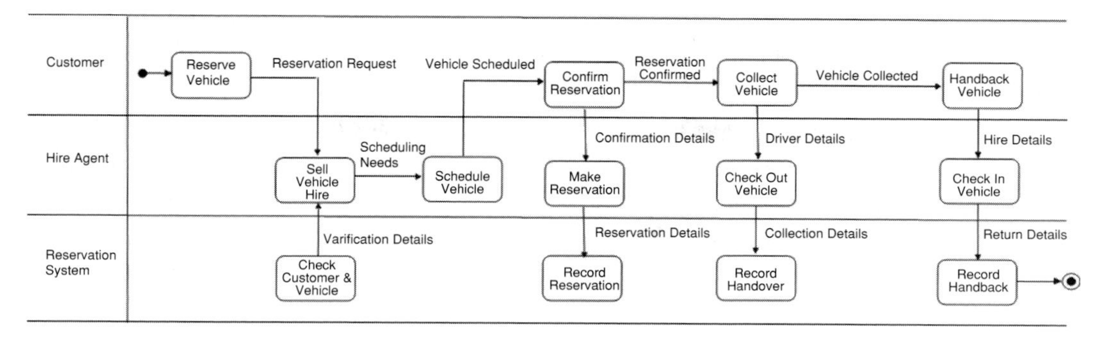

Figure 3.1 Example of a swim-lane diagram

through the diagram. Each activity is positioned in the swim-lane depicting the role responsible for it, as illustrated in figure 3.1.

Other features covered by business process modeling tools include diagramming of locations, information, business goals, inhibitors, organization charts, and various forms of matrix that allow different elements to be correlated (for example, location and role). However, despite the fact that it is possible to find many different kinds of model, there is one that comes up again and again and that dominates the domain: the swim-lane diagram.

3.2.3 Bringing workflow and business process modeling together

Let us look a little more at the technology aspects of these different tools. Increasingly, workflow tools began to include business modeling capabilities. Conversely, many of the modeling products began to include workflow facilities. These newer combined tools provided a relatively simple and higher-level way of defining processes and integrating applications to perform workflow.

However, these products tended to be built on top of proprietary transport layers. The two platforms that dominated this area were IBM's MQ Series and J2EE, on the one hand, and Microsoft's MSMQ and COM+, on the other. The underlying component models for J2EE and COM+ provided great potential for exposing common software services across processes within the boundaries of the individual organization.

Of course, problems occurred in tackling consistency within organizations that have mixes of applications based on both platforms. However by exposing these applications as software services (specifically using Web services standards) it became possible to achieve a consistent approach to the automation of processes. The term "business process orchestration" is generally used to refer to the coordination of software services to perform complex processes within the confines of the organization firewall.

3.3 Elements of BPM

While business process orchestration is a major step forward, problems emerge once again when the process stretches outside of the organization firewall. Even where partners share the same platform and component model there are often niggling inconsistencies that must be resolved – for example, different versions of J2EE, each with different vendor extensions. The upshot is that these technologies essentially are stretched to support workflow beyond the traditional organization boundaries. There is a need for something more.

In addition, as we seek further and further business improvements so the need to regulate and optimize the process becomes critical. Just as effective SEM is required, so effective BPM of run-time processes (in harmony with SEM) also becomes critical. In this section, we outline the requirements for BPM.

3.3.1 Business process choreography

The term "business process choreography" stretches the idea of orchestration outside of the organization firewall. Of course, things get much more complex where processes cross organizational boundaries. A particular issue is that of ensuring consistency for those processes built around integrating autonomous and independent applications, such as those provided by your business partners. The required transactions are "loosely coupled" and do not benefit from the consistency and integrity that comes with the support for global transactions provided by tightly coupled technologies such as COM+ and J2EE. This traditional transaction support guarantees that either all the actions making up a transaction successfully complete (on commit), or that none of the actions appears to have ever taken place (on abort). This guarantee ensures that any changes made by a transaction are not visible until it commits and is normally implemented using database locks. System administrators are not normally overly happy about granting database locks to unknown application code running on some unknown machine somewhere on the Internet. This lack of trust means that some other mechanisms have to be found to ensure overall system integrity.

Compensating operations are the most common approach used today to help maintain consistency in loosely coupled systems. Compensators semantically undo actions that have already been committed, perhaps in a nested transaction. For example, the compensating action to reserving an airline seat is canceling the reservation. The main problem with compensators is that they do not guarantee isolation, so changes made by a workflow can be seen by concurrent tasks before it commits. This problem, and others, often makes compensators difficult to implement, sometimes more difficult than the action they are undoing! A key requirement of BPM, as opposed to workflow, is that it must effectively address these kinds of cross-organizational choreography problems in the manner of loosely coupled systems.

3.3.2 Enabling standards

Immature standards have hampered our ability to deal with the kinds of issues discussed above. However, standards for choreography of Web services across multiple partners in a process are evolving rapidly, with some signs of convergence in what has been a somewhat fragmented area of activity.

For example, in May 2004 the Business Process Management Initiative (BPMI.org) announced the release of the Business Process Modeling Notation (BPMN 1.0) providing a graphical, royalty-free notation designed for both business process design and business process implementation. Earlier, in August 2002, OASIS published the Business Process Execution Language for Web Services (BPEL4WS) specification. This work is continuing with the aim of specifying the common concepts for a business process execution language. At the time of writing (July 2005) a new Web Services Business Process Execution Language (WSBPEL) is being developed at OASIS to continue this work. The intention is for WSBPEL to complement BPMN.

In parallel however, since January 2003 the Worldwide Web Consortium (W3C) have been developing a Web Services Choreography Description language (WS-CDL), which appears to be a higher-level language than BPEL4WS. These various developments can be tracked using the Web references supplied at the back of this book.

3.3.3 The BPM solution

BPM is focused on describing a process and then executing the process by sending tasks to employee terminals and by calling software services. The employees and the software applications are not part of the BPM capability, although they are necessary to create a functional BPM system.

There are three core elements of a BPM solution as shown in figure 3.2

- Business modeling: A component that includes *techniques* ranging from swim-lane diagrams to service dependency diagrams that are linked into the SOA; we describe the requirements in detail in chapters 4 and 5.
- Process engine: A component that generates executable code from the business process model. This code *choreographs* the Web services that it invokes from the service realization layer.
- Service realization layer: A component that offers sets of Web services, and that provides mechanisms to realize these services as different types of run-time software units (for example, legacy software systems or modules). The sets of Web services are offered in the form of *service buses* (described in more detail in chapter 9).

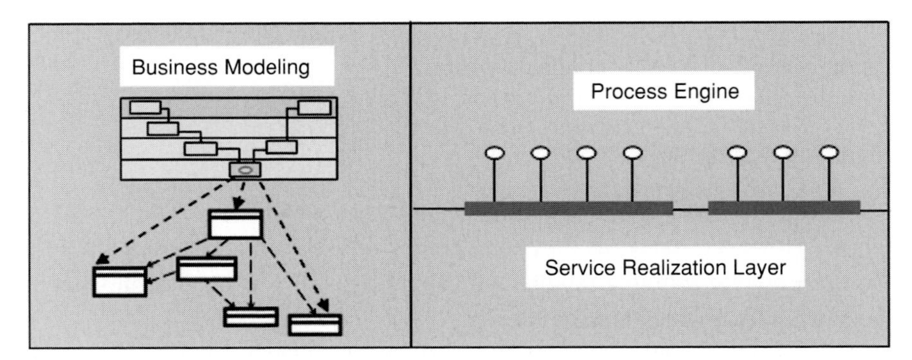

Figure 3.2 Core elements of BPM

A business analyst should be able to change the business process model and automatically change the way in which employee activities or software applications are called or activated.

Three further components complete the BPM capability, as shown in figure 3.3.

- Business activity manager (BAM): A component that assesses and controls the performance of the process engine in real time. The BAM component should be enabled with connections to SEM in order to automatically monitor and control QoS attributes, BLAs, and event patterns. It must also connect with the user interface component to allow user intervention in controlling BLAs.
- Business rules manager (BRM): A component that allows users to define BLAs (business rules and goals) and event patterns, via the user interface component. Business rules and events are a key part of the SOA discussed in chapter 9. It is therefore necessary for the BRM to maintain BLAs and event patterns as part of the asset inventory that is used to store all SOA artefacts.
- User interface component: A component that enables users to interact with the process engine in their daily work ideally via a customized portal. This component should provide users with compelling real-time visualization for business rules, events, alarms, alerts, statuses, and threshold definitions for quality of service and BLAs.

3.3.4 BPM as multi-faceted

BPM, in the true sense of the term, represents a coming together of many of the threads that we have briefly explored so far. Let us look at three examples.

The first point to note is a human one. As I have mentioned already, throughout the 1990s most large companies embraced the use of document workflow systems, and the use of ERP and EAI systems for at least some processes, and many used business process modeling tools to explore and document their processes. Overall, however, the workflow and business process modeling efforts were often divorced: the former tended

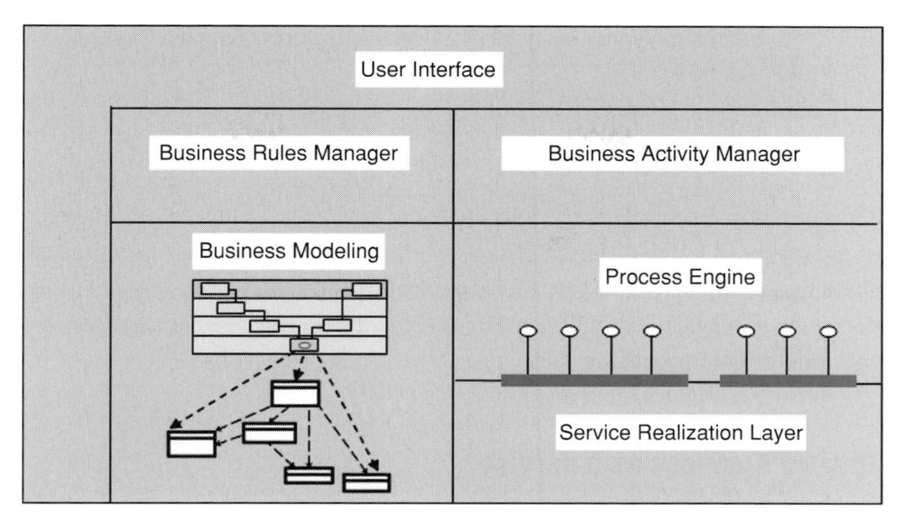

Figure 3.3 Business process management components

to be dominated by software developers, the latter by business managers. Perhaps the most telling feature of BPM is that it requires some serious coming together of these often disparate groups.

The second point concerns techniques. A much more innovative, diverse, and business-focused set of techniques for modeling is certainly needed and we shall come on to that point in chapter 4. Right now, we just want to draw attention to the increased integration of techniques that is required across tactical and strategic levels. Although the business process modeling tools were and still are usefully employed to document the human aspects of workflows, the techniques employed by the business process modelers often tend to be more sweeping, visionary, and strategic. In contrast, the use of document workflow and ERP systems represent a much more tactical step-at-a-time approach to process redesign. Of course – as we alluded to earlier – in some cases the introduction of an ERP solution caused the process itself to change to accommodate the constraints of the software. So it could be argued that some workflow tools caused wholesale process change. The problem in these cases, however, was that there simply was no true business process analysis or design!

Another way of looking at this is to consider that business process modeling tools lend themselves to abstraction from existing constraints and promote the "ideal model" view. In contrast, the workflow and ERP approaches focus on automating existing processes, and replacing existing, departmentally-focused, legacy systems with new software modules that are designed to work together. These systems are (with the exception of the wholesale ERP solution) narrowly focused. All rely heavily on IT people to put them in place. Effective BPM requires that we can model processes strategically but in a way that allows us to "drill down" and quickly provide executable support for these processes and relate the results back to the strategic goals (Kiepuszewski, Paluskiewicz, and Stokalski, 2004).

The third point concerns the confluence of capabilities of tools for BPM. As previously mentioned, vendors have entered the BPM market from a variety of different market segments – such as workflow, ERP, EAI, visual modeling, and now process choreography. Each offers specific experience with the management of different types of processes and with different dimensions of these processes. BPM represents a major challenge for these vendors to offer support for all process types. For example, vendors that focus on EAI today will need to develop capabilities in managing more complex and longer-lived processes.

3.4 Toward business as a service

It should be becoming clear by now that BPM encompasses both the modeling of services (as well as processes) and the choreography of software services. Looking at figure 3.4,[1] we can see that the business modeling component of BPM is in fact just one part of the much broader business architecture, a subject that we shall focus on in part 2 of this book. The broad principle that we established in chapter 1, and expanded in chapter 2, is that business architecture, SOA, and SEM must be approached in a unified way. Our approach to BPM reflects this principle.

However, while the right architectural and modeling concepts are essential, they are all for naught unless the difference between services and processes is fully understood and reflected in those architectures and models. In this section, we therefore focus on the differences between processes and services, and upon the overall gear shifts that are required in business modeling techniques in order to cope with these differences.

3.4.1 Business process improvement

The major focus of business process re-engineering (BPR) in the early 1990s was on removal of latency and duplication. Whole departments that did little more than shunt data from one department to another were cut away as organizations strove to become more efficient. Layers of middle management were removed as part of the downsizing and empowerment movements of the 1990s. However, it is well known that many of these sweeping and ambitious initiatives failed for one reason or another. Sometimes, the political will was just not there as departmental managers waged turf wars in protection of their own selfish interests. In other cases, the plans were simply on too grand a scale. More often than not the supporting technologies – the early workflow and ERP

[1] The abstract swim-lane diagram is shown in relation to a service dependency diagram, with a service-enabled process invoking three different services. This is just one illustration of how services can be handled in business architecture.

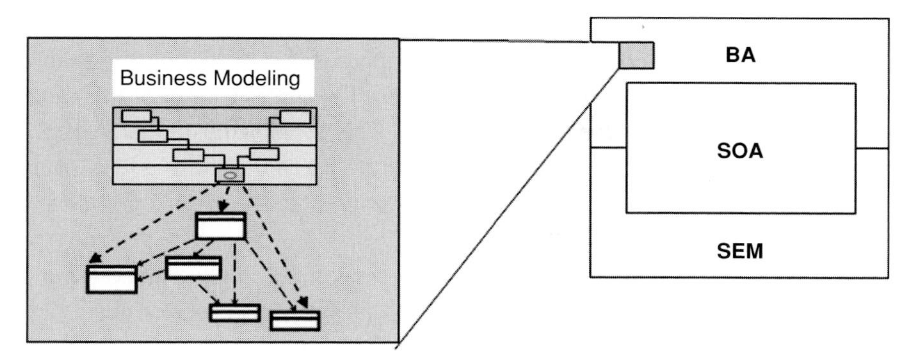

Figure 3.4 Business modeling in context

systems – were simply not up to the job. The upshot was that most companies settled for much less ambitious initiatives, under the title of business process improvement that sought a much more realistic step-by-step incremental route to better processes.

Nowadays, the accent of process redesign in most companies is on small-scale changes that lead to better alignment of processes with business strategy and increased operational efficiency. These redesign efforts are easier and quicker to implement, more acceptable to employees, and generally less costly to complete than traditional BPR initiatives. As companies wrestle with constant change in their business and technical environments, a continuous and incremental approach to business process improvement poses the least risk.

In parallel with these efforts, a further motivation for business process improvement is to identify and specify processes that are badly defined, as part of the call for corporate governance that was borne out of the corporate financial scandals of the early years of the new millennium.

At the same time, the very nature of BPR and improvement is changing as a result of changes in commercial practice toward service orientation, in which partners work together in flexible supply chains that are not constrained by the traditional organizational barriers that delimited earlier re-engineering efforts. Collaboration, choreography, and agility are the metaphors upon which BPM technology is founded.

It is important to note that a good part of the required strategy for change concerns managerial or organizational changes that lie beyond the scope of our current discussion. What is interesting from our current viewpoint is that part of the approach that impacts the SOA.

3.4.2 The manufacturing paradigm

There is a wealth of literature on business process analysis and design that contains much excellent advice that has built up over the years, from Porter's value chain analysis

(Porter, 1980) to the process diagrams of Rummler and Brache (1995). The dominant modeling technique common to all these approaches centers on the manufacturing paradigm of an organization as a product line. The original form of this model was to transform raw material into finished goods. This manufacturing paradigm is directly applied to service industries that transform raw data into "finished" information and knowledge. There are several interesting things to note about this dominant view of an organization:

- Processes are decomposed into *sub-processes* with an emphasis on different levels of granularity, but very much from the point of view of a particular chain of processes and on the flow of work. Despite the best of intentions, commonality of process sometimes tends to be relegated to a secondary role.
- The processes are essentially conceived as belonging to a *single organization*, with little thought for delegation of work to different organizations, according to core competencies, along the production line.
- The emphasis is on *internal process efficiency*, with little thought about how the organization collaborates in market context. In particular, opportunities to sell processes (or parts of processes) to partners are not leveraged.
- The *initiating inputs and resultant outputs* are essentially given. There is an emphasis on how best to transform the material or data, not on what the inputs and outputs should be in the first place, and how these might change.
- The processes are *repeatable*. Repeatable processes are essentially production systems in which the inputs can be contained within specified boundaries. Repeatable processes are input-driven and they repeatedly produce the same outputs, if, and only if, the inputs remain within tolerances. Little thought is given to creativity and change, to collaboration and exploration.

In immediate contrast and direct correspondence, five of the most important features of service orientation are:

- Commonality of process
- Delegation of work outside of the organization
- Collaboration with partners
- Agility to meet changing business requirements
- Unpredictability of process.

3.4.3 Introducing service-oriented viewpoints

The two trends that we discussed earlier in this chapter – the steady evolution of the technologies and the business move toward service orientation – are complementary. Each fuels the other. However, there is no free lunch. Most large companies have huge swathes of legacy systems that though vital to the running of the company do not lend themselves easily to service orientation. Effective execution strategies are needed to cut a path out of the legacy jungle.

However, the biggest challenge is much less a technical one and much more a cultural and methodological one: it is making the shift from the *manufacturing paradigm* to a *service perspective*. It is useful to consider this perspective in terms of seven service-oriented viewpoints ("SOV7" for short):

- *Transparence*: The smoothness of the customer's experience in using the service; includes consistency of information.
- *Customer fit*: The ability to use core competencies to provide customers with excellent products and experiences; an important aspect of this is to be able to tailor offerings to variations in customer needs.
- *Partner connectivity*:
  - The ability to use third parties to perform commodity services; examples range from ubiquitous package courier services to sales contact management software services through salesforce.com.
  - The ability to offer a service to different partners to streamline a business process, improve business relationships, or to generate revenue; for example, in the last case a large bank offering its accounting services to smaller banks for a charge.
- *Adaptation*: The ability to gracefully adapt the process to changes in the marketplace (Haeckel, 1999).
- *Multi-channel capability*:
  - The ability to support the customer end-to-end though the process, using different channels to achieve continuity; for example, buying an air ticket over the Internet and checking in through a kiosk.
  - The ability to offer the same service through different channels; for example, buying an air ticket over the Internet or buying an air ticket over the counter.
- *Optimization*: The ability to offer services in real time at high performance levels.
- *One-stop experience*: The ability to cater to different needs of the customer through one set of services, typically offered through one channel at one time (often via a portal); for example, Amazon.com offers a wide variety of third-party services, in addition to its original book purchasing, through its web site.

We will examine these viewpoints in detail in chapter 5. For now, we simply note that business process improvement involves new service-oriented viewpoints that stretch way beyond the bounds of traditional production line thinking. Even manufacturing industries, those that are traditionally conceived of in terms of product lines, are changing. Drug companies, which once pored over designing compounds, now generate millions of compounds and then test them using ultra-sophisticated, ultra-speedy mass spectrometers. Toyota employs set-based design in its automobile design process, maintaining multiple design options on components until late in the development process. Boeing designed the 777 in software (using sophisticated simulation programs) before building physical components. From materials research to drugs to airplanes, companies are relentlessly driving the *cost of change* out of their new product development processes. These service-oriented style processes increase experimentation and diversity of paths to foster innovation in the form of "exploration" projects that severely challenge the traditional production-oriented process view that attempts to predict paths, and conform to detail plans.

This is not to say that processes are no longer valid of, course. However, what we are saying is that in order to realize the benefits and opportunities of service orientation it is important to understand the concept of a service and to apply it in business process analysis and design.

3.5 Where to next?

The approach to BPM outlined in this chapter must be integrated with effective tools and technologies for both SOA and SEM, as part of a holistic approach to service orientation. However, all this technology is for naught unless we address some fundamental gear shifts in how we do business modeling.

The business drivers behind service orientation are much less to do with optimizing production lines and more to do with factors such as improving the customer experience, collaborating effectively with partners, and achieving agility over different channels. In part 2 of this book we introduce techniques that are designed to help tease out these kinds of characteristics in your business models.

These techniques are designed to help achieve a better set of business models and SOA. In fact, we shall see that the business models and the SOA are very much different sides of the same coin.

Part 2 Business architecture

Most organizations are not able to meet the challenges of service orientation with a "clean slate" – they do not change overnight. In part 2, we therefore provide an approach to business architecture (BA) that is based on evolution of best practices and in tune with the need to provide useful business process improvements quickly. We develop an example that continues through the remainder of the book and that illustrates integration of the BA with both SOA and SOM. The main focus is on the following aspects of the BA:

- business process models
- service policy
- domain and service models.

Chapter 4 provides foundation concepts for understanding services in the business context. In particular, we provide a service-oriented business process redesign pattern, and discuss how sourcing and usage strategy influences our approach to redesigning processes.

Chapter 5 focuses on the business process model. We provide a simple example of tactical first steps in moving toward service orientation based on incremental business process redesign given an existing legacy portfolio. We also consider the concept of service-oriented viewpoints in more detail, and show how to apply it to achieve a better business process redesign.

Chapter 6 discusses service policy, domain models, and service models. We address the longer-term, strategic view of identifying and designing services. In particular, our focus is on exploring how services can be used in different business contexts. We wrap things up by discussing the key aspect of surveying and cataloging services as corporate assets.

4 Service-oriented process redesign

4.1 A stepwise approach

The great majority of organizations are not in a situation where they have the luxury of designing their processes from scratch. These companies must adjust and develop their existing processes to meet the challenges of service orientation described in chapter 3. For example, with a service-oriented approach, the emphasis is much less on decomposition and value chain flow, much more on collaboration and reuse of business capability.

In addition, the business models of these organizations, whether explicitly stated in the form of a business process model or whether implicit in the way the company operates, are dominated by the notion of the *end-to-end value chain*. However, it is unrealistic to expect that these business models and modes of operation can be replaced overnight. As we have repeatedly said, service orientation is much more a matter of *evolution of best practices*.

For all but a few companies, able to start out with brand new business models, the need is for *stepwise process redesign*. The need here is for delivering "quick wins" that increase return on investment in terms of increased business value or reduced business costs. At the same time, the idea is to evolve a set of services which are offered to the processes, as part of the longer-term goal of business agility. This evolutionary approach is a keynote of both our case studies: Queensland Transport (detailed in chapter 14) and Credit Suisse (detailed in chapter 15). So a balance is needed. In this chapter, we lay out the groundplan for achieving this balance in terms of a service-oriented redesign pattern and introduce the core concepts that are used in that pattern.

4.1.1 Evolving process redesign

Despite the common tendency to think in terms of production lines, there are a great many useful established business process modeling techniques already in use in many organizations. The intention is not to sweep these techniques aside but rather to augment them with guidelines for combating production line thinking and for addressing the challenges of service orientation. In particular, we need to address the seven service-oriented process viewpoints (SOV7 for short) that were introduced in 3.4.3.

It is not our intention to retread established ground in business process modeling and design, or related areas such as business metrics, already covered by such a large domain of literature. In particular, *Business Process Change: A Manager's Guide*, by Paul Harmon (Harmon, 2003) provides a seminal account that is rich in business modeling techniques and guidelines on business metrics. Harmon introduces the notion of *business process redesign patterns*. A pattern is *reusable knowledge* that describes a problem and its recommended solution. A good pattern is concise: small, but packing a heavy punch. This helps an experienced practitioner recognize a pattern, including its core concepts, so that it can be adapted to the specific requirement. Although software design patterns have received most attention, the concept applies to any problem-solving domain.

A business process redesign pattern provides an overall approach that helps to define the overall goal of the redesign, regardless of the specific methodology you are using. In this chapter, we introduce a service-oriented process redesign pattern to help address the gear shifts in business process redesign that are required for service orientation.

4.1.2 Recasting existing software resources

We have emphasized that as well as fostering the trend toward federated processes a service-oriented approach must also support low-risk migration and integration of the legacy software portfolio to achieve ease of maintenance, flexibility, and responsive solution delivery. The SOA is usefully pictured as the "meat in a sandwich" between the available software resources (including the company's legacy software portfolio) and its processes. The SOA consumes services from a range of sources, including legacy systems, and offers these services to processes.

The foundation concepts and techniques for identifying and modeling these services must not be too purist. They must be pragmatic enough to encourage identification and reuse of the existing software services. Similarly, our approach to process redesign must be pragmatic enough to cope with legacy systems.

At the same time, because existing software resources are employed in new ways and in new contexts, it is more appropriate to think in terms of *recasting* the legacy software rather than simply reusing it. It will often be necessary to tighten up or to tune the software in such a way that it can be offered as services, in much the same way that an actor can be recast in various plays by learning lines and rehearsing.

4.1.3 Business architecture

The key elements of the BA, from a service-oriented viewpoint, are shown in figure 4.1, though this is not meant as an exhaustive list; the reader is referred to Harmon (2003) for more comprehensive treatment. In this chapter, we focus on some

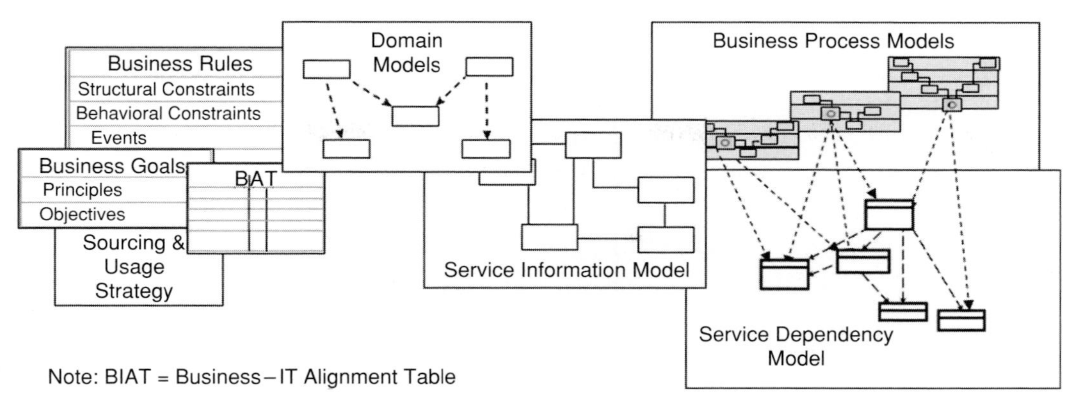

Figure 4.1 Key elements of BA

of the key concepts involved in understanding services. In particular, we provide a service-oriented business process redesign pattern and discuss the sourcing and usage strategy, which influences our approach to redesigning processes. Armed with these concepts, in chapter 5 we focus on the business process model and elaborate our techniques for service-oriented business process redesign. Then, in chapter 6, we move on to examine the remaining elements of the BA shown in figure 4.1.

4.1.4 The software service as a unifying thread

Earlier BPR and improvement efforts tended to be exclusively business- or IT-driven. Business-driven approaches often resulted in business models that were completely divorced from any real IT solution. The resulting BA became shelf-ware. Conversely the IT-driven approaches tended toward first-generation workflow. For example, as we saw previously, the wholesale adoption of off-the-shelf ERP solutions often caused organizations to forfeit the uniqueness of their processes.

An SOA transcends the traditional boundary between business and IT, as illustrated in figure 4.2. In this figure software services are depicted as offered by interfaces. Various types of software units realize the interfaces.

These terms are discussed in much more detail in chapter 9. The point to realize now is that it is *software services* that provide the bridgehead between the BA and the SOA. The concept of software service provides the new thread that can unite these disciplines. This is a thread that runs all the way to live execution as we shall see in part 4 of this book. At the same time, the SOA is very much a *software* architecture, in that it is software services that are modeled in the SOA. Software services may be offered by one or more interfaces and a service may also use other services, as illustrated in figure 4.3.

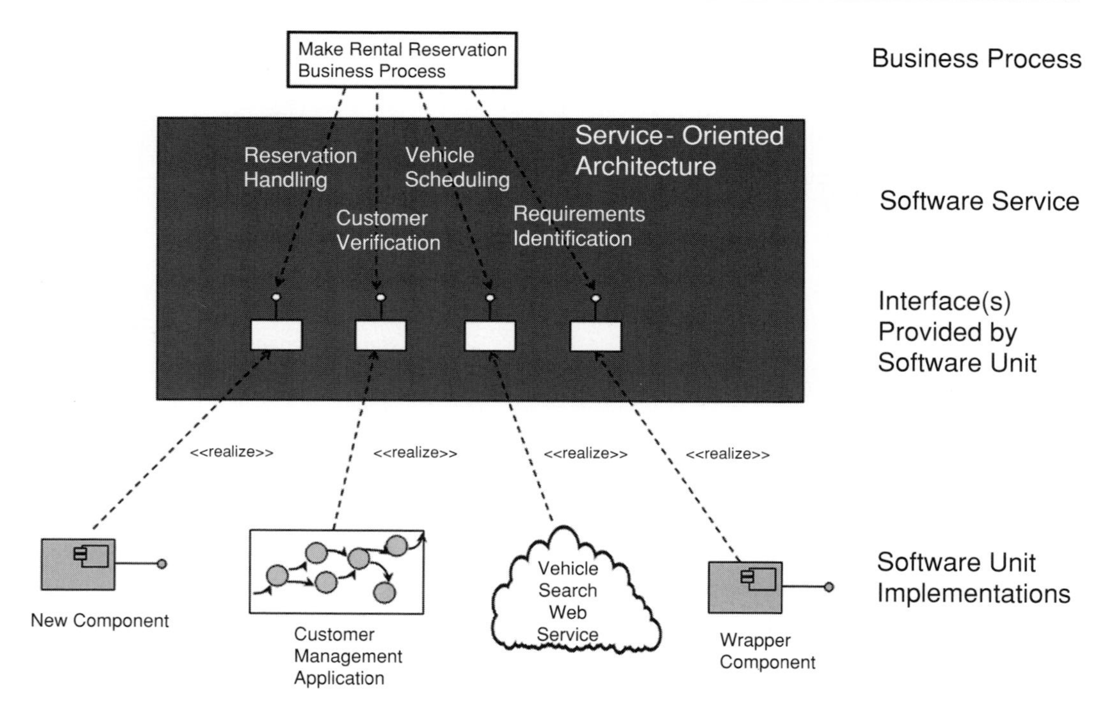

Figure 4.2 SOA in overall context

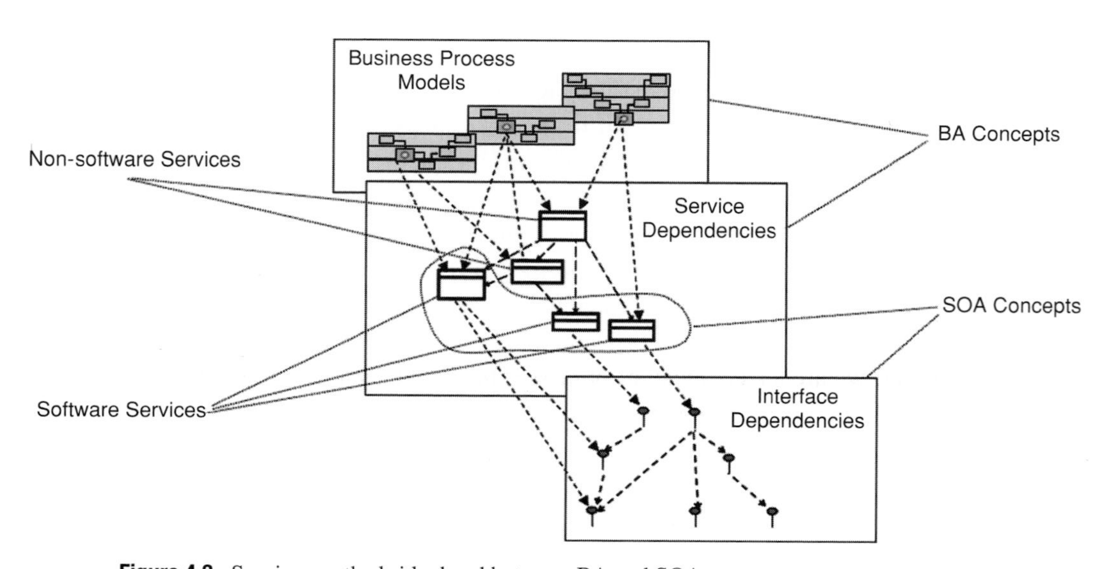

Figure 4.3 Services as the bridgehead between BA and SOA

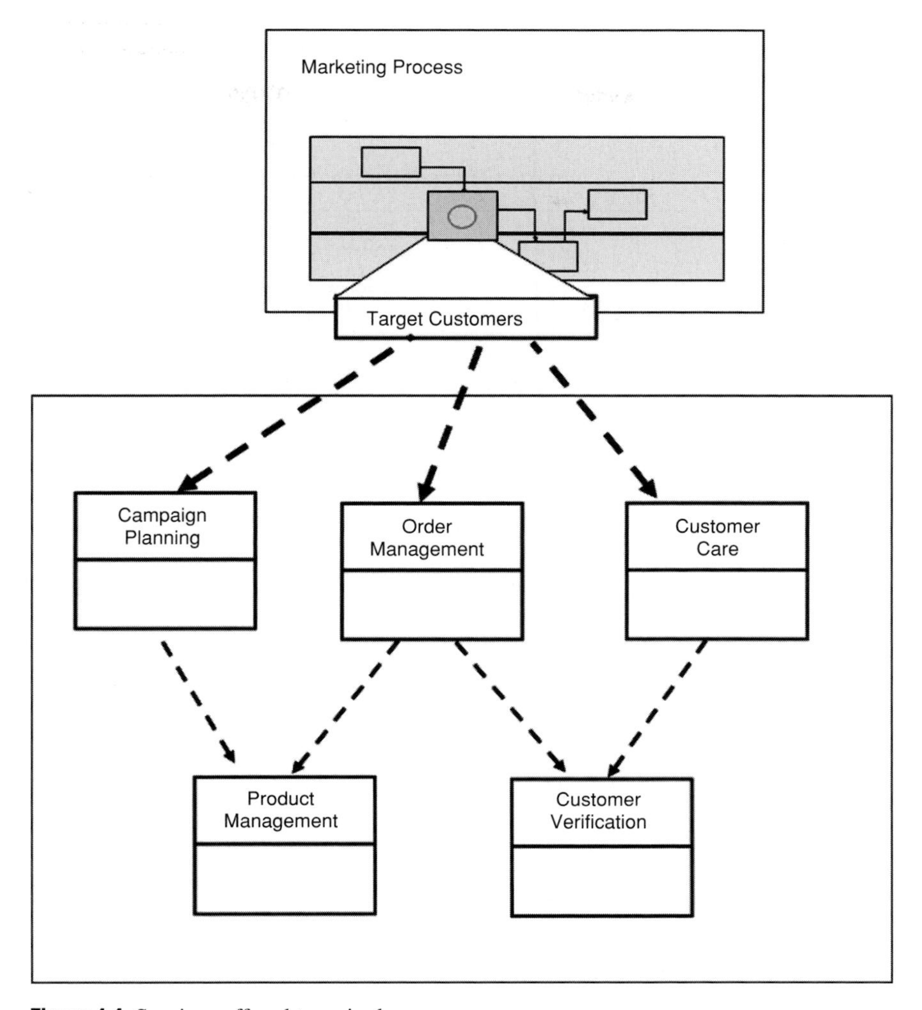

Figure 4.4 Services offered to a single process

4.1.5 Focusing on services

However, we are a step ahead of ourselves here. We cannot just assume that software units and interfaces are somehow magically "given," as is unfortunately sometimes the case. One of the key points about a service-oriented approach is to focus on the services as significant in their own right. We need to explore services in relation to *business needs*. That will greatly expedite the key task of identifying interfaces when we get on to designing the SOA (in chapter 9). Of course, it may be that we have little choice over some services, where we are reusing services supplied by existing systems. However the choice must not be assumed – part of the purpose of a BA is to help make those choices rationally. So we picture the services working to support the processes, as shown in figure 4.4. It is the software services that form the glue that

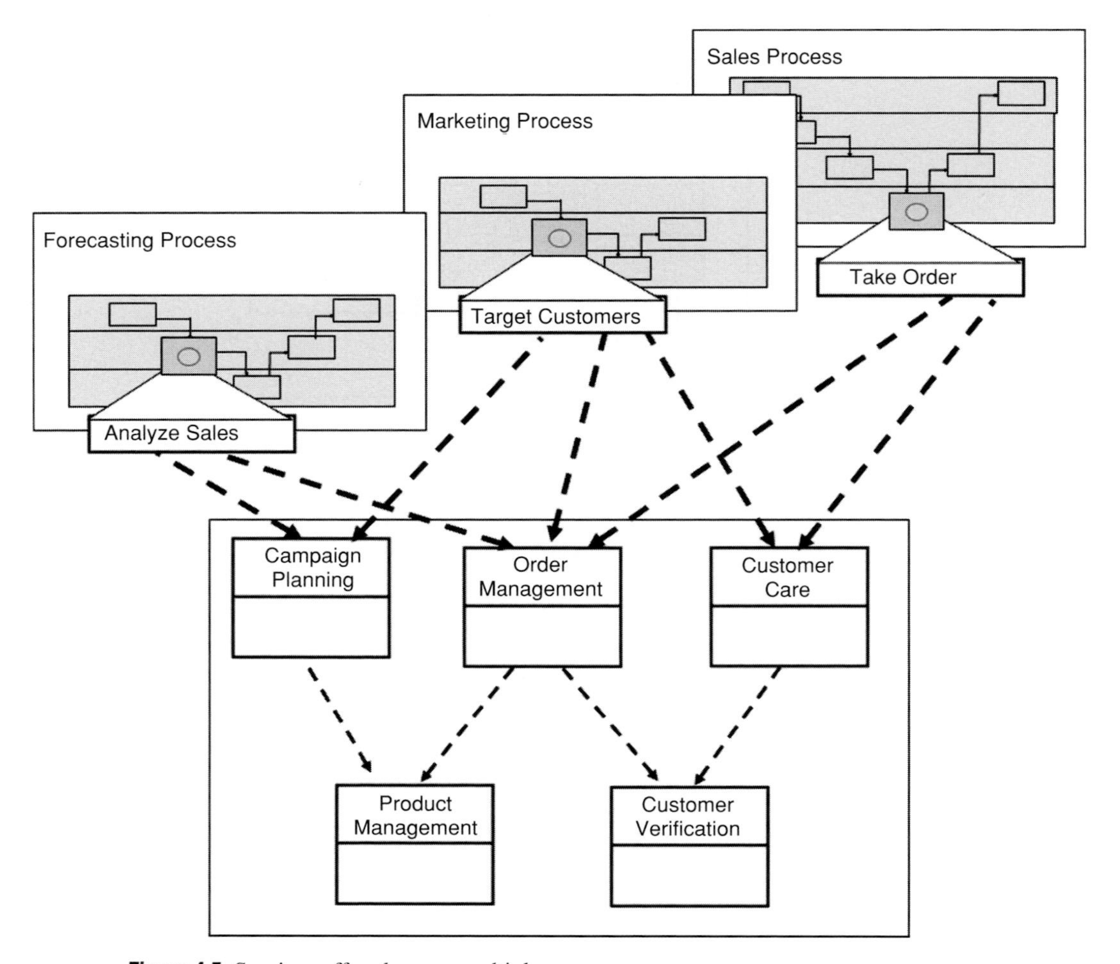

Figure 4.5 Services offered across multiple processes

links your processes with the software resources that are used to support and perform them.

Enhancing individual process designs with techniques to help achieve improved processes that are more service-oriented is a useful step forward. However, without the right techniques for concentrating on the services this approach has severe limitations. The longer-term aim is to achieve a good set of agile services that can be used across different processes, as illustrated in figure 4.5. Each service is akin to a reusable chunk of a process that can be mixed and matched with other services. The challenge of process change is met by reconfiguring well-designed services rather than having to unpick and restitch the processes together. In the following sections we introduce a service-oriented process design pattern that is designed to help evolve this set of services in parallel with individual process improvements.

4.2 Process redesign patterns

Patterns are approaches or solutions that have often worked in the past. Paul Harmon (Harmon, 2003) applies this idea to the redesign of processes. *Process redesign patterns* are determined by the overall goals for the process redesign. One or more patterns may be used, depending on these goals. In this section, we build on Harmon's work by introducing a process redesign pattern tailored to the needs of service orientation.

4.2.1 Types of process redesign patterns

The patterns are divided into two sets: *basic process redesign patterns* and *specialized process redesign patterns*. Harmon identifies four basic patterns – reengineering, simplification, value-added analysis, and gaps and disconnects – that are described in table 4.1.

Specialized patterns extend the basic patterns or solve specialized problems. This group includes management alignment, a set of software automation patterns, and patterns tailored for supply-chain design and for human performance improvement. The software automation patterns include an XML process language pattern. This pattern is particularly appropriate for introduction of BPM, as described in chapter 3. BPM is playing an increasingly important role for the implementation of service-oriented processes. This pattern will be particularly useful alongside the service-oriented process redesign pattern that we are about to describe.

4.2.2 A service-oriented process redesign pattern

Our service-oriented process redesign pattern is described in table 4.2. The pattern seeks to achieve a balance between incremental process improvements aimed at increasing business value, and the longer-term evolution of a set of services to achieve business agility. There is a major emphasis on getting more value out of what an organization already owns, on the recasting of existing assets, including legacy software systems, as services. The new pattern focuses on recasting existing assets and contracting others to do what a company has to do just to get by, and harnessing the company's own resources to do what it does best.

There is some overlap with some of the original basic patterns. For example, value-added analysis seeks to concentrate on developing a company's customer-facing processes. Simplification and gaps and disconnects both seek to streamline existing processes in different ways. Similarly the service-oriented process redesign pattern seeks to increase customer value and streamline processes.

However, there are some significant differences. The driver of the service-oriented process redesign pattern is to reduce costs of market participation, releasing the

Table 4.1 *Basic business process redesign patterns (after Harmon, 2003)*

Pattern	Driver	Approach	Time Required	Impact and Problems
Reengineering	Major reorganization desired. Major changes or new technology is to be introduced.	Start from a clean slate. Question all assumptions. Design process from ground up using best practices.	Major effort Considerable time required.	Can achieve major breakthroughs in productivity and efficiency. Potential for disruption and risk of failure proportionately high.
Simplification	Eliminate redundancies and duplicated effort in processes.	Model IS process and ask at each step: "do we really need to do this?" Focus especially on similar processes and ask if they can be combined.	Usually a mid-sized effort.	Usually results in a modest to major increase in efficiency and productivity, largely dependent on the amount of redundancy in the process.
Value-added analysis	Eliminate non-value-adding activities.	Model IS process and ask at each step: "does this activity add value or enable a value-adding activity?"	Usually a mid-sized effort.	Usually results in a modest to major increase in efficiency and productivity, largely dependent on the amount of non-value-adding work in the process.
Gaps and disconnects	Problems occur when information or materials are passed between departments or functional groups.	Model IS process and ask at each point: "when information or material pass between a department or function, what happens and what needs to happen?" Requires a process sponsor and a matrix organization.	Usually a mid-sized effort.	Usually results in a modest to major increase in efficiency and productivity, largely dependent on the amount of problems between departments. Depends on a strong process sponsor.

company's own energy to increase business value. It is the balance between reducing costs and increasing value that is the true hallmark of the new pattern. Moreover, although the business value *could* be to increase direct customer value or streamline a process, it is only by attention to the service-oriented viewpoints that the business value is determined.

Table 4.2 *The service-oriented business process redesign pattern*

Pattern	Driver	Approach	Time required	Impact and problems
Service-oriented	Reduce costs of market participation, releasing company's own energy (IT and HR) to increase business value through the use of services.	Start from a set of existing resources. Deliver measurable business value and process improvements in small chunks. At the same time evolve set of services aimed at long-term business agility.	Two levels: Short-term: process improvements. Long-term: services.	Need to manage expectations of senior management as this is not a silver bullet! There is a potential to become bogged down in academic debate. A certain pragmatism and opportunism is called for.

Finally, note that the service-oriented process redesign pattern does not religiously mandate that all functionality be offered as services. It may well be, especially in the early days, that there are actually very few services resulting from the process redesign efforts.

4.2.3 Context for service-oriented process redesign

As with any approach to business process redesign, the *business strategy* should provide a context for the effort. Similarly a pattern is applicable in certain contexts, and it must have a driver or a motivation for use.

The service sourcing and usage strategy has a significant influence on the service-oriented process redesign pattern, as indicated in figure 4.6. The driver is to "Reduce costs of market participation, releasing own energy to increase business value through the use of services." The sourcing and usage policy should emphasize these factors. If it does not, then a service-oriented approach may well be not appropriate. Remember this is not a religion!

If the sourcing policy tends toward outsourcing of processes, then the accent will be on searching for external services, whereas an insourcing strategy will tend toward reuse of existing processes or parts of processes (including legacy software systems). In regard to usage policy, if services are to be offered only to selected internal users then the quality of services may not be a big issue. At the other end of the spectrum, if services are to be offered to the general external market then the quality of services and the use of SLAs to manage expectations are very likely to be a big issue. Once we have laid out the conceptual groundwork, we return to look at sourcing and usage policy in more detail in 4.6.

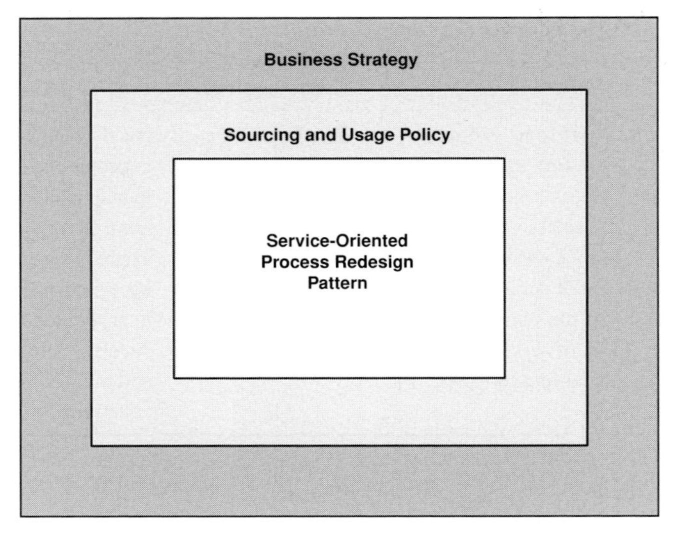

Figure 4.6 The service-oriented process redesign context

4.2.4 Overall approach

Our approach to service-oriented process redesign is iterative and incremental, and can be broadly characterized as follows:

- *Scope process redesigns*, with measurable goals, in line with SOV7
- *Redesign the processes* focusing on using existing assets or external services
- Plan and implement *incremental process improvements*.

The idea is to seek opportunities to glean business value in small chunks. Underpinning the redesign effort are three parallel ongoing activities that are aimed at achieving long-term business agility:

- Establish *service policy*[1] in relation to business goals
- Understand *domains* and identify services and their *dependencies*
- Survey and catalog *assets*.

The idea is to improve selected processes by making these processes more service-oriented and recasting existing assets wherever possible. This primarily involves examining possible service usage (with the accent on available resources, including legacy systems) by existing processes. In early iterations, it may well be that few if any services are employed in the redesigned processes. As more incremental improvements are gained so the accent may switch to achieving business agility by producing services (usually by "servicizing" existing assets; see 4.3.1).

It is important to note that what we are describing here is a *pattern* and not a *process* (or set of procedures). The pattern is as much about the concepts (described in the

[1] Service policy includes QoS levels, business rules, and sourcing and usage policy.

following sections) as the process, and is not prescriptive. In fact, there are many possible ways to construct processes in support of the pattern. What we have done is to characterize service-oriented process redesign in terms of its overall shape. While we will be using a rigorous approach to modeling of services it is not the intention in this book to prescribe a process by *diktat*. We much prefer to use illustrations and provide guidance, as most clearly exemplified in the case study chapters 14 and 15.

4.2.5 Scoping the service-oriented process redesign

Because of the problems of sheer scale of potential business process improvement in large organizations it is necessary to focus on those areas that are likely to be most fruitful. A traditional approach, common to the manufacturing paradigm, is to divide the organization along functional lines. The problem here is that a functional partitioning of effort tends be self-defeating in that it tends to lead to the replication of existing organizational inadequacies.

Another approach is to scope your business process improvement effort using a business theme that dominates company direction, such as the idea of focal points offered by Kalakota and Robinson (2003). Focal points define what your business is all about and have crisp themes that direct their process improvement efforts. They describe ten generic focal points:

1. *Easy to do business with*: Walking in the customer's shoes and eliminating hassles by integrating channels – for example, Staples' efforts with an integrated multi-channel strategy linking office supply stores, e-commerce, catalog operations, and contract stationery businesses.
2. *Customer-centric integration*: Integrating internal applications and infrastructure to allow a shift from product silos to customers.
3. *Low cost*: Offering customers "every day low price" by creating low-cost or low-inventory supply chains – for example, Dell's ability to execute a build-to-order, customer self-service strategy.
4. *Lowest overhead*: Minimizing overhead in areas such as HR, logistics, and contact centers via either consolidation or outsourcing.
5. *Zero-defect quality*: Producing premium quality goods and services with very few defects – for example, Toyota's success in aligning changing customer priorities with its production process, allowing it to take a new car from design to production in less than a year (compared to up to three years for its competitors).
6. *Productivity multiplier*: Maximizing employee productivity using technology such as self-service portals or mobile computing.
7. *Fast service*: Promising customers the fastest service by constantly speeding up the delivery and fulfillment of commodity products.

8. *Product innovation*: Enabling product innovation so that companies may leap from laggard to fast follower or market leader.
9. *Evolving business model*: Edging into new businesses or markets (particularly relevant for companies with customers that have changing needs).
10. *Real-time business*: Responding and adapting in real time to market and customer demands.

While the focal points provide a refreshing antidote to functional partitioning, the concept is rather open-ended if applied in isolation. However, as we shall see later in this chapter, they can be used to help shape our approach. Also, as described in chapter 6, if applied in combination with domain partitioning focal points can work well as means of prioritizing areas for improvement.

4.3 Identifying services

As we saw in chapter 3, *services* are distinct from *processes*. A service offers a coherent family of functions that work toward a defined business purpose, and that are described in terms of the contracts that obtain between customers and providers of the service. In particular, a software service offers a coherent family of operations. A software service is therefore akin to the concept of the *interface* in component-based design.

In this section, we offer some general guidelines for identifying services, and outline some criteria for achieving a good level of granularity of services.

4.3.1 Guidelines

Whereas a process is something a company does, a service is something that a company offers or requires. The focus is on searching for the *business capabilities* offered by or required by existing processes. There is an onus on identifying possible parts of processes that might be recast and tightened up so that they can be offered as services; we call this "servicization." A useful technique is to apply SOV7 to existing processes and ask how value can be increased in this way.

In balance with this "process view," services are also identified by considering the overall business domain and SOA policy, including business goals and rules.

At the same time, services are identified from an analysis of existing resources, including those provided by partners and third-party service providers, as well as existing IT systems and software packages. A major goal of a service-oriented approach is to harvest existing assets that can provide value, in reasonably quick time and at low cost, in the manner of "low-hanging fruit." In these cases, the emphasis will be on taking suitable assets and servicizing them.

In a nutshell, the process of identifying services is one of balancing what the business would like to do with what it is currently capable of doing. This is a *holistic process*, in contrast to the traditional approach of creating elaborate "as is" and "to be" models and mapping one to the other. The balanced holistic approach works in an evolutionary and incremental fashion.

Further guidelines on what makes a good service depend on the type of service as described in 4.4.

4.3.2 Business process granularity

Before considering service granularity it is useful to consider the concept of *process granularity*, as this is likely to be a consideration for many organizations that have followed previous-generation approaches. In other words, "To what level of detail do we model processes?" The term "activity" is commonly used to refer to the lowest level of process that is modeled. "Processes can be subdivided into smaller and smaller units or subprocesses. An activity is the smallest subprocess that a given business process team decides to illustrate on their process diagrams" (Harmon, 2003, p. 457). Reversing this, we might say that a process is made up of one or more activities. However, as Harmon points out, an activity could vary from a single step (placing an arm on a chair) to a whole series of steps (assembling a chair). Further guidance is needed. The concept of *elementary process* can be useful here.

> An *elementary process* is a process that is triggered by a single *business event* and that does not require further events to occur in order to complete an execution. The work of an elementary process is performed by one organization unit in one location, continuously until it is done.

The underlying idea here is to decompose each process into its constituent elementary processes, and examine the value chains of these elementary processes, to achieve an optimum management of the process.

4.3.3 Service granularity

Similarly with a service-oriented approach, it is necessary to use a consistent and optimum level of granularity for services. Vagueness results from too high a level of service granularity. At the same time, too low a level of granularity and it is difficult to see the wood for the trees. Services are defined in terms of their sets of functions – or, in the case of a software service, in terms of their sets of operations. It is possible to group services into higher-level services that consist of lower-level services. What we are primarily interested in is the "lowest level" of service. Here are some guidelines for achieving an optimum level of granularity for a lowest level service:

- It should be possible to describe the service in terms of functions, information,[2] goals, and rules; but *not* in terms of groups of other services. At the same time, the service may be dependent upon other services to help achieve its goals.
- The function set of a service should operate as a family unit that offers *business capability*; for example, an inventory operations service might consist of functions that maintain inventory, determine inventory status (for example, quantity on hand of an item), or handle special situations (such as back orders). An inventory information service might provide various levels of management information (such as finding expected dates of arrival for certain types of part).
- A *single role* should take responsibility for the service. However, the actual execution of the functions of the service may be distributed across several roles, which may be human or automated (operations) or both.
- The service should be as *self-contained* as possible. Ideally it should be autonomous. However, as noted above, it may have dependencies on other services.

The above guidelines are offered as ideals to aim at. However trade offs must inevitably be made in different organizational situations. The design of services must consider the degree of agility required, as well as the different contexts in which the service must be offered; see chapter 9. In particular, we shall need to consider the various QoS requirements and balance these against each other and against requirements for agility of service; see chapter 10.

4.4 Types of service

There are many ways in which services (and processes) may be categorized. In this section, we discuss three key types of service – commodity, value-add and territory – that are particularly relevant to service-oriented process redesign. We also consider some other characteristics of services that are particularly useful in assisting in the identification and organization of services.

4.4.1 Three key types of service

It is useful to distinguish three key types of service:

- *Commodity services* are stable treadmill functionality that simply represents the cost of market participation; examples are payroll and insurance claims. Commodity services are ubiquitous and sufficiently established to be used at low risk of failure or poor performance. These services should be provided and used with maximum reuse in mind. They are often outsourced or used on a subscription basis.

[2] This includes both information for which the service is responsible for maintaining and information that the service needs to access to perform its work.

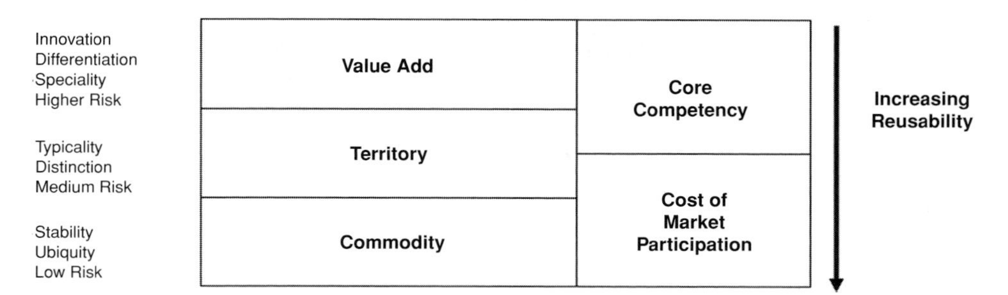

Figure 4.7 Characteristics of service types

- *Value-add services* are functionality that reflects the special value that an organization brings to market. They differentiate the company from its rivals, conferring competitive advantage and are often highly innovative; examples are product marketing and new channel sales. It is important to note that value-added functionality is often much better not designed as a service, certainly in its first iterations. This is because the innovation that comes with such functionality is best exploited quickly, without incurring the overheads inherent in servicization. There is a greater risk involved that brings greater potential rewards (DeMarco and Lister, 2003). Later, however, it may be appropriate to servicize the functionality especially if offering to a wide audience or where QoS requirements are important. Servicizing the value-add functionality can be seen as a way of managing the risk associated with it.
- *Territory services* are functionality that represents an essential characteristic of the business. They "come with the territory" and are central to business success. Territory services are likely to be volatile because they are subject to a high degree of change. So there is a risk that therefore requires careful management; examples are retail pricing and insurance product rules. These services should be as self-contained as possible, with minimal dependencies on other services, but at the same time often as reusable as possible.

The above discussion is summarized in figure 4.7.

Remember that not all functionality needs to be offered as services. In fact, at the outset of the process redesign efforts, there are likely to be few if any services. Certain processes or sub-processes will typically evolve into services. The above classification of services is also usefully applied to processes and sub-processes in terms of commodity processes, territory processes, and value-add processes.

The idea of distinguishing service types is comparable to previous techniques that center on classifying processes and IT systems. These kinds of approach can be usefully employed to facilitate the above technique. For example, a technique proposed some years ago, the "McFarlan Grid" (Cash, McFarlan, and McKenney, 1992), provides a course view for assessing a company's use of IT in terms of four quadrants. This technique is discussed in chapter 14 as part of the Queensland Transport case study.

4.4.2 Relativity of types of service

The three types of services are relative to the organization. For example, I quoted insurance claims as a commodity service. This is true of my own automobile insurer who outsources claims to a specialist claims settlement company. However, in the case of the claims settlement company insurance claims is probably a territory service, or a value-add service if some unique feature is offered. This is largely because that company will be competing with other claims companies to offer the best service.

4.4.3 Identifying different types of service

Value-add and commodity services are often fairly self-evident from a consideration of the company's business strategy. In the case of a vehicle sales company, for example, promoting the brand and negotiating a sales deal may be services that the company wants to resource and nurture itself. They are value-add services that leverage the organization's specialized knowledge and capability to innovate.

In contrast, administering a warranty is a commodity service that represents only the cost of market participation and might be outsourced. Commodity services are usually common to several processes and highly reusable. Software services from third parties provide further possibilities for commodity services. To extend the example, it may be that ordering of parts is achieved via a Web service offered by a parts distributor and perhaps a Web service enabled package is purchased to provide human resources management.

Territory services are more difficult to assess. These services are central to the very fabric of the business. They are critical to business success, often highly reusable but at the same time volatile in that they are subject to most change. It is critical that these services are agile enough to support ongoing change without disruption. For example, the vehicle hire company might see the setting of pricing policy as a territory service because although it does not provide differentiation or direct value, it is nevertheless necessary for the organization to keep control over pricing, rather than treat it as a commodity service. This might be because prices can frequently and quickly change according to market conditions and because prices are reused in many different business contexts. The effects of a change in price can have a serious knock-on effect, and the volatility of price changes poses a key business risk.

4.4.4 Service variation

Some services are more stable than others. These more stable services tend to be associated with a fixed number of highly predictable inputs and outputs. Their business logic consists of a few well-trodden paths. Examples include currency conversion and warehouse distribution. The former converts a single defined input to a single defined output using a conversion table, depending on currency required. The latter might actually consist of a series of highly complex algorithms which operate upon different combinations of product, quantity, availability source, and destination. However the complexity is well defined and predictable, and the usage of the service not context-dependent.

In contrast, other services are variable in the sense that they provide large numbers of possible options to their users that allow the users to customize the service to suit their needs through evolving sets of rules and requirements. Examples include customizations of products such as laptops and automobiles. The usage of these variable services is highly context-dependent.

Fixed services tend to be of the commodity service type, and are often candidates for outsourcing or subscription. Variable services tend to be of the value-added service type, and are usually best performed as an in-house core competence. At the same time, these are not hard and fast rules, but guidelines to assist in service type identification and sourcing strategy.

4.4.5 Service genericity

Service genericity is very similar but subtly different to service variation. Services may also vary according to *how widely they are used*. For example, a service such as product purchasing may be highly generic as it is reusable in many different contexts – say, inventory management, product planning, sales, and so on. On the other hand, a service such as product configuration may be much more specific to use in a specific context – say, product engineering.

Generic services tend to be of the commodity service type, and are often candidates for outsourcing or subscription. Specific services tend to be of the value-added service type, and are usually best performed as a core competence.

4.5 The line of commoditization

Understanding and managing dependencies between services is a critical theme of service orientation. Service design offers techniques that assist in the management of dependencies between software services, described in part 3 of this book. However, first and foremost, service dependencies must be addressed from a business view.

A challenge that many companies face today is that because their processes are not understood and organized in a service-oriented way – using the service types described above – they find it increasingly difficult to cope with the service-oriented business landscape in which they find themselves. The company needs to ensure that its services are broadly in line with its business strategy. In this section, we therefore introduce the idea of the "line of commoditization" as a means of framing this problem in business terms, in preparation for mapping out service dependencies as described in chapter 6.

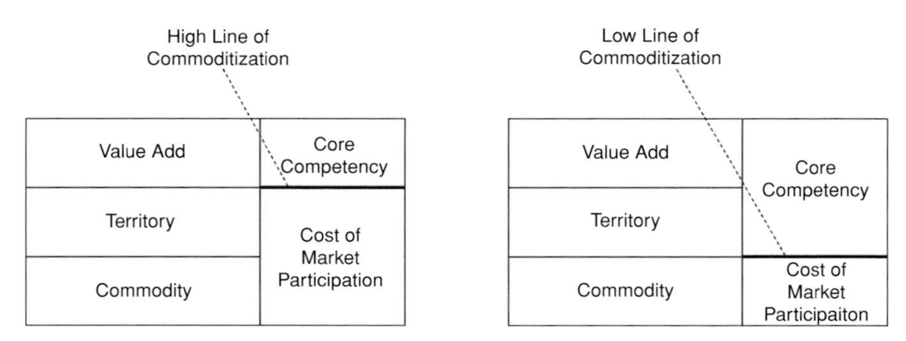

Figure 4.8 High versus low lines of commoditization

4.5.1 Organizing services

As a broad generalization, organizations that compete on very high turnaround at very tight margins might require a majority of highly efficient services that represent the cost of market participation – for example, certain food retailers or airlines. These organizations have a high "line of commoditization." At the other end of the spectrum, organizations that provide specialized knowledge or unique products might well require a majority of core competency services – for example, careers counselling or travel planning can be services that are highly tailored to customer needs. These organizations have a low line of commoditization. These differences are illustrated in figure 4.8. In order to manage its agility, an organization must understand and address its line of commoditization.

For purposes of illustration the situation is greatly simplified in figure 4.8. However, do not make the mistake of assuming that the line of commoditization is a nice and simple straight line that clearly cuts through the organization for all to see. Finding the line of commoditization is a challenge in its own right. We have already seen that the same high-level process can be supported by an even spread of all three types of service. Sometimes territory services present a challenge, in that they can be treated as a commodity to be outsourced or a core competency that the organization wants to nurture. Again, in order to fully address these issues we must form a *sourcing policy*, which we discuss briefly at the end of this chapter and in more detail in chapter 6.

4.5.2 Organizing software services

The problem that many organizations are faced with today is that because their software is not organized according to whether it is commodity, value-add, or territory they have, to some degree, lost control of their software. Worse still, because of their increased dependence on software, they have lost control of their business! These organizations lack a clearly defined line of commoditization that allows them to make the right

Table 4.3 *Focal points versus service types*

Focal Point	Value-add service type	Territory service type	Commodity service type
Easy to do business with	✓	✓	
Customer-centric		✓	
Integration			
Low cost		✓	✓
Lowest overhead			✓
Zero-defect quality	✓		
Productivity multiplier		✓	
Fast service			✓
Product innovation	✓		
Evolving business model		✓	
Real-time business	✓		

decisions on service provision. In some cases where economic pressures reach boiling point, it is all too easy to outsource whole processes: "Act in haste, repent at leisure" is a maxim that comes to mind.

The same problem – lack of awareness of the line of commoditization – also manifests itself in the converse way. IT human resources are commonly the most expensive IT resource. Clearly it is desirable to apply this resource where it provides most value and avoid wasting it on creating and maintaining commodity software services. It is not uncommon, however, to find valuable IT human resources tied up in such activities.

4.5.3 Focal points and the line of commoditization

The focal points introduced earlier can be used as a rough guide to plotting the line of commoditization. By addressing the dominant focal points (Kalakota and Robinson, 2003) it is possible to gauge the likely emphasis on service types, as illustrated in table 4.3.

The idea of focal points is akin to previous techniques used to distil the company's unique value that it can bring to market. For example, Treacy and Wiersema's (1995) value-discipline model has proven useful and is described in chapter 14 as part of the Queensland Transport case study.

4.5.4 A moving target

Thus far, for the purposes of explanation, we have assumed a static picture of the organization. We must also consider how organizations change over time to respond to changing business conditions and to developments in technology. The competitive advantage of a value-add service often lies in the opportunities to use IT innovatively in ways that competitors cannot easily follow. In time, the strategic advantage ebbs away as competitors catch up. Unless carefully monitored and developed value-add services do not endure.

A classical example is American Hospital Supply (AHS), who in 1976 introduced an innovative system called Analytic Systems Automated Purchasing (APS) that allowed hospital purchasing agents to order goods electronically via terminals at their sites. This was very attractive to the hospitals because it allowed them to reduce costs and streamline processes in a way that was simply not possible before. APS was a proprietary mainframe system that effectively shut out the competition and provided considerable competitive advantage for several years. Within a decade, however, the emergence of networking standards, PCs and packaged software began to erode the benefits of heavily proprietary systems. APS was not sufficiently nurtured to keep up with these developments. Nor was it replaced by its owners with a better alternative. APS therefore turned from what was initially an asset into a liability as customers found it uneconomical and inflexible compared with the emerging alternatives.

Without careful monitoring and good awareness of market trends it is all too easy for services, particularly value-add ones, to degenerate. Other organizations, however, have been much more successful at sustaining advantage through value-add services. American Airlines' SABRE reservation system, for example, has shown extraordinary robustness.

4.6 Sourcing and usage of services

Service orientation involves major strategic *business* decisions on the sourcing and usage of services. Sourcing concerns how services should be provided – for example, whether services should be outsourced or performed internally. Usage centers on the constituency to which services should be offered – for example, whether to offer services only internally, to trusted partners, or outside of the organization to the wider market. Both decisions are significant board-level decisions that should lie at the very heart of the company's business strategy.

The sourcing and usage policy forms a context for the application of the service-oriented process redesign pattern. At the same time, application of the pattern should help to inform the business strategy, offering help and guidance in the making of the decisions on sourcing and usage. The service-oriented process redesign pattern helps to expedite and articulate these decisions by exposing the line of commoditization.

It is important to note that we are talking *services* in this context, not software services. Decisions on whether to outsource or offer to market software services are secondary to the decisions on process sourcing and are dealt with as part of the SOA. At the same time, consideration of the current software portfolio will often impact the process sourcing strategy. For example, if there are legacy software systems that "do a job," and it is not cost-justified to replace them, then it may well be expedient to exploit those systems in support of the process sourcing strategy.

In this section, we include some overall guidelines and pointers on sourcing and usage of services. It is recognized that each company is different and will have its own unique policies on when (and *when not*) to outsource a service, and on when (and *when not*) to take a service to market. At one extreme are companies with a high line of commoditization that adopt a very aggressive attitude toward outsourcing based upon the lowest-cost focal point. At the other extreme are companies with a low line of commoditization operating in niche markets where there is a need for customer quality regardless of cost. This policy must be stated explicitly, in order to provide a sound basis for process redesign.

4.6.1 Outsourcing services

Commodity processes and their associated services are usually the best candidates for outsourcing. Outsourcing should be considered where it is likely to achieve the business goals of the process better, usually through greater efficiency and cost-effectiveness. Further criteria that support the argument for outsourcing are that the process is:

- Fixed, as opposed to variable
- Generic, as opposed to specific
- Frequently used.

The process definition should be compared with the market services offered by providers using gap analysis. The process definition acts as the contract definition to an outsourcing provider, although of course the gap analysis could reveal that outsourcing of the process is not a viable strategy!

The candidate process may comprise a number of services, some or all of which are software services. Unless the whole process is outsourced as a set of services, it will be necessary to decide the individual sourcing strategy for each of the services. Sample options might include preparing legacy systems for use as Web services, subscribing to a Web service, purchasing a software package, or outsourcing the running or development of the service to an external software provider. These decisions must be made explicit as part of the SOA (discussed in chapter 7).

By understanding a process in terms of its constituent services we can help to mitigate risk by sourcing a process in smaller lower-risk chunks, each of which should be easier to manage. Each service should, where appropriate, be clearly specified in the form of an SLA that governs the contractual relationship between the customer and the provider; guidelines on when (and *when not*), to use SLAs are given in 13.5. SLAs help to mitigate risk further by clearly stating the responsibilities of customers and providers, thus managing expectations and helping to avoid possible contractual disputes.

4.6.2 Insourcing services

"Insourcing" of a process refers to the running of a process in-house. In addition, a variant of insourcing, known as "backsourcing," refers to the situation where a process that is currently being outsourced is brought in-house.

Insourcing should be considered where it is likely to achieve the business goals of the process best, usually either through greater differentiation and customer value, or greater efficiency and cost-effectiveness. Territory and value-add processes (including sub-processes) are usually the best candidates for insourcing. Further criteria that support the argument for insourcing are that the process is:

- Variable, as opposed to fixed
- Specific, as opposed to generic.

Backsourcing should be considered in the case of poor-performing providers. The candidate process should be examined to see where the provider is falling short of expectations, using gap analysis comparing performance against the process definition. It is important to realize that gap analysis could reveal that bringing the process back in-house is not a viable strategy – for example, where suitable in-house resources are not available. In that case, it may be that "re-sourcing" to a different provider is appropriate.

As with outsourcing, splitting a process into a set of potential services can help mitigate risk through insourcing in smaller chunks and through clearly specified SLAs. It may, for example, be that an organization decides to insource part of a previously outsourced process. However, much will depend here on the constraints of the original contract between the customer and the provider.

4.6.3 Using services

Processes may be organized into sets of services that are offered to consumers. Identifying and deciding which services to offer is dependent not only on the service type, but also on the consumer context in which the services are to be offered. Again, each company will have its own attitude toward taking services to market, depending on its business strategy, attitude to risk, and other factors such as company culture. The following are simply offered as some pointers on offering services.

Services may be offered in three ways: for internal use only, to trusted partners, or to the external market as a whole.

Internal context

In one sense, all services are offered internally; it is just part of the nature of a service. After all, the whole point of servicizing a process is to ensure that it is good enough for anyone to use. However, there is a subtler angle here. Initially it is likely that any service will be used in one context only. Only once it has been tried and tested, found useful, and evolved is it likely to be reused in different internal contexts. The more contexts in which it is used, the more it will need to be well designed, documented and implemented, and offer long-term stability. Generally speaking, therefore, the order of play for offering services internally is commodity, territory, value-add.

Trusted partner context

Services offered to trusted partners are likely to be territory services that can operate within the context of cross-organizational processes, with the objective of streamlining these processes – for example, automatic ordering of parts offered by a manufacturer of vehicle components to the vehicle manufacturer. Offering services for trusted partner use is generally more challenging than for internal use. Legal implications and the consequences of changing or withdrawing a trusted partner service are a consideration, though they are likely to be covered by broader arrangements between the participating companies. The QoS requirements are also likely to be much more stringent than for internal services.

Open market context

Services offered to the external market as a whole are likely to be either territory services or value-add services. The reason for taking the service to market may be to make money from selling it. For example, a company may have developed a powerful customer verification capability that it can sell to the open market on a usage basis.[3] However, the reason does not have to be concerned with immediate revenue generation – for example, many insurers offer free quotation services with the longer-term objective of persuading customers to buy their products. In the case of value-add services, the benefits of exposure to the open market must be weighed against the loss of any competitive advantage gained by keeping the service restricted internally or to selected partners or customers. Offering services for external use is generally more challenging than for internal or partner use. In particular, legal implications and the consequences of changing or withdrawing an open market service must be very carefully considered. The QoS requirements also likely to be much more stringent than for internal or trusted partner services.

4.7 Where to next?

There is no one "magic way" to service orientation. In this chapter, we have introduced some concepts that are intended to help at the business level of process redesign.

The reader may well be asking at this point, "I thought service orientation was all about Web services – how does all this stuff relate to software?" After all, in the great majority of cases it is likely that software is employed as part of the redesigned process.

[3] The "URU" service, provided by British Telecom, provides an example.

First and foremost, service-oriented process redesign is concerned with services in the general sense of the term. Also, the ethos of our approach (as mentioned in 4.1.4) is that SOA transcends the traditional boundaries between business and IT. This approach allows processes to be modeled as simply as possible, without making dangerous assumptions about technology, and understood in a language that business people can understand.

At the same time the SOA *is* a type of software architecture that emphasizes the importance of dependencies between services. Our approach to service-oriented process redesign is aimed at ensuring a good business context for the SOA.

Incremental process redesign[4] works "top-down," identifying and using services defined in the SOA. In chapter 5 we look at how to glean business value through these incremental process redesigns.

At the same time, to achieve long-term business agility, we must work "outside-in" to identify services through consideration of business policy and domains that stretch across individual process redesigns. There also is an emphasis on using existing assets by these redesigned processes. Hence we also work "bottom-up" on surveying, cataloging, and using these assets. In chapter 6, we consider the areas of business policy, domains, and assets.[5]

[4] Incremental process improvements involve software solution design and implementation. Various approaches, including prototyping incorporating object-oriented design, may be used here. Agile development methods are most appropriate for this type of work. However, it is important to apply these approaches with an emphasis on assembly of software solutions from pre-existing software services. BPM software plays an increasingly important part in this area, so care is needed.

[5] Longer-term evolution of software services depicted in the SOA involves software service design and implementation. The emphasis here is on the SOA as a basis for specification and implementation of services. There are also important links to service management that will need to be catered for. We discuss SOA and service specification and management in part 3 of the book.

5 Gleaning business value

5.1 Gleaning early value

As mentioned earlier in chapter 2, the first moves toward service orientation are likely to involve stepwise tactical exposure of legacy systems to glean early business value. Recall the approach described in our service-oriented process redesign pattern:

- Scope process redesigns, with measurable goals, in line with SOV7
- Redesign the processes focusing on using existing assets or external services
- Plan and implement incremental process improvements.

We begin by focusing on process redesigns likely to yield "quick wins" with minimal effort, pinpointing legacy functionality that can be offered as software services to the redesigned process. Our main focus in this chapter will be on business process models, as depicted in relation to the BA in figure 5.1. In early iterations it is unlikely that much time, if any, will be spent on the three underpinning activities of the pattern, aimed at achieving long-term business agility:

- Establish service policy[1] in relation to business goals
- Understand domains and identify services and their dependencies
- Survey and catalog assets.

However as time progresses so these longer-term activities (described in chapter 6) will come increasingly into play.

In this chapter we shall provide a simple example of tactical first steps in moving toward service orientation based on incremental business process redesign given an existing legacy portfolio. We also consider the concept of SOV7 in more detail, and show how to apply it to achieve a better business process redesign.

The example, which continues into the following chapters, emphasizes that the transition to service orientation is likely to be gradual in most organizations, with legacy systems likely to play a crucial role for many years. It is also important not to restrict attention only to home-grown systems. Bear in mind that often a major component of legacy systems is software packages and outsourced IT services. In addition, it may be that commercial arrangements dictate that certain suppliers will be chosen to provide certain services for some time to come.

[1] Service policy includes quality policy, business rules, and sourcing and usage policy.

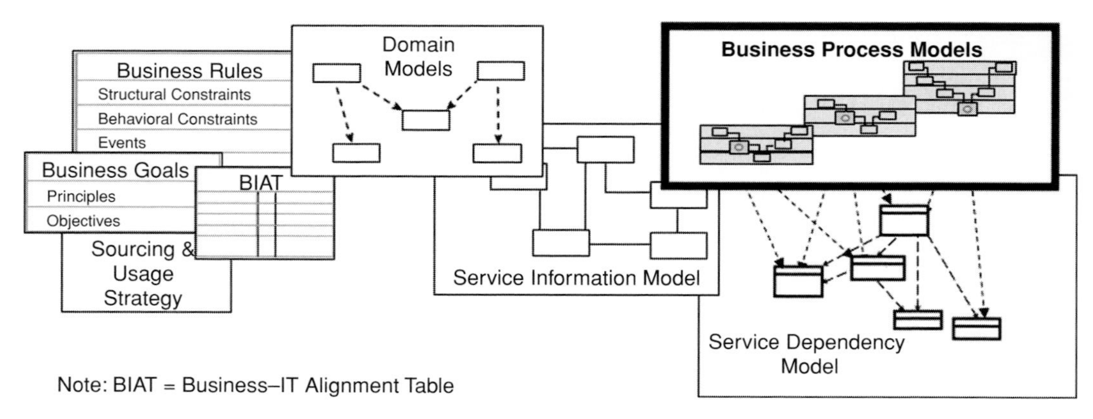

Figure 5.1 The focus of chapter 5

5.1.1 Spirit of approach

The approach described in this chapter is more about *attitude* and *results* than about detailed modeling notation. The modeling aspects of the approach must be kept simple and as familiar as possible if they are to work in practice. For example, we continue to use the swim-lane diagram to model processes even though it is evocative of the manufacturing paradigm. It is a ubiquitous notation, so we stick with it but use it in a rather different way through techniques such as SOV7.

Although a certain degree of planning is necessary the emphasis, at least in the early stages, is also on *exploration* and *iteration* as opposed to advanced planning, following the concept of emergent design (Austin and Devin, 2003) and also of agile software development.[2] The topics of emergent design and agile development are well covered in the literature and are therefore not repeated in this book. However, both are highly appropriate for the development of the software services that are required to enable the kinds of business process improvement we are aiming for. At the same time, an emergent (or agile) approach to gleaning business value should not be used as an excuse to avoid clear specification and management of services[3] (Allen, 2001a). It is also interesting to note that both of our case studies (chapters 14 and 15) bear this out.

It is also vital to appreciate that an emergent approach is dependent upon *low-cost iteration*, which in turn requires enabling technologies that reduce reconfiguration and exploration cost, and make industrial application realistic. We have already discussed BPM in some detail: the BPM software should provide flexibility of process orchestration and choreography. At the same time, the software services used by the BPM

[2] Agile software development techniques include eXtreme programming (xP) (Beck, 2000), Scrum (Schwaber, 2004), and the dynamic systems development methodology (DSDM) (Stapleton, 1997).

[3] Contrary to popular misconception agile techniques are not incompatible with a service-oriented approach.

software should be stored in a version control system so that it is relatively simple to rewind changes. As the processes start to scale up, so execution management and security software play an increasingly significant role.

5.1.2 Use of Web services

As we saw in chapter 3, BPM is playing an increasingly significant role for the implementation of service-oriented processes. Throughout this book we shall be referring to software services, rather than Web services, unless the context specifically demands it. This is in line with our approach of arming readers with an approach to service orientation that is as resilient as possible to changing technology standards.

However, it is useful at this point to just remind ourselves that Web services are the underlying enabler behind the standardization of software services and the steady evolution of the BPM technologies. As our emergent process redesigns grow more service-oriented so XML business process languages (such as BPEL4WS) become increasingly necessary to automate the choreography of our growing portfolio of Web services.

Recall that specialized process redesign patterns extend the basic patterns or solve specialized problems. These specialized patterns include an XML business process language pattern. This pattern is particularly appropriate for introduction of BPM (Harmon, 2003).

5.2 Starting the redesign effort

We begin by sketching out an example that is illustrative of the situation that many companies find themselves in today: a need to compete with agility in an increasingly challenging market but faced with a portfolio of legacy systems. There is no clean white slate. The example also recognizes that previous generations of methodology – structured, object-oriented (OO), and CBD – have often been tried with some success. There is a need to build on hard-won expertise as well as on legacy systems.

5.2.1 A case for business process improvement

As an illustration, consider a vehicle hire company. Make Reservation is an activity within the Vehicle Hire process. Currently, this process is slow because of inefficiencies in the Make Reservation activity. Customers book vehicles over the phone with a hire agent. Originally, several databases and applications had to be checked before customer details could be confirmed and vehicle requirements established.

The company has a complex set of existing systems with a long history. Several attempts have been made to integrate these systems using EAI techniques with some

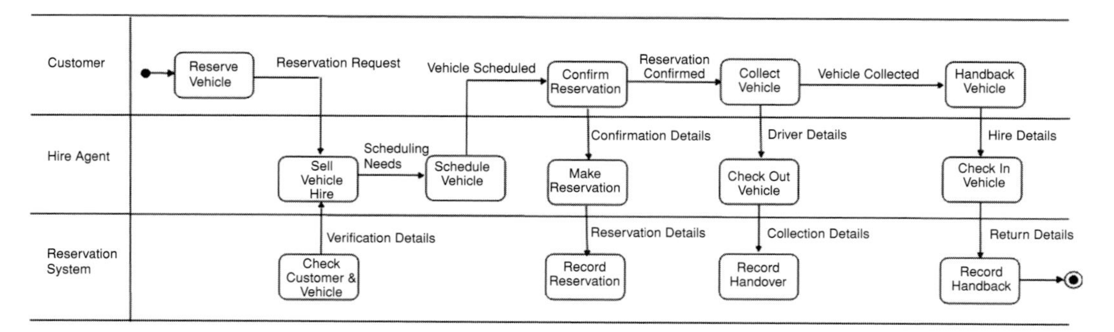

Figure 5.2 Existing vehicle hire process

success, though progress has been slow because of the proprietary nature of the middleware products used to support these initiatives. The various initiatives have been somewhat piecemeal, driven as they have been by departmental managers rather than as a coordinated enterprise plan. The company has also tried out some OO techniques and also made some early forays into CBD, with some limited success.

Despite all the technical progress the hire process is still being hampered: the integration software is slow, unreliable, and difficult to maintain. Moreover, once the customer and vehicle type have been identified the hire agent has to check manually through a wall chart in order to actually schedule a particular vehicle. Customers often hang up in frustration at the long waits. Clearly, there is a case for business process improvement. The Vehicle Hire process is modeled using a business process model to identify activities and the flow of information, as illustrated in figure 5.2; in this case, a swim-lane diagram has been used. Note the use of the Reservation system (as a separate swim-lane) to provide the required software support.

5.2.2 Scoping the process redesign

Recall the first step in our service-oriented process redesign pattern:

- *Scope process redesigns, with measurable goals, in line with SOV7*

The manual scheduling is causing delays to the Reservations part of the process. This will form our scope, which is in tune with the dominant focal points: *Easy to do business with* and *Low cost*. The Check Customer and Vehicle functionality provided by the Reservation system is in fact slow, unreliable, and difficult to maintain, causing customer frustration. Also, by the time that the reservation is finally confirmed vehicles are sometimes no longer available.

Following extensive discussions with the customer (see chapter 13 for role definitions), using participative brainstorming in line with the spirit of our approach (see 5.1.1), goals for the redesigned process are established as follows:

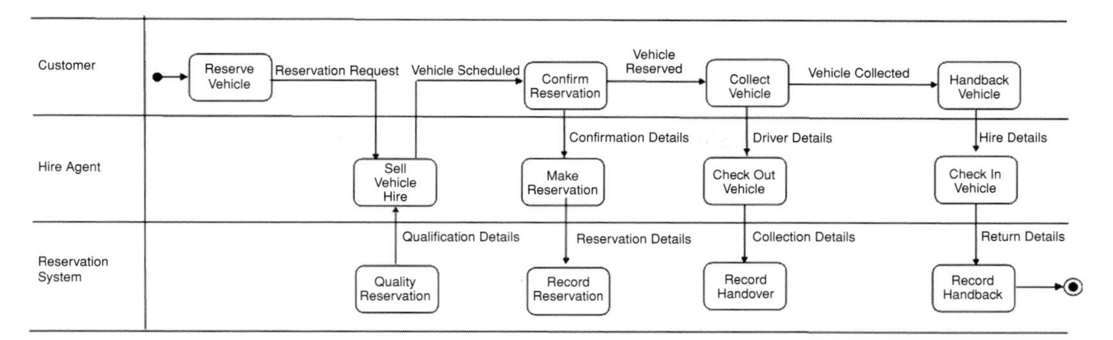

Figure 5.3 Redesigned vehicle hire process

- Average turnround time from customer enquiry to confirmed reservation reduced from 1 hour to 5 minutes
- 20% increase in number of hires
- 20% increase in productivity of hire agent.

5.2.3 The first increment

The next step in our service-oriented process redesign pattern is:

- *Redesign the processes focusing on using existing assets or external ervices*

A proposed process redesign, for the first increment, is shown in figure 5.3. Scheduling of vehicles is now to be automated and integrated with checking of customers and vehicles under a consolidated activity, Qualify Reservation. At this stage, there is nothing particularly service-oriented about the redesign. However, we check consistency with SOV7 and simply note that the first increment is largely about increasing customer transparence. Later in the chapter we will apply SOV7 much more strongly.

The new automated activity Qualify Reservation will involve an enhancement to the Reservations application. In addition to the new scheduling functionality, our approach needs to leverage the existing legacy systems while proving a realistic migration path to service orientation.

We want to avoid a piecemeal approach in which we just build a whole new generation of legacy systems. Ideally, we would like to discard inefficient and problematic systems but retain useful and valuable ones. We therefore need to survey the existing portfolio of legacy systems. Pressure for business results means that at this early stage it is not realistic to spend a great deal of time on that. As our redesign emerges so we shall be able to conduct a more complete survey and start to compile an inventory of existing assets, as described in chapter 6.

A small team is therefore tasked with a day or two's investigation of the customers and the vehicles integration software. Let us suppose that two EAI projects have resulted in ICustomers and IVehicles interfaces that provide consistent customer and vehicle information. The customer information is accessed and filtered from existing legacy applications written in COBOL. ICustomers is offered by an EJB component wrapper. The IVehicles interface is offered by a COM+ component; this might have been developed by a separate team.

5.2.4 Limitations of the first increment

Because the technologies used to implement the ICustomers and IVehicles interfaces are mixed, the Reservations application will need to resolve any implementation differences. In addition, it will need to contain business logic to secure the reservation – date, type of vehicle, rate, special features, and so on. Of course, we could write this business logic as a separate Reservations component, but that would take time and effort. The temptation to fold this business logic – as a "quick fix" – into the application is too often easy to succumb to in today's high-pressure business environment. In addition, the calls to ICustomers, IVehicles become hardwired into the Reservations application. This feature severely compromises reuse and replaceability of these functions by introducing recompilation and testing overheads every time changes are required to them. However in the short term the decision is made to go with modifications to the existing Reservations application in the interests of minimizing development time and money spent.

The upshot is that we are part of the way toward a good flexible design that helps to streamline the process. At the same time, compromises have inevitably been made under the pressure of tight deadlines and these now constrain the adaptability of our solution. And we have not even touched on the adaptability of the components themselves, interface dependency management, and the key area of SOA. However, the fact is that this situation is fairly typical of many process improvement projects today. So deal with it we must.

5.2.5 Tackling the second increment with Web services

The next step in our service-oriented process redesign pattern is:

- *Plan and implement incremental process improvements*

At this point, we have only a limited plan mainly addressed at achieving the goals previously identified. A slightly more detailed plan will emerge a little latter (see 5.4.6), once we have seen how our early process improvements have worked out in practice.

The overall goals for the redesigned process are partially met by the first increment. However these goals are compromised by problems in the Reservation application discussed in 5.2.2. In addition, constant changes to hiring rules mean that the application has to be opened up frequently, with consequent fixes and downtime. Some of these changes have to be dealt with through manual adjustment in the interim. So we fall short of the original goals:

- Average turnround time from customer enquiry to confirmed reservation reduced from 1 hour to 30 minutes – not the intended 5 minutes
- 10% increase in number of hires – not the intended 20% increase
- 5% increase in productivity of hire agent – not the 20% increase

The company recognizes that Web services standards offer the potential of solving many of the internal integration problems that are at the root of these shortfalls. Web services are therefore used to provide a technically neutral layer over the EJB and COM+ components in the example, obviating the need for application code to sort out technology mismatches. At this stage, no thought has been given as to what services are required. Basic access, create, amend, and delete operations have been wrapped over the top of existing functionality.

5.2.6 Stepwise migration to new services

In a nutshell, the strategy is to hide all access to existing systems, such as the customer and vehicle applications, so that the company can make a graceful transition to service orientation. Start out by building a thin shell of Web services that employ legacy systems to do the actual computing work. Gradually, on an opportunistic basis as solutions are designed and delivered (ideally in small increments), functionality is migrated out of the legacy systems and into new Web services. Over time, the shell grows thicker and the legacy core shrinks.

After each solution is developed, preferably as much as possible by assembly of pre-existing Web services, the asset inventory is revisited, re-evaluated and gradually built up; we discuss the asset inventory in more detail in chapter 6. If the functions of a legacy system end up being packaged together as large-grained software units that offer Web services to the solution, then keeping the legacy system intact becomes a more attractive option. On the other hand, if legacy functionality is spread across several small-grained software units that involve complex interdependencies then it becomes difficult to preserve the existing system. There is then a much stronger case for either re-engineering parts of the legacy systems or maybe abandoning altogether in favour of subscription or packaged services.

5.2.7 Challenges and limitations of this approach

In one sense, we might argue that our vehicle hire company is now in a position where a BPM system can be used to orchestrate the services created thus far and provide flow routing to and from other activities in the overall process. However, there are many challenges to be overcome. Seven examples are typical:

- It is one thing to use a few isolated operations offered as Web services to glean some early business value. It is quite another thing to apply the idea of software

services – that is, closely related sets of operations – to achieve long-term business agility. Our thinking needs to address the *analysis and design of services*.
- Efforts must be consistent with the *process sourcing strategy* (see 4.6). If company policy is strongly in favour of outsourcing services then there may be no mandate to recast existing internal systems.
- The current state of *legacy systems* in most companies is often a challenge in its own right. For example, it is not uncommon for large companies to have upward of twenty different sales order systems with their own customer databases.
- The prevalent approaches to business process redesign have centered almost exclusively on the manufacturing paradigm. Our thinking needs to broaden to cater for *SOV7*.
- Scaling up early successes demands major attention to *QoS* requirements and to the *quality of information* provided by the services.
- *Standard specifications* for Web services are still at an early stage. Scaling up to meet much tougher demands for security and transactions at the *interorganizational level* is a challenge that the Web services standards community is addressing as I write.
- Scaling up early successes to cater for *larger and larger transaction volumes*, and for more *complex and critical processes*, distributed over more and more *execution nodes* is a challenge that cannot be underestimated. As we saw in chapter 2, effective execution management of Web services is a critical enabler of such industrial applications.

At the same time, because something is difficult that does not mean that it is necessarily a good idea to ignore it. Business is changing toward service orientation for a variety of reasons. Sometimes the change may be a relatively low-key one, as in the vehicle hire example discussed above. In other cases, the effect may be much more dramatic as a company changes its business scope in order to compete or just to survive. Either way, it makes good sense to have techniques in your bag that help to bring out the changes in your business model.

In the next section we examine service-oriented viewpoints in more detail and provide examples of how these might be applied to the vehicle hire scenario. In chapter 6, we then move once again to our roadmap to consider the impact of policy, domains, and existing assets.

5.3 Service-oriented viewpoints

In this section we recap SOV7 and describe some simple questions that need to be asked of the process redesigns. We then discuss each of the viewpoints in more detail with some guidelines on the kinds of technique that are useful for each one. In particular, we focus on how each viewpoint raises issues that are not adequately covered in traditional process re-engineering and improvement approaches.

5.3.1 In a nutshell

To recap, SOV7 addresses the following:

- *Transparence*: The smoothness of the customer's experience in using the service; includes consistency of information.
- *Customer fit*: The ability to use core competencies to provide customers with excellent products and experiences; an important aspect of this is to be able to tailor offerings to variations in customer needs.
- *Partner connectivity*:
  - The ability to use third parties to perform commodity services; examples range from ubiquitous package courier services to sales contact management software services through salesforce.com.
  - The ability to offer a territory service to different partners to streamline a business process, improve business relationships, or to generate revenue; for example, in the last case a large bank offering its accounting services to smaller banks for a charge.
- *Adaptation*: The ability gracefully to adapt the process to changes in the marketplace.
- *Multi-channel capability*:
  - The ability to support the customer end-to-end though the process using different channels to achieve continuity; for example buying an air ticket over the Internet and checking in through a kiosk.
  - The ability to offer the same service through different channels; for example, buying an air ticket over the Internet or buying an air ticket over the counter.
- *Optimization*: The ability to offer services in real time at high performance levels.
- *One-stop experience*: The ability to cater for different needs of the customer through one set of services, typically offered through one channel at one time (often via a portal); for example, Amazon.com offers a wide variety of third-party services, in addition to its original book purchasing, through its web site.

These viewpoints are applied in the modeling of processes within the chosen process redesign scope.

5.3.2 Some questions to ask

Our viewpoints can be applied in various ways simply by asking questions in relation to each one and adjusting the process design depending on responses to the questions, as indicated below (table 5.1).

5.3.3 Transparence

Earlier process re-engineering and improvement efforts focused very much "inside-out," on getting the production line working like clockwork. Clearly efficiency of

Table 5.1 *Example viewpoint questions*

Viewpoint	Questions
Transparence	Is the customer experience as hassle-free as possible? Are there customer interactions that are unnecessary or wasteful of customer time?
Customer fit	Can the customer experience be enriched? Do customers have options to customize offerings according to their particular personal needs?
Partner connectivity	Have partners been considered for supply of certain parts of the process? Can money or time be saved by using these partners? Have opportunities been considered for offering parts of the process to partners?
Adaptation	Can the process easily accommodate business change (for example changes in business rules)? Can the process be easily changed to provide competitive advantage or insulation from competitor threats?
Multi-channel capability	Can the process be easily extended to cater to further channels? Can a user employ different channels at different points in the process?
Optimization	Can the process be improved by providing immediate access or updating of information? Can real time alerts help leverage business opportunities or avoid problems? Can wastage of valuable resources at quiet times be avoided? Can speed and efficiency be maintained at busy times?
One-stop experience	Does the process fit into the wider customer picture by easily linking to other services commonly associated with the process?

processes is, and always will be, a significant concern. The drivers for cost reduction do not go away.

Today, however, the demand for faster performance and cost reduction is external as well as internal. It is embodied in the maxim: "Time is money." Customers are increasingly intolerant of spending precious time listening to so-called "courtesy music" and being transferred from one agent to another when they call with an enquiry. Customers grow frustrated with interruptions in their dealings with companies. They do not want to repeatedly enter their names and addresses every time they make an order. They expect a smooth and consistent experience in which inner company workings are transparent.

Redesigning for transparence involves taking the customer's perspective and asking how the process can be made easier and quicker. Simple techniques such as walking through possible customer scenarios using storyboards are very useful. This technique also overlaps with the customer value analysis redesign pattern described in (Harmon, 2003).

5.3.4 Customer fit

Earlier process re-engineering and improvement efforts often gave little thought to enhancing the customer experience by fully engaging the customer. The result was poor "customer fit" of the process.

What we are seeing today, however, in many business areas is a move toward *economies of scope* in addition to the economies of scale of earlier times. In Henry Ford's day you could have any Ford automobile as long as it was black and no different to anyone else's. Processes that supported tuning of the production line were dominant. Henry Ford understood that better than any of his competitors, which was a major reason behind the success of Ford. Today, car manufacturers offer a huge diversity of options around which to customize the vehicle to suit the customer. The ability to be able to analyze and design processes that can cater easily for this kind of diversity is therefore increasingly critical.

Redesigning for "customer fit" involves examining the quality of the customer experience and asking how this can be enhanced by the process. The question of customer fit assumes increasing importance with the degree of *service variation* (see 4.4.4). This technique also overlaps with the idea of *variation analysis*, in which subtle variations of services usage are explored (Lycett, 2001). The technique also relies on the ability to readily distinguish between so-called "mass customization processes" (for example, personalizing the automobile) and more stable processes (for example, building the automobile infrastructure). Consideration of value-add service types, introduced in chapter 4, is used to help identify mass customisation processes. The value-add services must often be designed to be customizable.

Analysis of how QoS requirements may vary, depending on the type of customer, is also relevant to determining "customer fit." For example, a VIP customer might have 24/7 access to a service, in contrast to other customers who may be restricted to access during certain periods. Other criteria, such as service response times, may also be varied with different performance thresholds set for different types of customer.

5.3.5 Partner connectivity

Earlier process re-engineering and improvement efforts focused on internal processes. Today's processes are becoming increasingly *externalised* in linking not only customers but partners, providing services such as logistics, marketing, and customer relationship management, as well as more obvious basic commodity services such as HR and accounting. Business partners increasingly need to share knowledge and work together to create innovative products and services. Connecting all process participants, whether they are humans or applications, is fast becoming one of the key characteristics of successful processes.

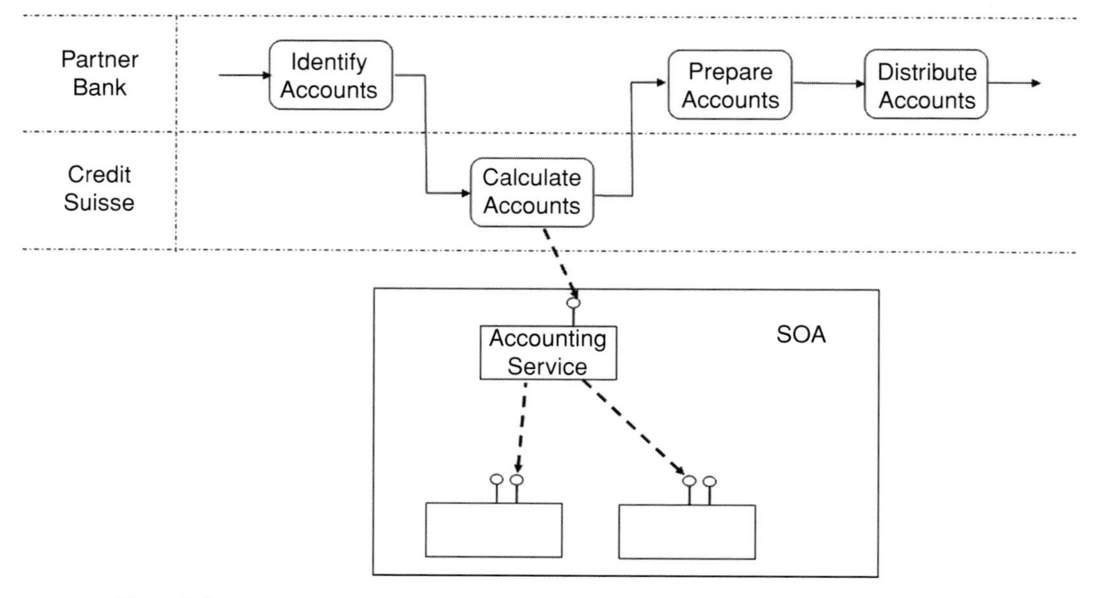

Figure 5.4 Partner connectivity at Credit Suisse

There are two main aspects of *partner connectivity*, the first of which centers on the use of third parties to perform commodity services. While this is not a new idea, as witnessed by the fact that we have used postal services to transport our mail for many years, there are ongoing developments in the areas of commodity software services – such as sales administrations services provided by salesforce.com – that it is important to stay abreast of.

The second aspect of partner connectivity concerns the ability to offer services to partners to streamline a process, enhance a business relationship, or simply make money. For example, QT offer their vehicle registration software service to motor dealers, but would ultimately like to offer it to the vehicle manufacturers. This would streamline the process and enhance the business relationship with the manufacturers. In this way, regulatory arrangements would be applied "at source ," thus opening up further possibilities such as automatic systems for in-car road-use charging.

In addition, large companies with established and proven territory software services may choose to make these available on a subscription basis to smaller partners who are unable to afford the cost of building their own software, or even buying a package. For example, Credit Suisse offer an accounting service to other banks in their business ecosystem, on a subscription basis over the Internet, as illustrated in figure 5.4.

The astute reader will note that many of the characteristics of the partner connectivity viewpoint overlap with the multi-channel viewpoint. That is fine. The viewpoints are not meant to be mutually exclusive categories that are applied in the manner of a taxonomy. They are simply offered as a means to expedite service-oriented business process redesign.

Redesigning for partner connectivity involves a range of considerations that focus on the line of commoditization introduced in chapter 4. One of the most useful is responsibility, authority, expertise, work (RAEW) analysis. This is a simple but powerful approach, more traditionally used for pinpointing process malfunctions but highly applicable for service orientation; for examples see JISC infoNet (2004).

Partner connectivity is likely to adopt a high profile where outsourcing is part of the sourcing strategy (see 4.6.1), or conversely in situations where services are to be offered to partners as part of the usage strategy (see 4.6.3).

5.3.6 Adaptation

Earlier process re-engineering and improvement efforts focused on linear production line-style processes. However, today's processes are increasingly not simply sets of static, deterministic relationships or contracts. Instead, they are characterised by organic change and adaptation.

Redesigning for adaptation mainly emphasizes the structural techniques – such as domain information modeling and policy setting – introduced in chapter 6. However the impact of changes to processes can be addressed in many ways. Change scenarios can be brainstormed – for example, we might consider how new business opportunities arise as a result of the growth of new technologies. Compelling events can be envisaged and discussed – for example, we might consider the effect of a merger or takeover, or the effects of a disaster such as a terrorist attack.

There is also a temporal aspect to the categorization of services that must not be overlooked. Roles frequently change; for example, the same entity can switch from customer to competitor to supplier to enabler depending on the business context. These changes of role reflect changes in the market and organizational strategy. Companies can now play multiple roles in relationship to each other (Coplien, 2004).

Business process frameworks are particularly useful where *adaptation* is important. These are models of processes common to specific industry sectors and that are developed by industry groups. One such industry consortium is the TeleManagement Forum (http://www.tmforum.org). eTOM defines all of the processes that are essential to a telecom company. The Supply Chain Council's Supply Chain Operations Reference (SCOR) provides another example of a business process framework that gives supply chain teams a way to quickly define a supply chain process using a common vocabulary (http://www.supply-chain.org.).

These frameworks provide an increasingly useful technique for helping to identify and evaluate a firm's territory and commodity services. Existing or future business process models can be compared and mapped to the appropriate framework, or the

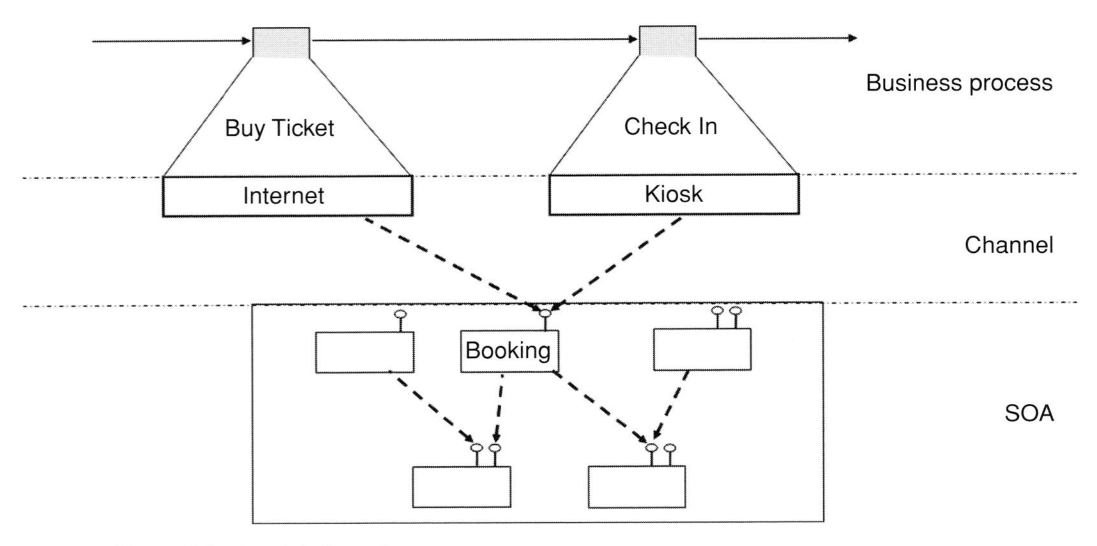

Figure 5.5 A multi-channel process

framework can be used as a starting point for developing a future business process model. Value-add services can then be layered over the top of the model.

5.3.7 Multi-channel capability

Earlier process re-engineering and improvement efforts tended to assume that the process was conducted through a single route to market. In the early days the interaction with the customer was manual, with internal staff performing tasks through their company desktops. Now, not only do customers increasingly deal with companies direct, but they are able to do this through an increasing variety of channels. Different channels can be used within a single process; the same service can be offered through multiple channels. Therefore it will be potentially rewarding to identify services that are used in different channels as in the airline ticket scenario mentioned earlier and illustrated in figure 5.5.

Redesigning for multi-channel capability involves exploring processes in the context of further channels. For example, a portal or a kiosk can be added to a swim-lane diagram (Harmon, 2003). We can also think about multi-channel capability in a slightly different way. Once candidate services have been identified they can be mapped to possible channels, as illustrated in figure 5.6, which is based on a transport provider business.

In a nutshell, on identifying a candidate service, bear in mind that it could be used in different contexts, and try to design the service so that it is independent of process and user interface technology.

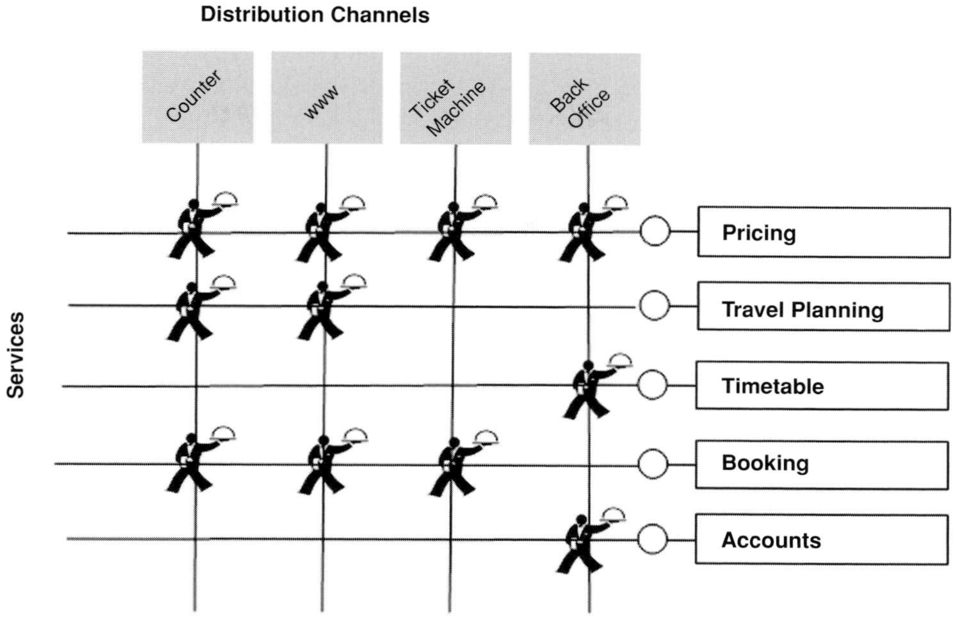

Figure 5.6 Multi-channel services

5.3.8 Optimization

Earlier business process re-engineering and improvement efforts focused on cutting costs by streamlining internal processes. The primary objective was to reduce operational costs, thereby reducing the total cost of the end product. With service orientation the need to trim costs continues unabated with a focus on optimization of services. The movement toward utility computing is evidence of the broader need for business process optimization. In this dynamic situation, the abilities to monitor, understand, and respond, based upon clearly defined QoS requirements, are critical and raise a need for effective real time process monitoring, control and query mechanisms provided by BAM software (see 3.3.3).

Business people also want the assurance that if anything goes wrong with their processes, they can quickly make changes while the process is running and not lose time by having to start over from the beginning. Other process execution facilities include work load balancing in order to iron out bottle-necks in the most efficient manner, and audit trails for legal or contingency purposes.

Yet more significantly, the ability of software to react promptly and appropriately to time critical situations extends beyond pure efficiency to changes that impact the process itself. These situations are characterized by a need to respond to events in real time. For example, a shopper has his credit card declined by a retailer because it has gone over the limit. The traditional situation would be to use letters and/or call centers over a period of several days to raise the limit where the customer indicates that he or she is ready for an increase. The event-driven situation would be that, on sensing the limit will be exceeded, the credit card system sends the customer a message to his/her cellphone with an option to raise the limit before the threatened transaction is aborted.

Techniques such as state charts, as used in real time analysis, can be useful for modeling these kinds of situation. Another technique is to look for patterns of events that show certain symptoms that require action. Business rules analysis might also be useful where an event is not an external event (like customer orders product) but conditional (like credit limit is exceeded); business rules are covered in chapter 6.

5.3.9 One-stop experience

Earlier process re-engineering and improvement efforts were based on functional lines. The advent of increasingly sophisticated portal technology continues to increase opportunities to aggregate services in "shop windows" targeted at customer needs across traditional functional boundaries. Amazon.com is the classic example. However, this is not just a portal phenomenon – it is a feature of everyday life. For example, supermarkets that once concentrated on retail domestic goods now offer pharmacies, petrol stations, and insurance.

Redesigning for the one-stop experience involves ensuring that the process is open enough to enable partner services to be "plugged in," through the shop window if you will. Context and collaboration modeling (Allen, 2000) are both useful here.

5.4 Applying service-oriented viewpoints

Once some progress has been made in delivering some value to the business through early iterations of our service-oriented process redesign pattern we can turn much more fully to SOV7. In this section, we look at how SOV7 might be applied to the vehicle hire scenario.

5.4.1 Assessing the goals

We now return to the first step in the next iteration in our service-oriented process redesign pattern:

- *Scope process redesigns, with measurable goals, in line with SOV7.*

Goals for the redesigned process are revisited and checked for consistency with the dominant focal points (*Easy to do business with* and *Low cost*). As a result the goals are restated as follows:

- 50% decrease in numbers of phone reservations
- Target of 1,600 online reservations per day
- 20% increase in number of hires
- 20% increase in productivity of hire agent
- 50% increase in customer satisfaction rating.

5.4.2 Assessing the viewpoints

So what might the next steps be for the vehicle company? To provide an answer to this question we return to our service-oriented viewpoints and brainstorm the possibilities:

- *Transparence*: Introduce reservations over the Web as well as by phone. Streamline the customer's experience by automating verification checks for risk and creditworthiness.
- *Customer fit*: Introduce special offers and options for Gold customers ranging from hotel deals to route planning. Personalize the reservation by using the history of customer's dislikes and preferences. Introduce delivery of vehicle to customer's location.
- *Partner connectivity*: Work (as a consumer) with vehicle broker to provide access to further vehicles, where company is unable to fulfill customer's hiring requirement. Work (as a supplier) with garages who need to supply their customers with temporary vehicles while their vehicles are being repaired or serviced.
- *Adaptation*: Anticipate and support possible changes:
  - Organization changes such as mergers and takeovers
  - Business process changes such as delivery of vehicles to customer's location; see also *Customer fit*
  - Information changes such as pricing structures for bulk reservations
  - Business rules changes such as changes to hiring rules
- *Multi-channel capability*:
  - Introduce reservations over the Web as well as by phone; see also *Transparence*.
  - Support reservation process using different channels consistently through the process; for example reserve over Web, enquire by cell-phone.
- *Optimization*: Offer reservations of predetermined vehicle types in real time by cell-phone,[4] and real time alerting of vehicle incidents (from faults to thefts).
- *One-stop experience*: Introduce a Web site incorporating reservations as one piece of complete travel package with links to hotels, ferries, restaurants, and leisure outlets such as theaters and cinemas.

Weighing up the goals in relation to SOV7, the scope now widens from the reservations activity to embrace the whole hire process.

[4] Note that this is a new channel, and therefore there is some overlap with the multi-channel viewpoint.

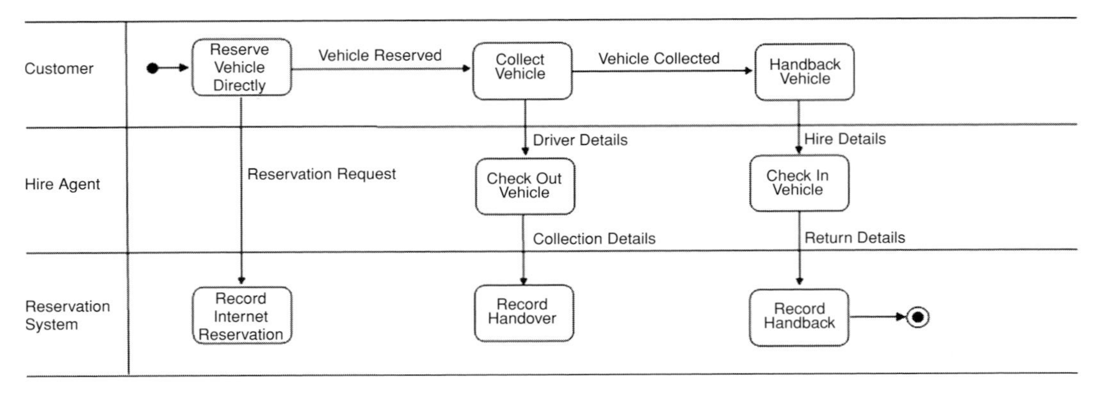

Figure 5.7 Internet vehicle hire process

5.4.3 Redesigning the process

The next step in our service-oriented process redesign pattern is:

- *Redesign the processes focusing on using existing assets or external services.*

A streamlined process based upon Internet reservations, shown in figure 5.7, is proposed to run alongside the existing phone-initiated process. Record Internet Reservation is an automated activity that will subsume all previous elements of the reservations activity, allowing the customer to be verified, select vehicle types, and secure her reservation online.

5.4.4 Looking for possible external services

We should also investigate opportunities to reuse external services. There is a significant business problem to be overcome that concerns verification of customer details. Customers become frustrated with having to give personal details over the phone and waiting for the hire agent to check out whether the customer is on a bad risks file and to then make a call to the credit supplier before they can continue with their transactions. This same situation causes further problems with respect to an Internet scenario. Although it is possible to capture the reservation online, information would need to be transmitted to the reservations assistant to make the checks offline and then get back to the customer by email.

By incorporating a customer verification Web service provided by a finance house and invoking this directly from the component implementing the Customers Web service we can make some real inroads into improving the customer's experience. If customers don't have to hang on the phone, or wait for an email, they are going to be much happier!

A feature of service orientation is the emergence and growth of companies who act as brokers between providers of services and customers of services. It's necessary for organizations to keep abreast of these developments. Suppose, for example, that a vehicle broker emerges in this business space. The next step might be to improve the process by partnering with this broker, thus extending the range of vehicles supplied and gaining more sales, without actually increasing the company's own stock of vehicles. A search for vehicles external to the company's own fleet can be introduced by invoking a Web service offered by the vehicle broker.

5.4.5 Assessing priorities

The next step in our service-oriented process redesign pattern is:

- *Plan and implement incremental process improvements.*

We use focal points (see 4.2.5), alongside our SOV7 assessment (see 5.4.2) to help select and prioritize the next features to be included. Recall that *Easy to do business with* and *Low cost* are the dominant focal points. The following order of priorities is agreed:

1. Introduce reservations over the Web as well as by phone
2. Streamline the customer's experience by automating verification checks for risk and creditworthiness
3. Introduce delivery of vehicle to customer's location
4. Provide access to further vehicles, where company is unable to fulfill the customer's hiring requirement.

5.4.6 Revising priorities

So how can we begin to improve the vehicle hire process, at minimal risk and without incurring too many overheads, from a service-oriented perspective? The ideal strategy is to incorporate one or two relatively independent and isolated software services that add some business value in line with the priorities in our plan. The top priority is to introduce vehicle reservations over the company's Web site. This is technically feasible based upon the Web services developed thus far, though these will need to be scaled up and stress-tested to meet more stringent QoS requirements for use over the Internet.

The upshot is that management decide to offer the limited Internet reservations capture as an interim measure while the customer verification and vehicle brokering Web services are investigated and then introduced as part of the phone-initiated vehicle hire process.

Once these new features are bedded down, the company will move to full-blown reservations over the Web and to the other priorities listed in the plan, as indicated below:

1. Introduce interim (limited) reservations over the Web
2. Streamline the customer's experience by automating verification checks for risk and creditworthiness (by phone)

3. Provide access to further vehicles, where company is unable to fulfill customer's hiring requirement (by phone)
4. Introduce full reservations over the Web
5. Introduce delivery of vehicle to customer's location.

5.5 Where to next?

In order to make this further progress it will be necessary to address the information and policy drivers of the BA more fully in chapter 6. This will take us naturally into the discipline of SOA in part 3 of the book.

6 Achieving business agility

6.1 Introduction

We now turn toward scaling up the early efforts at business process redesign. In chapter 5, we sketched out some tactical scenarios that illustrated the use of existing assets to start to redesign existing processes. Web services were used in opportunistic fashion to leverage some of these assets through wrapping. The Web services were little more than single operations[1] applied to very specific business requirements.

In this chapter, we move to the longer-term, strategic view to identifying and designing services that can meet the wider SOV7 requirements. This will involve considering the aspects of the BA shown in figure 6.1. We will move from left to right across this picture, starting with the service policy aspects, then considering domain models, and finally considering the two service models on the right-hand side. In particular, we focus on exploring how services can be used in different business contexts. These considerations in turn will cause the processes to be re-evaluated and developed so that they become more service-oriented.

A major part of this work concerns surveying and cataloging of corporate assets, of which services are a part. At the end of this chapter, we therefore also consider the place of the asset inventory in relation to this work.

6.2 Service policy

Service policy covers business goals and rules, the business–IT alignment table (BIAT), and sourcing and usage policy: the four items shown on the left of figure 6.1. The service policy constrains and governs the services at various levels of detail, from a policy governing any service enterprise-wide to a policy applicable to a particular group or type of service, and eventually to a policy governing one particular service.

[1] Moreover, these operations were essentially simple synchronous transactions. Later, as part of SOA, we shall be looking to architect software services as groups of cohesive operations that may be complex and asynchronous.

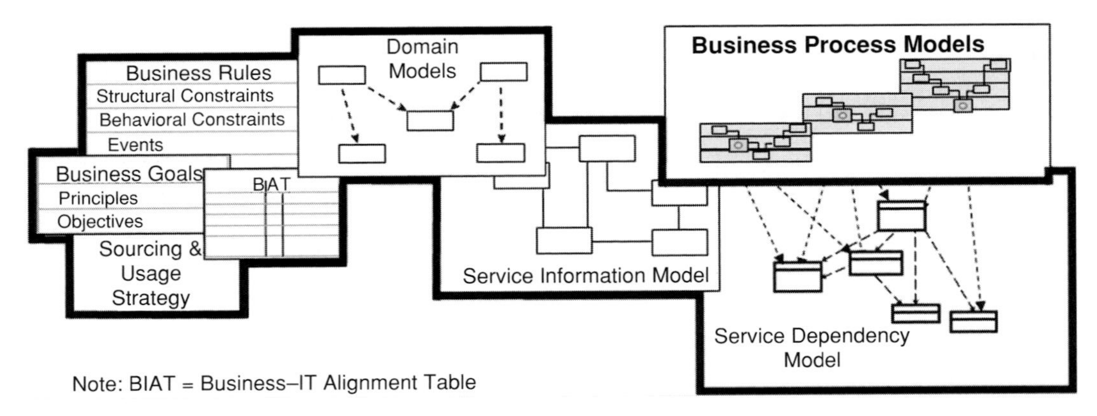

Figure 6.1 The focus of chapter 6

6.2.1 Process, goal, and rule

A process strives to achieve one or more business goals ("goals," for short). A goal is a *desired state* that must be *measurable*. The goal may be defined in quantifiable terms, such as profit, time, or quality. On the other hand, a goal may be defined in qualitative terms such as "this company will be regarded highly" or "this company will attract more reliable staff."

Goals may be very high-level, and act more as general principles than goals *per se*. An example might be "Better management of technology changes" – this might be measurable in terms of reduced cost and time to respond to these changes. A more specific goal, or "target" might be "Product defect rate must be less than 5%" or "The defect detection service must be available 24 hours a day." Targets also have times by which they must be achieved such as "Product defect rate must be less than 5% by November 1, 2006." However, such times are very often implicit in that they are specified for SLAs of which the individual service targets are part; see chapter 12.

Goals and targets describe *how well* the business must perform. High-level goals help to shape the SOA that will underlie our redesigned processes. Lower-level goals may both help to define QoS requirements as part of the SOA, and also form a significant part of service specifications and SLAs.

A goal may require the levying of a number of business rules ("rules" for short) that govern various aspects of the process. For example, "Product defect rate must be less than 5%" might be a goal of the product manufacturing process, which requires the rules "Finished products must be authorized by the production manager" and "If product dimensions do not match specified product dimensions the product is rejected."

Rules describe *what* conditions govern the business. Rules will form an important part of service specifications.

6.2.2 Assessing the business–IT alignment

The *raison d'être* for a service-oriented approach is to improve the business so that it is fit to compete in the increasingly service-oriented commercial climate. However, this can be difficult to get across to top management, and it can often be difficult to obtain buy-in from business leaders and business unit managers. It is therefore necessary to communicate the *business value* of a service-oriented approach to the movers and shakers of the organization.

Many organizations have instituted *business metrics programs* such as balanced scorecards (Kaplan and Norton, 1996) or approaches based on Six Sigma (Smith and Fingar, 2003b) in order to pinpoint business value. Unfortunately sometimes these programs remain isolated from IT concerns. The business measures must be correlated with the software used to achieve these measures. Alignment is particularly significant for a service-oriented approach predicated on contracts that stipulate QoS requirements that providers must meet. Such QoS requirements must ultimately connect with a business metric of some form or other, such as a key performance indicator (KPI).

A useful approach is to create a BIAT. The BIAT provides baselines against which the business value of services can be assessed, and helps demonstrate why a service-oriented approach is necessary in business terms. It is also used to guide the evolution of the SOA in a way that is designed to meet the right business goals (as described in 8.2.1). A BIAT can be created at different levels of granularity. A high-level BIAT correlates goals with service-oriented features that can help to achieve the business goals.

Typically, the following three steps are involved in creating a high-level BIAT:

- *identify and weight goals* (concentrate on the primary goals – between ten and twenty as a rough guide) using focal points as a guide
- use *SOV 7* to help develop and classify the goals
- identify *service-orientation features* that are most relevant to the goal using a knowledge of service orientation in relation to an understanding of goals.

However, there are no hard and fast rules. Brainstorming, involving responsible business leaders, along with IT architects, is essential to develop the BIAT.

In 5.4.2 we identified goals for our process redesign work and classified them in terms of SOV7. We now need to identify the service-oriented features that help enable these goals. The resulting BIAT is shown in table 6.1.

6.2.3 Dealing with lower-level business goals

The BIAT can be further developed to a more detailed level of granularity by adding sub-goals and KPIs or other business metrics, as defined in balanced scorecards or Six Sigma models. These can all form useful input to the SOA as detailed in chapter 7,

Table 6.1 *BIAT for vehicle hire company*

High-level goal	Service-oriented feature
Transparence: Introduce reservations over the Web as well as by phone. Streamline the customer's experience by automating verification checks for risk and creditworthiness.	An SOA promotes identification of well-designed services that reflect business needs. Such services can provide a high level of automation and performance that can be tuned, using appropriate execution management software, to varying run-time constraints.
Customer fit: Introduce special offers and options for Gold customers ranging from hotel deals to route planning. Personalize the reservation by using history of customer's dislikes and preferences. Introduce delivery of vehicle to customer's location.	Exposing the line of commoditization allows the organization to focus on designing value-add services that ensure customer fit through interoperability with other service types.
Partner connectivity: Work (as a consumer) with vehicle broker to provide access to further vehicles, where company is unable to fulfill customer's hiring requirement. Work (as a supplier) with garages that need to supply their customers with temporary vehicles while their vehicles are being repaired or serviced.	Services are specified to a rigorous level that allows certification to validate stated availability and security requirements across organizational boundaries. This helps consumers to effectively manage the external service provider relationship, and helps suppliers to ensure that consumers are receiving the agreed level of service.
Adaptation: Anticipate and support possible changes: • Organization changes such as mergers and takeovers • Business process changes such as delivery of vehicles to customer's location; see also Customer fit • Information changes such as pricing structures for bulk reservations • Business rules changes such as changes to hiring rules.	Guidelines that structure services in a way that minimizes the potentially disruptive effects of change at all levels. This approach leverages the potential of services for: • *Adaptability*: Regarding business change, services can be extended in "plug and play" fashion in different business contexts • *Flexibility*: Services can be designed to smoothly accommodate change to information and rules • *Replaceability*: Regarding associated technology changes, a service can be smoothly replaced with another having different component implementation code.
Multi-channel capability: Introduce reservations over the Web as well as by phone; see also Transparence. Support reservation process using different channels consistently through the process – for example reserve over Web, enquire by cell-phone.	An SOA helps remove dependencies on specific technologies through the use of standardized interfaces. Services are designed that can exploit this capability thus allowing reuse in different channels and technology contexts.

Table 6.1 *(cont.)*

High-level goal	Service-oriented feature
Optimization: The ability to offer reservations of predetermined vehicle types in real time by cell-phone. Real-time alerting of vehicle incidents (from faults to thefts).	Services can be deployed using technology appropriate to varying run-time constraints and that provides the capability to respond to high transaction volumes in real time. Note that effective scaling up of the SOA depends upon effective SEM technologies that provide monitoring and control of software services.
One-stop experience: Introduce a Web site incorporating reservations as one piece of a complete travel package with links to hotels, ferries, restaurants, and leisure outlets such as theaters and cinemas.	An SOA fosters appropriate design of loosely coupled software services, allowing them be used together in different combinations in different business contexts. The usage across organizational boundaries requires potentially high levels of availability and security. This depends upon effective SEM technologies that provide monitoring and control of software services.

where we shall see that different service types may have different QoS requirements, with different priorities for each type of service. At the same time we must not "gild the lily" – the documentation should be useful: too much detail can be self-defeating.

Lower-level business goals are often attributable to specific processes. For example, "Product defect rate must be less than 5%" may be a goal of a product manufacturing process. These lower-level goals may result in further goals that are attributed to the services supporting these processes. For example "The defect detection service must be available 24 hours a day."

In chapter 9, we examine the relationship of goals to the SOA, and the allocation of business goals to services.

6.2.4 Establishing business rules

Much prominence has been given to business process modeling and, to a slightly lesser extent perhaps, business information modeling. Despite a wealth of literature, business rules have tended to be something of a poor relation when it comes to business process redesign.

This is unfortunate in that, while rules can be simply assigned to services in many cases, the separation of certain significant business rules (from process logic) in the form of *separate rules services* is a useful strategy to help achieve reuse and adaptability of business rules.

In this section, we introduce the basic types of business rule, with a view to moving on later to illustrating how rules can be used within a service-oriented approach. Again, there is a plethora of literature in the business rules area (Morgan, 2002; Chisholm, 2003; Ross, 2003).

6.2.5 Types of rule

There are many different ways of classifying rules. We follow Penker and Eriksson (2000), pp. 153–62, who provide a simple and pragmatic classification, using the following types of rule:

- *Structural constraints*: Conditions that regulate business policy, usually about information. For example:
  - a reservation must specify a type of vehicle
  - a contract must specify an actual vehicle
  - a contract must name a driver
  - only the hire manager shall award customers gold status
  - a deposit must never exceed 20% of the agreed reservation fee.
- *Behavioral constraints*: Conditions that constrain what must be true before or after an action is carried out. For example:
  - only verified customers may reserve vehicles
  - a contract can be closed only where a payment has been made.
- *Event constraints*: Stimulus/response rules that specify that certain actions should be carried out when certain business events occur. There are two types of event – direct and indirect.

  Direct events occur as a result of something happening in the business environment; for example:
  - If a vehicle breaks down then organize recovery and replacement of vehicle.

  Indirect events occur as a result of a state, such as a specified time, or condition occurring; for example:
  - If a vehicle is not returned by agreed return date then issue warning notice.

  Event constraints can be complex and involve relationships to other business rules; for example:
  - If excess vehicle mileage > 0 and the customer is not a Gold customer, then an excess premium shall be payable (see below).
- *Derivations*: Rules that describe how one piece of information can be derived based on another. There are two types of derivations. Inferences specify conclusions that can be made when certain facts are true; for example:
  - "Excess payable" is true if excess vehicle mileage > 0.

  Calculations are procedures or algorithms that specify how to derive one piece of information from other pieces of information; for example:
  - Hire mileage = recorded mileage – previous recorded mileage

– Excess vehicle mileage = hire mileage – (maximum day mileage × days hired)
– Excess premium is calculated according to a pre-stipulated formula.

Notice that the scope of rules can vary quite markedly. Some rules, such as structural rules, apply at a very general level – for example, "A contract must specify an actual vehicle." Others, such as some of the behavioral constraints, are much more specific – for example, "Only verified customers may reserve vehicles" is a condition that governs the making of reservations.

6.2.6 Capturing and documenting rules

Rules are discovered in many ways. One way is through inspection of processes and their associated goals and information requirements. A good technique is to develop process scenarios, using storyboards. Here are some questions to ask:

- *To uncover structural constraints*: What are the conditions that govern business policy?
- *To uncover behavioral constraints*: What are the conditions that constrain successful opening and closing of a process or activity?
- *To uncover event constraints*: What unusual or problematic situations can occur throughout a process, and what must be done to cope with them?
- *To uncover derivations*: What significant algorithms are required at different points of the process?

Because of the problem of sheer potential number of business rules, each business rule should have a *unique identifier*. The unique identifier enables you to catalog and refer easily to business rules.

The business rules will map to other artifacts, as discussed in chapter 7. We want to focus on significant rules that can be reused – the operating principles of the business if you will – that apply to different situations or contexts. More specifically, our aim, in the context of this book, is to identify rules that apply to *services*. In fact, there may well be certain services whose responsibility is rule management. This is often the case where rules govern the use of information that is central to the business and where these rules are complex enough to warrant separate management. A database of significant rules, integrated within the BPM tool, is a practical necessity in these cases in order to maintain the rules in harmony with the process, as described in chapter 3.

6.2.7 Capturing and documenting sourcing and usage policy

There are two further key questions that service policy must address:

- How are services to be implemented and provided?
- How are services to be used?

These questions are addressed with respect to the *overall sourcing and usage strategy*. For example, the first question involves defining which services are to be outsourced, which are to be performed internally, and which are to be a combination of both. The second question involves defining whether to offer services only internally, to trusted partners, or outside of the organization to the wider market.

Let us suppose that our vehicle hire company decides to outsource only services that form parts of its commodity processes in cases where there is genuine business benefit to be gained. Moreover, these services must be very well understood in terms of clear SLAs.

Our vehicle hire company also decides to offer most of its services both in-house and on the open market, where appropriate, but (at least for now) is not concerned with selling the services on a subscription basis or otherwise. The services are seen as part of its *business capability*, not as assets that are for sale.

6.3 Business domains

Business domains (domains for short) have a significant bearing on the identification and scoping of services. In particular, analyzing domains helps to achieve reuse of services across different processes.

6.3.1 Applying domain analysis for service orientation

Domain analysis was first introduced in the 1980s with the objective of identifying, capturing, and organizing information in order to make it reusable when creating new systems (Prieto-Diaz, 1990). Domain analysis focuses on supporting systematic and large-scale reuse (as opposed to opportunistic reuse, which suffers from the difficulty of adapting assets to fit new contexts) by capturing both the *commonalities* and the *variations* of systems within a domain to improve the efficiency of development and maintenance of those systems. The results of the analysis, collectively referred to as a "domain model," are captured for reuse in future development of similar systems and in maintenance planning of legacy systems (Foreman, 1996).

One of the difficulties with traditional approaches to domain analysis is the sheer scale and complexity of such undertakings, particularly in large organizations. There is the real danger that the domain model never gets finished – or, worse still, never gets used, because it is too far removed from the practicalities of everyday business or software development. The earlier approaches also fail to take into account the practicalities of service orientation – for example, partnering, outsourcing, focusing on core competencies, and offering services to market. They tend very much to the traditional, strictly bounded view of an enterprise.

Domain analysis must therefore be tuned pragmatically if it is to succeed as part of a service-oriented approach. For our purposes, the focus of domain analysis, applied within the service-oriented process redesign pattern, is on:

- Establishing the commonality of information *and functionality*[2] that is important for identifying services, especially commodity services
- Controlling complexity and managing scale by dividing a business into *sub-domains* that focus on particular fields of interest
- Examining *sub-domain dependencies* in preparation for creating an SOA that is not constrained by existing organizational structures
- Gaining a *strategic view* that helps to feed tactical incremental process redesign with services that can be used in other contexts.

6.3.2 Understanding business domains

There are many useful domain analysis techniques that can be applied in a service-oriented way. Perhaps the most useful of these techniques is to envisage the major business concepts in an enterprise and map these against high-level processes to form clusters of related concepts and processes (SOSA Team, 2004). Each cluster should suggest a *coherent group* of services.

For example, our vehicle hire company might have the following clusters: Vehicle Marketing, Vehicle Hire Sales, Customer Relationship Management (CRM), Vehicle Management, Financial Management, Warranty Administration, and Legal Management.

The clusters can be usefully modelled as *domains* using a simple dependency diagram, as illustrated in figure 6.2. The initial exercise of identifying domains should be broad-brush, and not take too long.

6.3.3 Partitioning domains by types of service

The enterprise domain dependency diagram can be used to engage the business leaders in establishing corporate attitudes toward the different business areas represented by the domains. Questions such as the following can be raised in whiteboard discussion:

- Is this domain important enough to be sourced internally, or can it be outsourced at low risk?
- Should the functionality associated with this domain be offered as a set of services; if so inside the organization only, to trusted partners, or to the market as a whole?

This kind of exercise is very useful for exposing assumptions and articulating sourcing and usage policy, for helping to facilitate key decisions, and for forming an overall idea of the organization's line of commoditization, as described in chapter 4.

The domains are considered in terms of the types of services (value-add, territory, or commodity) that they relate to. Where appropriate, a domain may be broken down into *sub-domains*, each of which provides a different type of service. For example, it may

[2] Traditional approaches tend to focus exclusively on information.

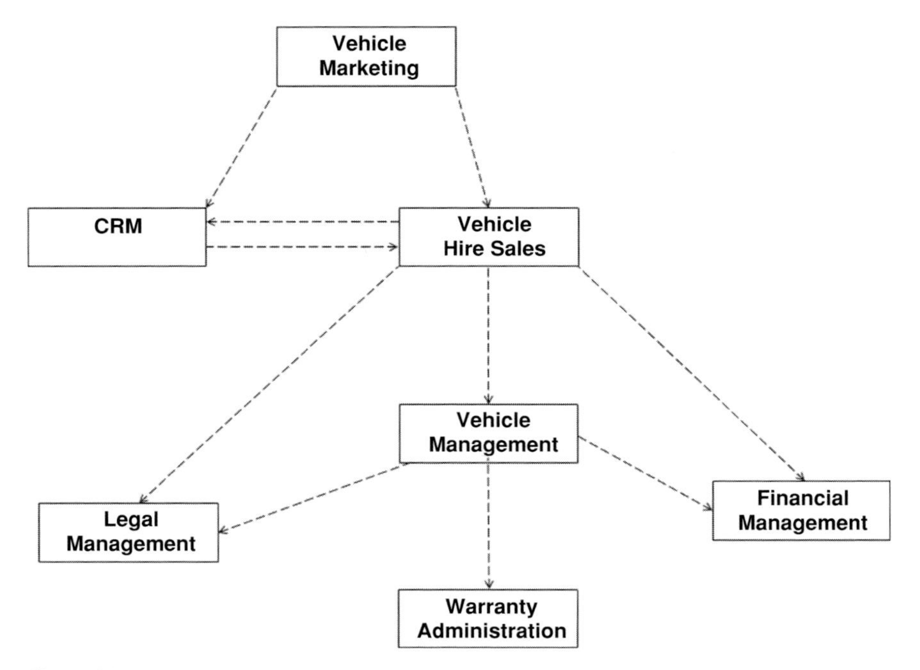

Figure 6.2 Enterprise-wide domain dependencies

be in the vehicle dealership that marketing of vehicles takes on a different significance according to whether we simply want to run a mail shot to existing customers, on the one hand, or to put together an innovative advertising plan around a distinctive brand on the other. The former looks very much like an activity that could be outsourced to a mail shot provider whereas the latter is something the hire company might want to either focus its specialist skills on, or to employ a marketing expert. In order to address these kinds of concern it is useful to further partition domains according to the type of service they provide. The different types of services are considered relative to domains using a *sub-domains table*; an example for our hire company is shown in table 6.2. Although service types run orthogonal to domains, some domains may line up very strongly with specific service types. For example the Warranty Administration, Legal Management, and Financial Management domains line up exclusively with commodity services. Other domains, such as Marketing, involve all three service types.

As we discussed in 4.5.1 organizations that compete on very high turnaround at very tight margins have a high "line of commoditization." At the other end of the spectrum organizations that provide specialized knowledge have a low line of commoditization. In order to manage its agility an organization must understand and address its line of commoditization. Techniques such as the sub-domains table can help to make a start.

Table 6.2 *Example sub-domains table*

Domain	Value-add sub-domains	Territory sub-domains	Commodity sub-domains
Vehicle Marketing	Brand Promotion	Campaigns	Prospects
Vehicle Hire Sales	Contracts		
CRM	Customer Care		Customers
Vehicle Management		Vehicles	Vehicle Maintenance
Financial Management			Audits Accounts
Warranty Administration			Warranties
Legal Management			Litigation Management

6.3.4 Examining sub-domain dependencies

Ideally, the domain and service type perspectives should coincide, as this will make service management much simpler. For example, Legal Management might be completely outsourced to a lawyer. On the other hand, as intimated earlier, Vehicle Marketing is much more complex from a sourcing point of view as there is a fairly even spread of different service types that might be sourced differently. For example, perhaps Promote Brand is sourced by a specialist advertising agency in collaboration with internal staff; Run Campaign is sourced internally; Run Mailshot is outsourced to a mailshot provider.

We must keep an eye on dependencies, especially between value-add and territory services on the one hand and commodity services on the other. Organizations that offer innovative products (such as some of the latest cell-phones) based on clever software designs, but are unable to bill their customers with efficiency, should take note!

6.3.5 Scoping using sub-domains

We have already emphasized the need to manage the potential scale and complexity of domain analysis. For practical purposes, however, we need to focus on the sub-domains that are likely to release the most value in the quickest time with respect to our plan for business process redesign. Recall that the business priorities for process redesign, identified in chapter 5, center on the hiring process. These are repeated for convenience below:

1. Introduce interim limited reservations over the Web
2. Streamline the customer's experience by automating verification checks for risk and creditworthiness (by phone)
3. Provide access to further vehicles, where company is unable to fulfill customer's hiring requirement (by phone)
4. Introduce full reservations over the Web
5. Introduce delivery of vehicle to customer's location

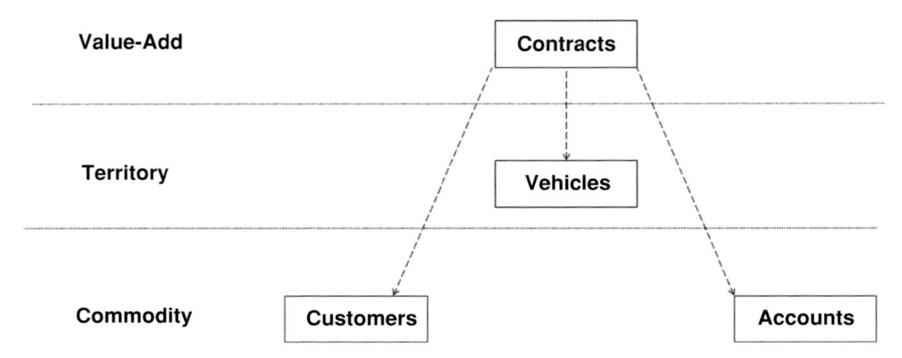

Figure 6.3 Sub-domain dependencies for the hire process

The hire process focuses on the Contracts sub-domain, but will also involve the use of services provided from other domains. We focus on the possible dependencies between sub-domains, as shown in figure 6.3.

Recall also that *Easy to do business with* and *Low cost* are the dominant focal points (see 4.2.5) that were used to assist our earlier scoping (5.4.5). This combination of focal points suggest a fairly even spread of value-add, territory, and commodity services, as reflected in figure 6.3.

6.4 Service models

The domain models are used alongside business process models and our knowledge of service policy to identify and structure services. In particular, with reference to our earlier roadmap (figure 6.1), we look in this section at techniques for developing the service information model and service dependency model.

6.4.1 Identifying services

A *candidate software service* is created for each of the sub-domains. Each software service must fit our earlier definition that a software service offers a coherent family of operations that work toward a defined business purpose. Operations are identified by considering business process models and, for each activity, asking what operations are required, as illustrated in figure 6.4.[3] Simple create, access, amend, and delete operations have been omitted from figure 6.4 for brevity.

[3] The reader should note that this is not a formal model. The idea is to consider possible software service usage by declaring these services and their operations in graphic relation to the activities that invoke them. By declaring services as boxes at the bottom of a swim-lane diagram brainstorming can be used to sketch out the invocations by drawing lines from the sub-processes to the invoked operations.

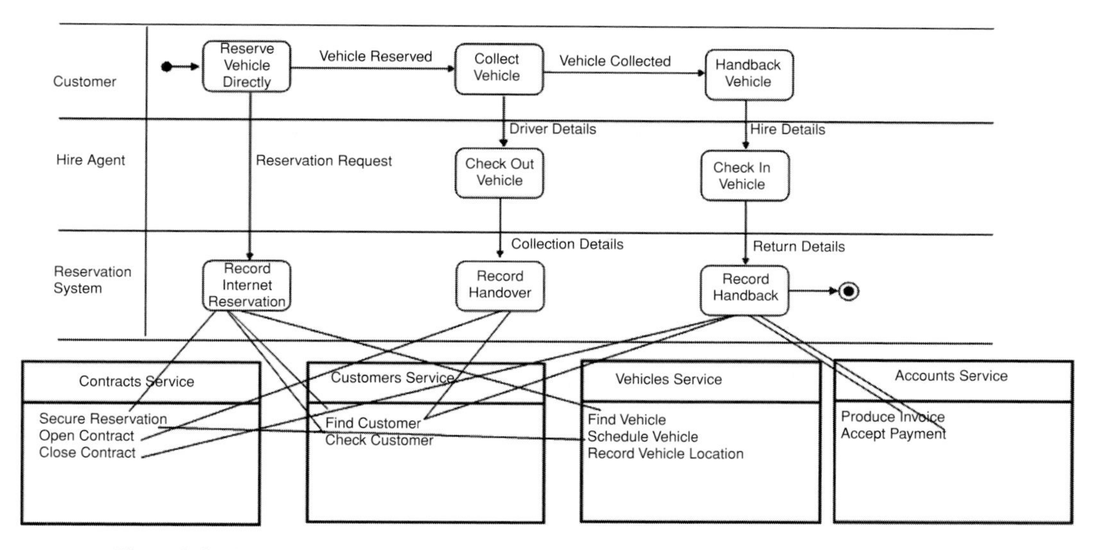

Figure 6.4 Candidate services for the Internet hire process

Of course, our collection of services represents a first-cut view. We must test out this view and develop the collection of services, by considering other processes, by exploring information needs, and by considering business rules. In doing that, we shall examine how the services collaborate together in different contexts and there are various techniques that can be used for that. For example, the idea of class responsibility collaborator (CRC) cards (Wilkinson, 1995) can be usefully adapted to services.

The operations of the services provide "hot buttons," so to speak, that are pressed at various points of the process. The services are akin to reusable chunks of process. However these chunks are offered as banks of defined functionality in the manner of virtual web sites, in contrast to a linear business flow of the process. Moreover, the service buttons (operations) are described in a nonprocedural way according to the contract, in contrast to the procedural description of the process. A BPM system is usefully employed to choreograph the pressing of these buttons according to the business rules that govern the process.

Though this has many of its roots in OO and CBD approaches, it is worth noting some of the differences. The main difference is that services are identified and fully explored in relation to business domains, processes, policy (such as goals and rules), and available existing and future assets, before thinking in detail about the structure of those services. There is sometimes a tendency in a CBD or OO approach to emphasize the structure of components or objects in the early stages of analysis. In addition, a service-oriented approach does not aim for an ideal pure analysis model, which is mapped to an implementation. Instead existing and planned use of pre-existing services is declared up front.

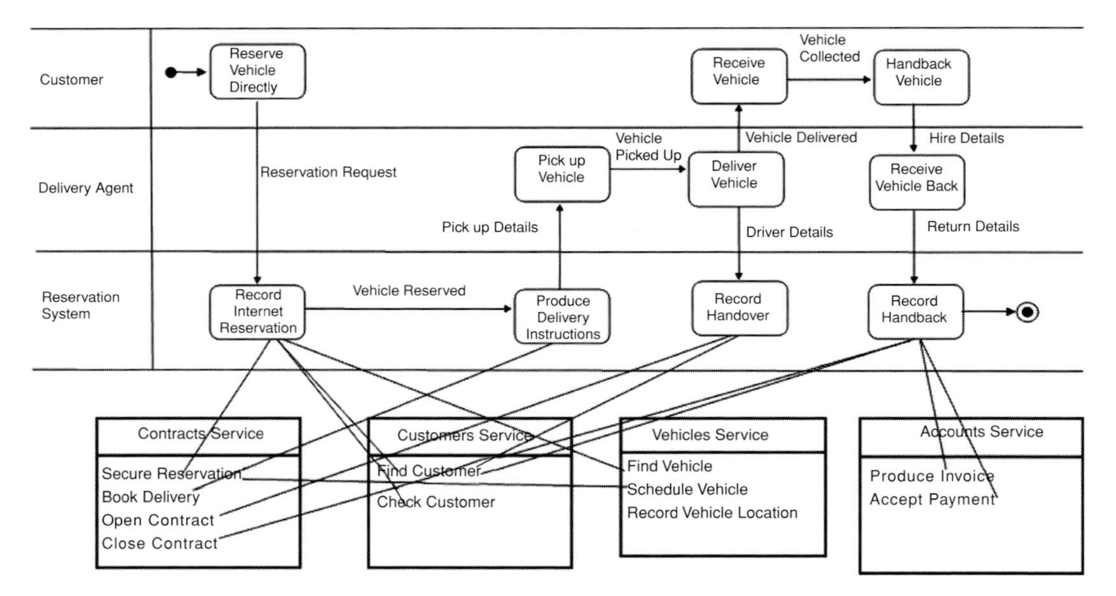

Figure 6.5 Candidate services for the enhanced Internet hire process

What is common to all three approaches – OO, CBD, and service–orientation – is the notion of *interface*. In chapter 7, we shall see that interfaces play a major role in SOA. However, we shall have prepared the ground at the business level by setting out the required services in a full consideration of business requirements first, in this chapter.

6.4.2 Examining different contexts for services

Our set of services should be as *reusable* as possible. We can apply the SOV7 guidelines to question the needs for transparence, customer fit, partner connectivity, adaptation, multi-channel capability, optimization, and one-stop experience to help here. At the same time, we must consider the process improvement plan:

1. Introduce interim limited reservations over the Web
2. Streamline the customer's experience by automating verification checks for risk and creditworthiness (by phone)
3. Provide access to further vehicles, where company is unable to fulfill customer's hiring requirement (by phone)
4. Introduce full reservations over the Web
5. Introduce delivery of vehicle to customer's location.

In particular, regarding point 5 above, a good step is to think about how the process might be extended to cater for delivery of vehicles direct to the customer with handover to qualifying delivery agents, as shown in figure 6.5.

6.4.3 Understanding service information

We focus on the sets of information that evolve with service usage and use business type models for this purpose.[4] For example, a customer interacts with the hire company through a typical life-cycle of searching, hiring, driving, and returning the vehicle. In first coming into contact with the company, the customer supplies personal details, which are captured as the first piece of information in the set. Types of vehicles in which the customer is interested are captured. The hiring process begins with the recording of a reservation. The customer then collects the vehicle with details captured on an entry sheet, before driving the vehicle away. On return of the vehicle, details are recorded on an exit sheet.

This kind of evolving collaboration between parties, in which a supporting information set builds up over time, is typical of service orientation. Other examples include order to cash, production planning, roll-up (hotel check in to check out), contract negotiation and applications (such as enrolment or employment).

We consider information within the context of the hire process, looking at each service in turn, starting with the Contracts service. It is important not to include too much detail at this stage. A good approach is to concentrate on the main business types (categories of information), such as Contract, Reservation, and Vehicle. We declare these to be *business types* and model the relevant associations between them, as indicated in figure 6.6. Notice that as far as attributes (specific data elements) are concerned we include only those that are the *responsibility*[5] of the service; in this case, these are the attributes of Contract and Reservation. Other types, such as Vehicle and Customer, may be required for the service to work but are not the responsibility of the service. These types are shown without attributes. Note that these types will appear with attributes within the service information models of the services that are responsible for them. We may well, of course, uncover many other attributes in the course of our investigations and record these, where appropriate, against their responsible business types in preparation for SOA modeling. However remember that right now we are concentrating on *business* matters, and do not want to muddy this activity with too much technical detail too soon.

6.4.4 Evolving the services

In practice, the business rules and the information models are developed alongside the process redesigns. As our understanding develops and the models unfold, so the process redesigns may change. It may also be that some services are removed or changed, and others added. For example, recall that we identified the following *event constraints*:

[4] The techniques used here are very much those of traditional information modeling applied to business types, relationships between types and attributes (Allen, 2000).

[5] A service that creates, maintains, or removes attributes or types is said to be *responsible* for those attributes or types.

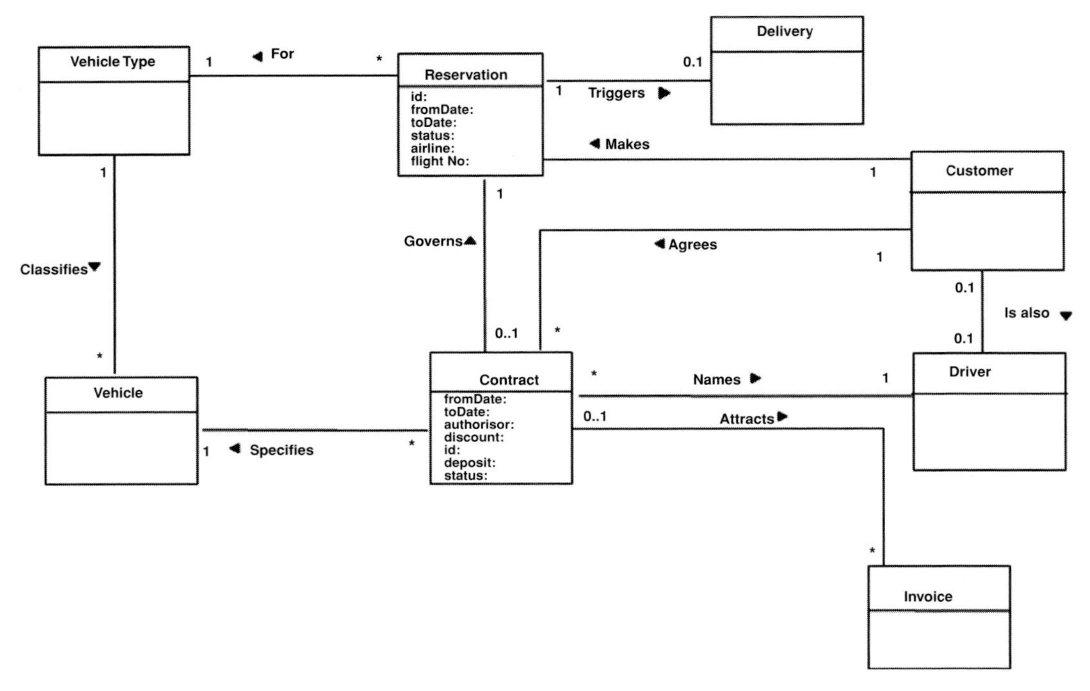

Figure 6.6 Information model for the Contracts service

- If a vehicle breaks down then organize recovery and replacement of vehicle (*direct*)
- If a vehicle is not returned by agreed return date then issue warning notice (*indirect*).

The breakdown event takes us back to consideration of the process itself. The business process redesign must be modified to take account of this event. (In fact, it is an event sensed in real time by a device in the vehicle and transmitted back to the hire office.) The Vehicles service needs to include operations to deal with recording the breakdown and recovery of vehicles. We will also need to change the information model to include recovery and replacement details. Clearly there could be some significant changes involved; at a minimum we will need to add a breakdown dates/times and recovery dates/times to the Vehicle type.

The failure to return event implies that our Contracts service needs to include an operation – Report Expiry – to sense expiry of return date and issue the warning.

We also previously identified the following *indirect events*:

- If excess vehicle mileage occurs and the customer is not a gold customer, then an excess premium shall be payable.

We also identified an *associated formula*:

- The excess premium is calculated according to a pre-stipulated formula.

We must decide how to handle calculating the excess premium. It might be that there are many different variables – for example, vehicle type, customer status and type,

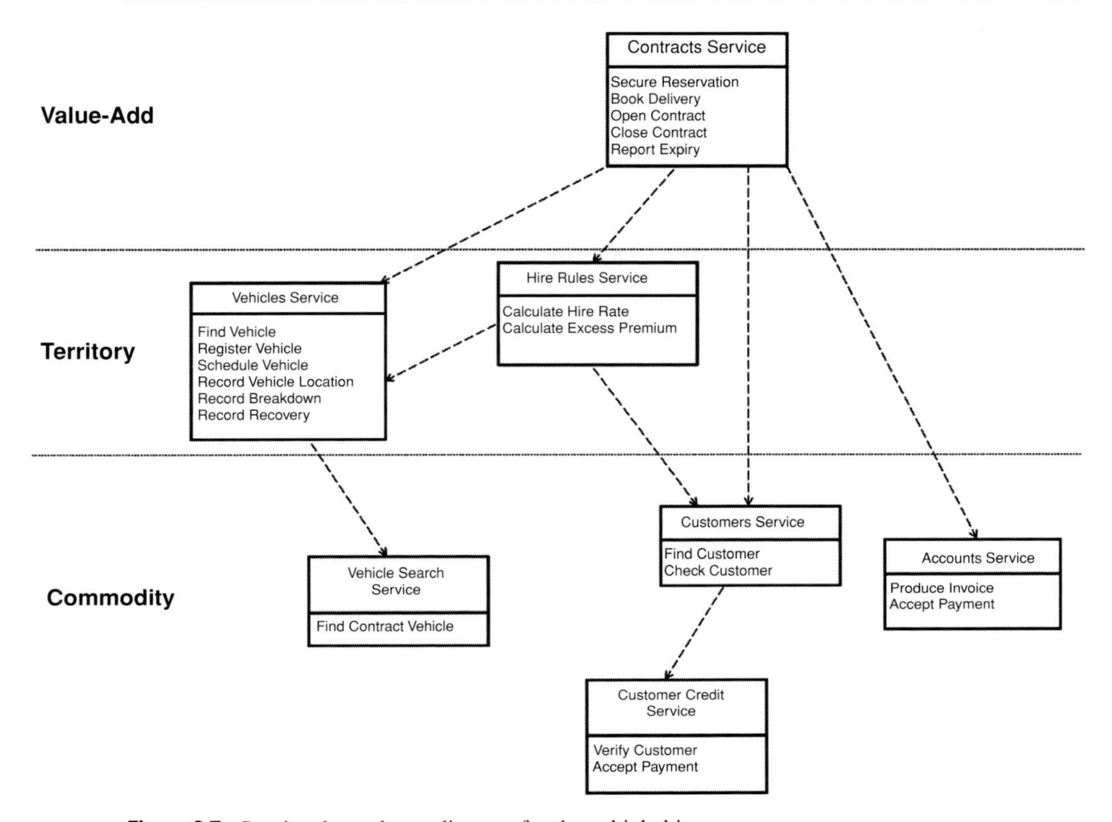

Figure 6.7 Service dependency diagram for the vehicle hire process

time of year, and region – that determine not only the excess premium, but also the hire rate itself.

This kind of example is typical in businesses with complex product rules. In these cases, it may well be appropriate to construct separate *rules management services*. In the hire example we choose to do just that and introduce a Hire Rules Service.

The resulting service dependency diagram is shown in figure 6.7.

6.5 Surveying and cataloging assets

Surveying and cataloging of assets is part of a much wider topic and therefore we focus here just on the aspects that are particularly relevant to service orientation. The shortness of this section is not an indication of the importance we attach to this activity: as well as being a vital cog in the wheel of service orientation it is an absolute necessity for effective corporate governance. In a nutshell, without an asset inventory it is impossible to manage services in any meaningful fashion, especially in an organization of any size.

As service-oriented process redesign unfolds so the various assets used by the processes should be surveyed and cataloged. The asset inventory should summarize both existing assets and possible future assets. These assets include all resources, not only services – including processes, business units and groups of individuals, as well as software such as databases, systems, and packages.

The asset inventory is used to support decisions on sourcing and usage. It has a key role to play in service design and in supporting the SOA. Further downstream at execution time, the asset inventory provides the foundation stone for SOM.[6]

6.5.1 Scope of asset inventory

The inventory covers all assets – both software and nonsoftware – enterprise-wide. In an organization of any size, however, the scope of an asset survey must be carefully managed if it is not to take on gigantic proportions. *Domains*, again, provide a very convenient scoping mechanism for helping to manage the scale of your survey.

6.5.2 Asset descriptions

The asset inventory should describe each asset in terms of its providers (internal or external) and customers; note that service descriptions are covered in chapter 9. The asset description should cover basic capabilities and rate the asset in terms of how well it meets business needs and how easily it can be modified to changing needs. Assets should be numbered to enable them to be easily identified and referenced.

Documentation of software assets should include only basic technical characteristics, such as the name and basic description of a service or software system. Further technical documentation can be elaborated as part of the SOA or the IT architecture. Such information might typically include number of users, usage rates, volumes of data, format and accessibility, degree of dependence on other systems, and types of application program interface (API) offered. At this point, the information should be just enough to expedite decisions on service sourcing and usage.

6.5.3 Evolving the asset inventory

Existing assets should be evaluated in terms of priority for replacement. Large, costly legacy software systems that are working well, are easy to modify and with clean interfaces to other systems are excellent candidates for servicization. Small, closed

[6] For example, we shall see later that the asset inventory has important connections to the configuration management database used in ITIL.

systems that are falling well short of meeting business needs are good early candidates for replacement. More careful judgment is needed for other systems which fall between these two extremes.

A useful technique (discussed in 14.3.3) is to assess the business value of each asset in relation to its current condition, and to decide whether it should be leveraged, maintained, replaced, or terminated. This can help expedite the sourcing strategy.

The asset inventory should also be used as a placeholder for possible future proposed services. As service-oriented process redesign unfolds, some existing assets may evolve into these new services.

6.6 Where to next?

We have now come as far as we can without investigating the structure and subsequent implementation of the software required to realize our plans for business agility. We need to address the SOA, the central topic of part 3 of this book.

In practice, the SOA and the business process redesign work unfold in an iterative fashion side by side. Software solution projects are triggered at various points to support the subsequent business improvements. SOA design, specification, and implementation projects are similarly triggered to ensure that the required services are in place where required.

Part 3 Service-oriented architecture

SOA has received its fair share of interpretations. It is a term that is commonly bandied about without clear definition. In fairness, the topic of SOA is probably as much about *cultural change* as it is about concepts and techniques. However, SOA is integral to the approach described in this book and therefore it is important to define our terms very clearly.

Part 3 of this book lays out the two major aspects of the SOA: *policy* and *structure*. We explain how to achieve an integrated approach to SOA that is driven top-down by the BA, and equally that is supported bottom-up by SOM.

Chapter 7 sets the context for the remainder of part 3 by providing an overview of SOA concepts and shows how the essentially static view of the SOA interconnects with the dynamic view of run-time services provided by the SOM. This paves the way for our discussion of SOM in part 4. We also provide an overview of the required changes in mind-set.

Chapter 8 deals with the policy aspect of SOA and provides techniques for the business–IT alignment of policy. We provide an overview of the four main areas of SOA policy – QoS, design, sourcing and usage, and technology – and preview the four key types of QoS requirement – agility, capacity, availability, and security.

Chapter 9 centers on agility, and provides a compendium of service design techniques for structuring the SOA and ensuring business–IT alignment. The topics covered range from modeling service dependencies, though interface design, to software unit architecture and modeling service buses. A keynote of the approach is the allocation of business goals and rules, identified as part of the BA, to the SOA.

Chapter 10 examines capacity, availability, and security in the context of SOA infrastructure design. We discuss how software services bring new challenges in these areas. The quality measurement and control bar must be raised! We describe a set of quality templates to help address these issues. We also discuss how the idea of an infrastructure service bus (ISB) can be used to help organize your approach to QoS infrastructure design.

7 Service-oriented architecture themes

7.1 Basic principles

We have emphasized that the SOA evolves over time in coordination with the business process redesign efforts. Now it is time to take a closer look at the concept of SOA itself. Unfortunately, SOA has become a much overused and abused term, subject to vague interpretations. It is important to clarify what exactly is involved in the idea of an SOA and how it differs from previous approaches to IT architecture. In this short chapter, we therefore provide an overview of the elements of SOA and outline the overall approach to evolving the SOA.

It is probably fair to say that the topic of SOA is as much about *attitudes* and *cultures* as it is about concepts. Apart from the shifts already discussed in the BA area, there are some other major shifts in mind-set required, that we discuss in this chapter.

While the SOA helps to plan strategies and specify services, the software has to work, and work extremely well in demanding conditions. Enter, therefore, SOM. In this chapter, we elaborate on the concept of SOM (introduced back in 2.3) and outline its relationship to the SOA. Later in this book, we shall build upon these introductory guidelines.

7.1.1 The elements of the SOA

What exactly, then, is the SOA? In this section, we provide the short answer in terms of a high-level bill of materials. There are two major aspects of the SOA: *policy* and *structure*.

The policy aspect of SOA falls into four parts:

- QoS
- Design
- Sourcing and usage
- Technology.

This chapter focuses on the policy aspect of SOA. We go into more detail on QoS in chapter 8.

The structural aspect of SOA is concerned with service design, and focuses on the following:

- Services[1] and their dependencies
- Service descriptions (which form the basis of service specifications; see 9.3.4)
- Interfaces and their dependencies
- Software units and their dependencies
- Software unit descriptions
- Service buses.

In chapter 9, we examine the structural aspect in some detail.

7.1.2 Overall approach

The overall approach to SOA is illustrated in figure 7.1. Think of the SOA as evolving piece by piece. Note that evolution of the SOA is a balance between "top-down" and "bottom-up" approaches. It is driven top-down in terms of the BA. At the same time it is driven bottom-up from consideration of the existing software assets, technology architecture, and SOM.

7.1.3 Project and enterprise SOA

It is usually necessary for practical reasons to break the SOA into different pieces, otherwise it becomes too big and complex to be useful. At least two levels of SOA (*project* and *enterprise*) are required in order to provide different levels of abstraction in an enterprise of any size

The distinction between the project SOA and the enterprise SOA is akin to the distinction between building or district architecture and city planning. While a project SOA provides an overview of the software services to be used by a particular project it also includes further details that are relevant to that project. The objective of the project architecture is to guide a particular project by declaring which existing services are to be reused and which are to be newly developed.

An enterprise SOA stretches across all the projects in an enterprise – or, more commonly, a significant sub-set of the enterprise. The objectives of the enterprise SOA are to set a vision for service orientation that transcends the traditional barriers between business and IT, taking a business-driven approach as indicated in chapters 4–6. Individual project SOAs should conform to the enterprise SOA.

For reasons of brevity we assume throughout this book that the SOA refers to the enterprise SOA unless the context demands otherwise.

[1] Note that, for the sake of brevity, unless specifically stated otherwise, the term "service" is used to mean "software service" in our subsequent discussions of SOA.

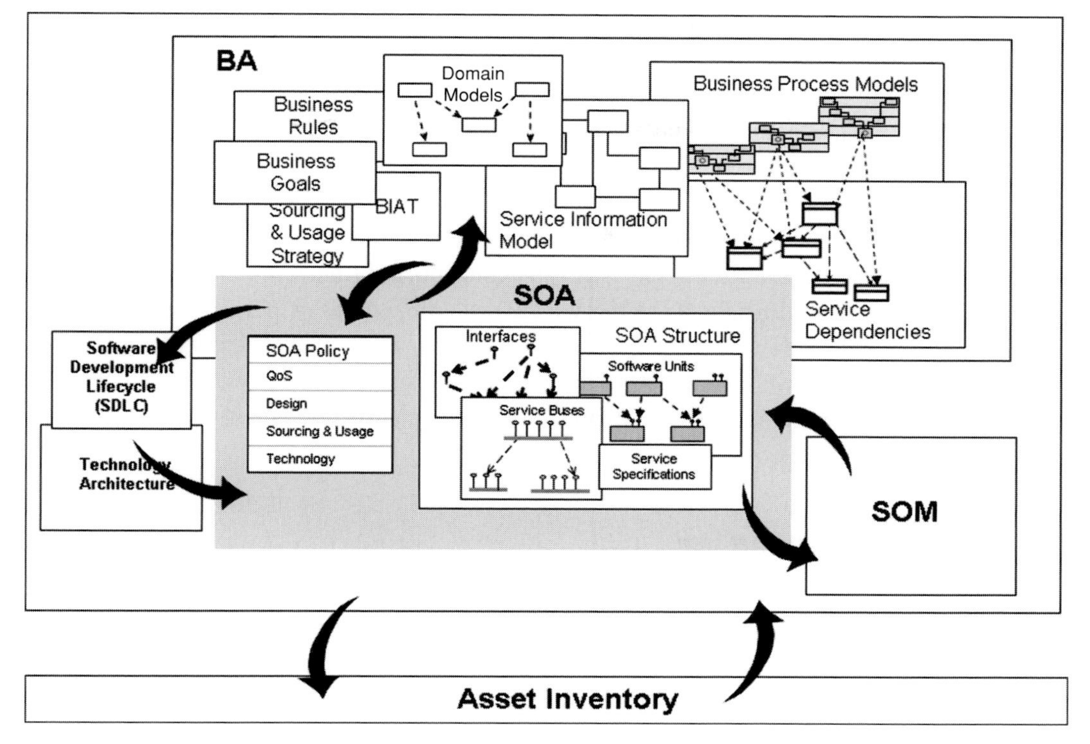

Figure 7.1 The SOA in a roadmap context

7.1.4 Provider and consumer viewpoints

The SOA has *consumer* (customer and user) and *provider* viewpoints. Consumers need to know what services are offered and how they can be put to work for the business. Providers need to know what services they are contracted to offer and also the dependencies on other services. Much of the work involved in the SOA is about balancing and trading off customer requirements for services against the available provider (including in-house) services and arriving at an effective overall software design. Similarly much of the work involved in SOM is about balancing and trading off user demands for run-time services against available supplier resources to actually execute the services, while ensuring service levels that have been agreed with customers. Moreover these demands can fluctuate and must therefore be handled in real time.

7.2 SOA perspectives

The SOA involves shifts in mind-set that do impinge upon some of the traditional assumptions of software development and IT architecture. It has to be said that this is potentially a very big topic! To manage scale and retain focus, we restrict attention to three particular key areas:

- Traditional IT architecture
- Risk management
- Operations management.

The intention is to provide some cultural underpinnings for chapters 8–12 that delve into the details of the SOA and execution management of the SOA. At the same time, SOA is not a revolutionary idea – many of the concepts have been around for a long time. We do not want the baby to be thrown out with the bathwater. Therefore, while being mindful that we do not want the baby to run the house (if you can bear with the analogy for a moment), we have emphasized the need to build on hard-won knowledge and things that work, while discarding the things that patently do not work in the brave new world of service orientation.

7.2.1 Traditional IT architecture

Until recently, books such as Shaw and Garlan (1996), Bass, Clements, and Kazman (2003), and the Zachman Framework (O'Rourke, Fishman, and Selkow 2003) dominated in the industry. Even though people still talk about these approaches today, to those involved in the latest enterprise architecture efforts the approaches seem very dated. In essence, Bass and Zachman are both focused on IT enterprise architectures. More important, they are focused on using the enterprise architecture as a *classification schema* that can be used to store information in databases. There is certainly some value in this approach, but it is losing its value as organizations try to become more agile and integrated.

Too often, the enterprise architecture is treated as little more than a collection of box diagrams. While there is nothing wrong with diagrams (indeed, they have an important role to play), an SOA puts much more of an emphasis on the capability of services to achieve business *agility*. This means joining up the business process redesign efforts with the *policies* that govern the QoS levels required to meet the business targets. We need to address policy in terms of the agility, capacity, availability, and security of services, as well as sourcing and usage of services. At the same time, this does not mean ignoring the design and technology approaches (topics usually covered quite well in traditional approaches) used to support the services. But it does mean a shift of emphasis. In the next section we consider one particular example: the increased profile of the SOA as a vehicle for managing risk.

7.2.2 Risk management

Risk has two components: the *cause* and the *effect*. Risk causes are possible future events that will lead to undesirable outcomes. Risk effects are the undesirable outcomes. Only the causes can be managed, but the justification for risk management in the first place is all about the possible outcomes. There are also many ways in which the concept of risk can be applied. As far as software is concerned it is important to distinguish *project* risk from *product* risk.

In a tremendously insightful account, DeMarco and Lister (2003) provide a simple yet very powerful approach to managing risk. The authors make the key point that most software people think in terms of *risk avoidance*. However, managing risk on software projects need not always be a defensive measure. Greater risk can, they argue, bring greater reward. A company that runs away from risk will soon find itself lagging behind its more adventurous competition. On the other hand, by ignoring the threat of negative outcomes, software managers can court software project disasters in the name of a positive "can-do" attitude. A *balanced approach* is needed in which threats are weighed up against opportunities. So, you ask, what has all this got to do with SOA?

The first point to note concerns product risk. Our approach to risk should match the kind of service that we want to offer the customer. The SOA helps to differentiate services according to whether they are simply commodities that are best bought or rented from providers, or whether they have a value-add that requires nurturing in-house. Value-add services are functionality that reflects the special value that an organization brings to market. They differentiate the company from its rivals, conferring competitive advantage and are often highly innovative: examples are Product Marketing and New Channel Sales. It is important to note that value-add functionality is often much better not designed as a service, certainly in its initial iterations.

This is because the innovation that comes with such functionality is usually best exploited quickly, without incurring the overheads inherent in engineering a service with demanding QoS requirements. DeMarco and Lister (2003) might well say that there is a greater risk involved that brings greater potential rewards. Later, however, it may be appropriate to upgrade the functionality to meet these QoS requirements (such as responsiveness and security) especially if, for example, we are offering to a wide audience. Upgrading or servicizing the value-add functionality can be seen as a way of managing the risk associated with it.

At the other end of the service spectrum are commodity services, such as currency converters, that have been tried and tested over and over again, thus ensuring high QoS levels. The larger the user base, the greater the "ripple effects" of quality problems. Commodity services simply have to work efficiently to manage the risk of the service degradation or failure. In a nutshell, an SOA helps to manage product risk by visualizing the dependencies and specifying the quality attributes of the services that are used and offered by the product.

The second point concerns project risk. The traditional software engineering methodologies that accompany IT enterprise architectures tend toward managing risks associated with the software development process. These software development methodology (SDM) approaches are perhaps most vividly exemplified by the Capability Maturity

Model® Integration (CMMI) (Ahern, Clouse, and Turhes, 2003; Chrissis, Kourad, and Shrum, 2003). However, as we have emphasized throughout this book, the IT world continues to move further and further away from software development and more and more toward integration and composition of solutions from disparate sources. As I have argued before (Allen, 2001b) this situation calls for a combination of lightweight processes and clear specifications, in contrast to the process-centric SDM approaches. The SOA is ideally placed to help manage the risks associated with this rapidly emerging world.

Rather than focus on controlling risks related to the software development process, DeMarco and Lister (2003) argue that it is more pertinent to focus on the actual *sources of uncertainty* and to ask how software projects can go about managing these factors. They provide a list of ten typical sources of uncertainty that do indeed characterize many of the challenges that organizations face today:

1. *Requirement*: What exactly is it that the system has to do?
2. *Match*: How will the system interact with its human operators and other peer systems?
3. *Changing environment*: How will needs and goals change during the period of development?
4. *Resources*: What key human skills will be available (when needed) as the project proceeds?
5. *Management*: Will management have sufficient talent to set up productive teams, maintain morale, keep turnover low, and coordinate complex sets of interrelated tasks?
6. *Supply chain*: Will other parties to the development perform as hoped?
7. *Politics*: What is the effect of using political power to trump reality and impose constraints that are inconsistent with end-project success?
8. *Conflict*: How do members of a diverse stakeholder community resolve their mutually incompatible goals?
9. *Innovation*: How will technologies and approaches unique to this product affect the eventual outcome?
10. *Scale*: How will scaling up volume and scope beyond past experience impact project performance?

Table 7.1 illustrates how traditional SDM and SOA approaches can help manage risk. We have kept the analysis deliberately simple and binary in that SDM and SOA are depicted as either supporting management of the risk or not. Of course, real life is much more complicated than that and there are degrees of support that might be identified in different circumstances, including types of organization and types of system. However the main intention is to keep things simple in order to illustrate the main point: that the SOA is an important tool for managing modern-day risks, in a way that traditional approaches are unable to do.

Notice that both SDM and SOA address the first two sources of uncertainty. Methodologies have always been reasonably good at addressing both functional requirements

Table 7.1 *Managing risk with traditional and SOA approaches*

Source of uncertainty	SDM	SOA
Requirement	☻	☻
Match	☻	☻
Changing environment		☻
Resources		
Management		
Supply chain		☻
Politics		
Conflict		
Innovation		☻
Scale		☻

and scoping and system interfacing issues through various modeling techniques. There is no shortage of techniques: everything from the dataflow diagrams and context diagrams of structured analysis approaches used in the 1980s to the different types of Unified Modeling Language (UML) diagram prevalent today. In addition, there is an established range of prototyping tools and techniques for separating out and understanding the human computer interface.

SOA provides capabilities for four more areas of uncertainty that are largely neglected by SDM. Let us take a brief look at each of these areas.

- *Changing environment*: SOA is expressly concerned with change by focusing attention on *agility*. In particular, SOA is based upon a clean separation of interface from implementation and focuses strongly on the management of interface dependencies.
- *Supply chain*: Service specification techniques provide a foundation for addressing the supply chain issue by expressing agreements between providers and customers in contractual fashion. The whole rationale of a service-oriented approach is based on unambiguous specification of *interfaces*.
- *Innovation*: SOA helps to visualize an organization's line of commoditization. Services are organized according to whether they are simply commodities or whether they represent unique core competencies, or whether they are somewhere between these extremes. This helps highlight areas where *innovation* is needed and also to make the right decisions on *service provision*.
- *Scale*: Service specification techniques address the need to measure *QoS* requirements head on and provide a foundation for addressing capacity, availability, and security management. This aspect of SOA plays a major role in addressing the need for scaleability but, at the same time, raises the bar for service management, as we discuss in the next section. The paradox is that there is an increased risk that comes with the scale and complexity of highly distributed systems enabled by software services.

Significantly there are four sources of uncertainty – resources, management, politics, and conflict – that are very largely untouched by either SOA or SDM. These are *human risks* that require softer approaches that relate to the culture and psychology of software projects and the organizations in which they exist. Neither SDM nor SOA are panaceas for all ills! Both have their limits, though it may be reasonably argued, of course, that the techniques associated with either SDM or SOA can play a part here. For example, both are often catalysts to the improvements in human communication that are required to handle these "softer" kinds of risk. At the same time, if not sensitively handled, both approaches – and SOA in particular – can actually increase these softer risks. Heavy-handed architects that seek to dictate demands to development teams are a case in point!

7.2.3 Operations management

We stated above that there is an increased risk that comes with the scale and complexity of highly distributed systems enabled by software services. Much better technologies and standards for interoperability are available than in the days when software development was largely confined within the single organization. However, the scope and complexity of the business problems that we are trying to solve has grown hugely in a way that transcends the organization boundaries of traditional operations management. Not only that but, as we try to keep pace with the speed of change, the standards are proliferating and becoming a challenge in their own right! While the SOA helps to plan strategies and specify services, more is needed to cope with the risks associated with this massive scaling up of business problems. Strategies must be executed. Clearly specified software services are all well and good, but the software has to work, and work extremely well in demanding conditions. Enter therefore SOM, which we discuss in the next section.

7.3 Integrating execution management

In this section, we provide an overview of the relationship between the SOA and SOM. This inevitably involves something of a quick preview of the elements of SOM, a subject that we return to in detail in part 4 of this book. It is also important to realize (as with the other themes discussed in this section) that what we a talking about here is not just a technical phenomenon: it is a shift in *culture* and *mind-set*; chapter 13 covers the key topic of roles. Enterprise IT architectures have traditionally been kept very separate from the approaches and techniques required to actually execute, monitor, and control run-time software. This is reflected in the time-honored split between the domains of systems development and operations. However, as we are about to see,

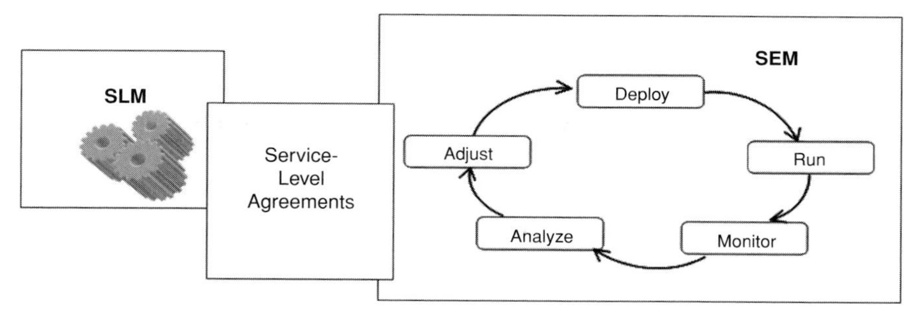

Figure 7.2 The elements of SOM

service orientation cuts right through this traditional divorce of disciplines in a way that calls for a much more *holistic* approach.

7.3.1 SOM in a nutshell

Recall that SOM refers to the combination of SEM with approaches to SLM enhanced to handle the challenges of service orientation. Effective SLM is in turn dependent upon good consistent SLAs that are used by SEM, as illustrated in figure 7.2. To underline a constant theme of this book: "You can't manage what you don't measure," but more than that "You can't measure what you don't specify."

At the same time, this is a two-way street: services must execute according to agreed QoS levels, and measures put in place to ensure that wherever possible the software implementing the service "self-corrects" with the minimum of disruption to users or to other services. SEM centers on the run-time execution of software services and consists of five main cyclic activities, depicted in figure 7.2, as follows:

Deploy: This activity moves a service from its acceptance-tested state into production.

Run: This activity covers the day-to-day operation of the deployed services, so that they remain available for users and consuming software to execute.

Monitor: This activity defines the tasks required to monitor the health of deployed production services. This includes setting "thresholds," observing system performance, monitoring the observations, and reporting on findings, in the form of alerts, online queries, dashboard displays, and periodic reports.

Analyze. This activity examines the monitoring data and attempts to diagnose faults and their possible remedies.

Adjust. This activity covers the actions taken to tune the technical infrastructure – hardware, network, and software platform for the services – to alleviate problems identified by analysis – or, better still, to prevent the future problems that can be foreseen by the monitoring and analysis.

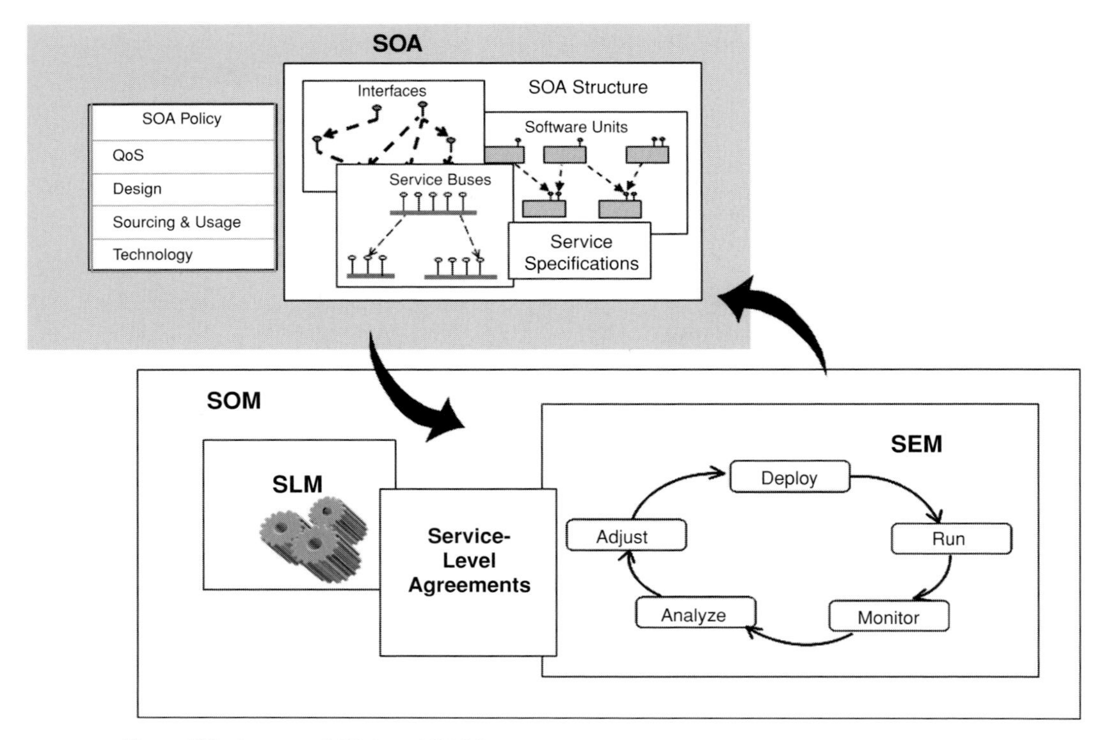

Figure 7.3 Integrated SOA and SOM

7.3.2 Integrated SOA and SOM

The monitor activity of SEM scrutinizes certain elements of the SLA, which reference QoS levels (such as agreed capacity, availability, and security levels) specified within the SOA. In summary, the SOA provides the specifications against which services are monitored. Conversely, the SOM provides input to the SOA design in the form of regular analyses of service behavior as well as requests to correct problems. Up until recently it may have been possible for SOA and SOM specialists to work in their own isolated comfort zones. Right now, though, the need is for a holistic approach to SOA and SOM, as illustrated in figure 7.3. This need can only get greater with increased uptake of software services.

7.3.3 The on-demand perspective

As software services usage scales up in terms of numbers of applications, each making numerous calls to software services on behalf of perhaps hundreds of users (involving increased amounts of information) across organizational and geographical boundaries, so agility, capacity, availability, and security of the software services become

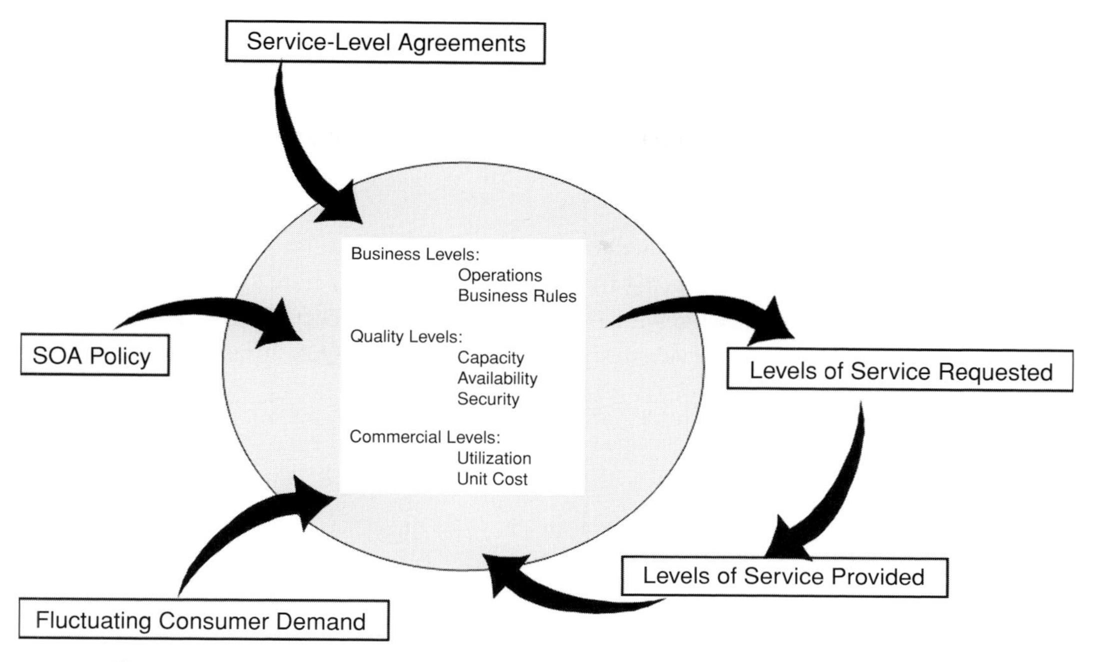

Figure 7.4 The context for on-demand software management

increasingly critical. These issues are magnified with increased uptake of complex asynchronous software services.

As the uptake grows, it becomes essential to alert key managers when capacity or availability falls below predetermined thresholds, or when security is compromised. In many of these situations, it is not realistic to wait for decisions to be made manually. Increasingly, corrective actions need to be automated in order to rectify problems in real time through effective control measures. This is the essence of on-demand management software.

If services are to adapt, in an on-demand fashion, to ensure that QoS requirements are met in ever changing run-time conditions, then SEM is absolutely key. Moreover the essence of the on-demand idea is that the service "self-corrects" in dynamic fashion with the minimum of disruption to users or to other services. Service supply must be balanced against service demand. Significantly, this also means that a service supplier can and must turn services "down," if it cannot meet the terms specified in the SLA.

In order to achieve this objective, measures must be put in place that enable the SEM software to make various trade offs, as illustrated in figure 7.4. The management of on-demand computing extends the traditional idea of gap analysis by adding the important dimensions of time and control. The SEM software continuously compares levels of service provided with requirements that are reflected in SLAs, which must be consistent with SOA policy. On this basis, the SEM software decides how much to adjust service levels up or down, ensuring value for money in terms of effective utilization – just enough computing resource for the demand.

7.4 Where to next?

In this chapter, we have outlined the elements of SOA and discussed some of the shifts in mind-set involved in moving from traditional IT architecture to SOA. In chapter 8, we build on this discussion with a survey of the policy aspect of SOA. This provides a context for discussing service design (in chapter 10) and QoS infrastructure design (in chapter 11).

8 Service-oriented architecture policy

8.1 Foundations of SOA policy

Recall from chapter 4 that business goals and rules, the BIAT, and sourcing and usage policy are captured as part of the BA policy. Traditional approaches tend to divorce business concerns from IT. In contrast, as we shall see in this chapter, a service-oriented approach involves a very close relationship between business and IT concerns, between BA policy and SOA policy.

Another problem with traditional approaches is that they tend to focus on business functionality and information, and to neglect the area of *policy*. In contrast, a service-oriented approach emphasizes policy as much as functionality and information requirements. In this chapter, we introduce the policy aspects of SOA along with techniques for capturing and developing that policy in alignment with business needs.

8.1.1 SOA policy aspects

The policy aspect of SOA, shown in overall context in figure 8.1, falls into four parts:

- QoS
- Design
- Sourcing and usage
- Technology.

While design, sourcing and usage, and technology all represent critical areas of SOA policy, our view is that progress in each of these areas has been up to now much better than in the case of QoS. To a large extent we can evolve what we already have learned regarding policy in the areas of design, sourcing and usage, and technology. Indeed, very often organizations have no choice but to evolve in a manner that seeks to maximize return on existing assets. Business leaders understandably will have it no other way. In addition, many of the required developments in design and technology are well under way.

Unfortunately, however, the same cannot be said of QoS. Part of our aim is to elevate the importance of this often neglected area and provide badly needed guidance in preparation for effective execution of services. While this chapter provides an overview,

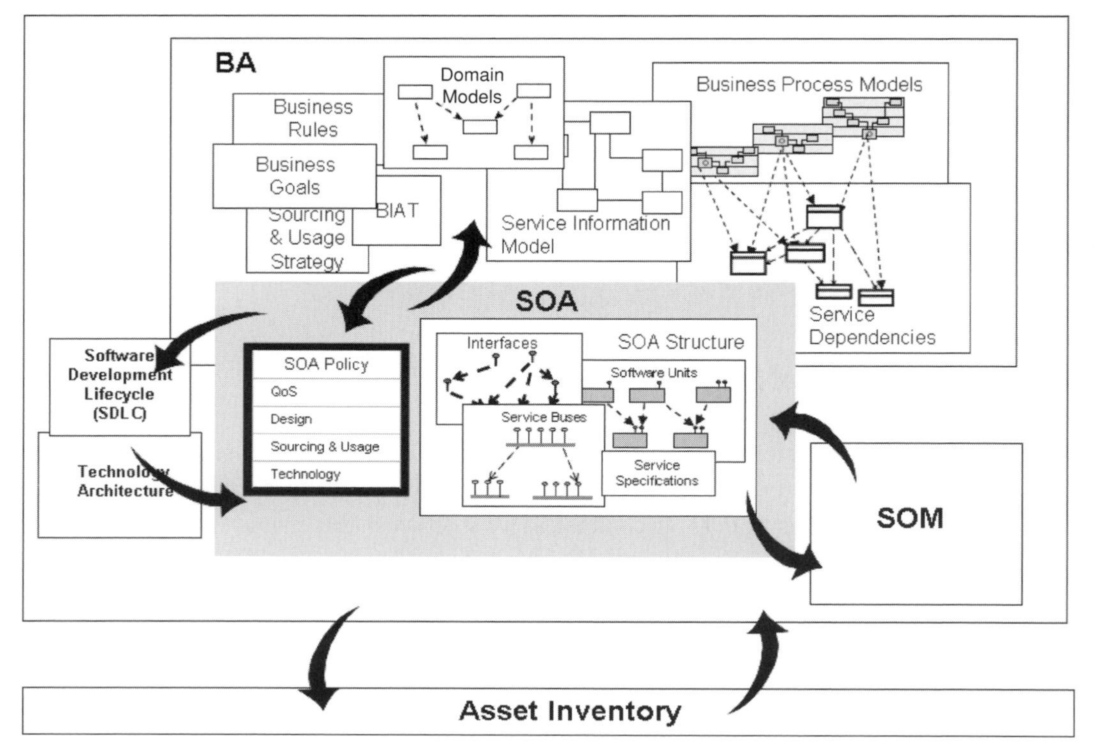

Figure 8.1 SOA policy in a roadmap context

with checklists, of each of the four aspects of SOA policy, we therefore go into more detail on QoS in chapters 9 and 10.

8.1.2 QoS types

QoS divides into four types:

- *Agility* is the ability to act quickly and with economy of effort in accurate response to change, and also to initiate change for business advantage.
- *Capacity* involves three inter-related capabilities of service – throughput, responsiveness, and storage capacity. In addition a fourth concept, scalability, focuses on how each of the three other capabilities is impacted by future demands and changes in technology.
- *Availability* is primarily concerned with keeping services free from failure within agreed limits and involves several parameters (that affect our trust in a service to do what it is intended to do), including reliability, maintainability, and resilience.
- *Security* focuses on the confidentiality and integrity of services, ensuring that the services are available when needed and free from denial of service attacks. In addition,

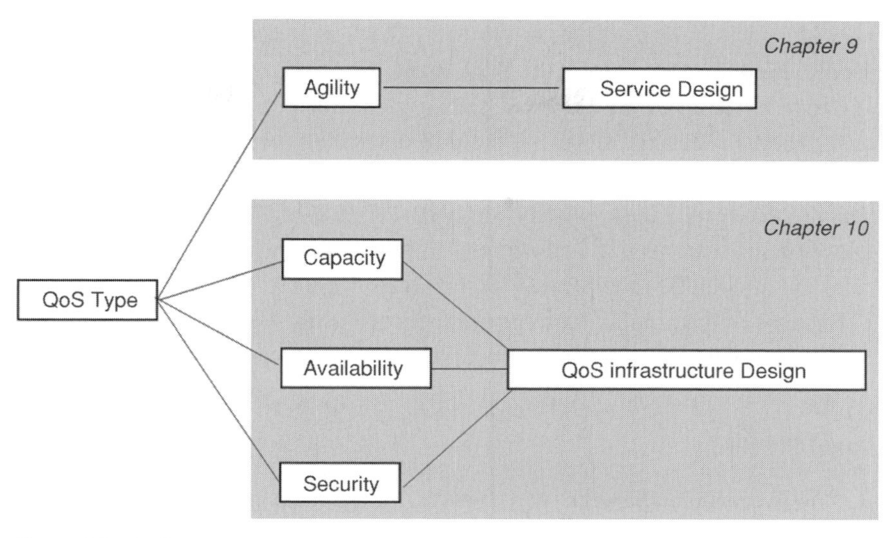

Figure 8.2 QoS types roadmap

security must also address three underpinning principles: authenticity, authorization, and nonrepudiation.

Each quality type comprises a number of quality attributes that we explore and define later in this book, as indicated in figure 8.2. A *quality attribute* is a specific measurable characteristic (such as responsiveness) that may take a specific value or QoS level (such as responsiveness = 20 msec).

The importance of specification of quality attributes in achieving software that meets business requirements was eloquently stated by Tom Gilb in a seminal account (Gilb, 1988). However it only seems to be quite recently (Allen, 2002) that we have woken up to the fact that they are a critical element of SOA. QoS types correspond to different aspects of SOA design.

Agility is an overarching characteristic that most strongly impacts service design in terms of the overall structure and shape of the SOA.

Capacity, availability, and security most strongly impact the technical infrastructure requirements of the SOA in terms of different supporting technologies. In addition, all three are intimately related to SOM, and may affect the specification of individual services and operations.

8.2 Business–IT alignment

In order to ensure a business-driven approach, our starting point is the overall policy as embodied in the BIAT, which relates service-oriented features with clearly defined business goals. While the BIAT is essentially a communication tool that helps demonstrate very clearly to business leaders why the SOA is necessary in business terms, it is also used to help drive and shape the SOA itself.

Other less high-level business goals will form measures of success for the business process redesign efforts and help to determine service policy in terms of quality attributes. Yet more detailed business goals, along with business rules, will relate to specific services. These must be supported by the SOA and allocated to elements of the SOA structure. A detailed account of how lower-level business goals and business rules are allocated to the SOA is included in chapter 9.

Because of the potentially huge numbers of business goals and rules in an enterprise of any size, we shall see that a selective and pragmatic approach is called for.

8.2.1 Using the BIAT

It is first worth recalling the overall goals for the redesigned hire process as follows:

- 50% decrease in numbers of phone reservations
- target of 200 online reservations per day
- 40% increase in number of hires
- 40% increase in productivity of hire agent
- 50% increase in customer satisfaction rating.

In our process redesign work (see 5.4.2) we interpreted these goals in terms of SOV7 and then mapped these goals to service-oriented features in the form of a BIAT (table 6.1, in 6.2.2). For each goal/service-oriented feature pair, we now consider which quality types are most relevant to the pair and add the quality types to the BIAT, as shown in table 8.1.

8.2.2 Developing the BIAT

The astute reader will observe that it is possible to relate each of the four quality types to each of the high-level goals of SOV7 so that a table can be constructed that provides twenty-eight rows, each corresponding to a pair. The row can then indicate the relevance of a particular quality type to a business goal. Indeed, that would be a valid strategy to follow, in effect the table being filled out interactively focusing on each of the twenty-eight rows in turn and asking how relevant the quality type actually is in regard of the high-level business goal.

In the example above, we have chosen to focus on the most critical of the quality types for each high-level goal in the context of the vehicle hire scenario. We can imagine that these results were arrived at following a brainstorming session with senior business executives and architects.

A further step is to prioritize each of the pairs of quality types and high-level goals; in the example, in the interests of simplicity, we have chosen not to do that.

Table 8.1 *BIAT for vehicle hire company*

High-level goal	Quality type	SOA feature
Transparence: Introduce reservations over the Web as well as by phone. Streamline the customer's experience by automating verification checks for risk and creditworthiness.	Capacity	An SOA promotes identification of well-designed services that reflect business needs. Such services can provide a high level of automation and performance that can be tuned, using appropriate execution management software, to varying run-time constraints.
Customer fit: Introduce special offers and options for Gold customers ranging from hotel deals to route planning. Personalize the reservation by using history of customer's dislikes and preferences. Introduce delivery of vehicle to customer's location.	Agility	Exposing the line of commoditization allows the organization to focus on designing value-add services that ensure customer fit through interoperability with other service types.
Partner connectivity: Work (as a customer) with vehicle broker to gain access to further vehicles, where company is unable to fulfill customer's hiring requirement. Work (as a provider) with garages that need to supply their customers with temporary vehicles while their vehicles are being repaired or serviced.	Availability Security	Services are specified to a rigorous level that allows certification to validate stated availability and security requirements across organizational boundaries. This helps customers to effectively manage the external service provider relationship, and helps providers to ensure that customers are receiving the agreed level of service.
Adaptation: Anticipate and support possible changes: • Organization changes such as mergers and takeovers • Business process changes such as delivery of vehicles to customer's location • Information changes such as pricing structures for bulk reservations • Business rules changes such as changes to hiring rules.	Agility	Guidelines that structure services in a way that minimizes the potentially disruptive effects of change at all levels. This approach leverages the potential of services for: • Adaptability: Regarding business change services can be extended in "plug and play" fashion in different business contexts • Flexibility: Services can be designed to smoothly accommodate change to information and rules • Replaceability: Regarding associated technology changes a service can be smoothly replaced with another having a different component implementation code.

(*cont.*)

Table 8.1 (*cont.*)

High-level goal	Quality type	SOA feature
Multi-channel capability: Introduce reservations over the Web as well as by phone; see also *Transparence*. Support reservation process using different channels consistently through the process; for example, reserve over Web, enquire by cell-phone.	Agility	An SOA helps remove dependencies on specific technologies through the use of standardized interfaces. Services are designed that can exploit this capability thus allowing reuse in different channels and technology contexts.
Optimization: The ability to offer reservations of predetermined vehicle types in real time by cell-phone. Real-time alerting of vehicle incidents (from faults to thefts).	Capacity	Services can be deployed using technology appropriate to varying run-time constraints and that provides the capability to respond to high transaction volumes in real time. Note that effective scaling up of the SOA depends upon effective SEM technologies that provide monitoring and control of software services.
One-stop experience: Introduce a Web site incorporating reservations as one piece of complete travel package with links to hotels, ferries, restaurants, and leisure outlets such as theaters and cinemas.	Availability Security	An SOA fosters appropriate design of loosely coupled software services, allowing them be used together in different combinations in different business contexts. The usage across organizational boundaries requires potentially high levels of availability and security. This depends upon effective SEM technologies that provide monitoring and control of software services.

Regardless of how much detail you choose to go into, the critical point is to glean an idea of *overall direction for SOA policy*. This applies to design, sourcing and usage, and technology policy, as well as QoS requirements.

8.3 QoS criteria

Despite the very best of intentions, software quality is usually relegated to a back seat role in traditional approaches to software engineering and IT architecture. This situation can be summed up no better than in the very expression "nonfunctional requirements" that is often used to characterize software quality and which conjures up a view of quality as being a cross between "nonimportant" and "dysfunctional." In the world of service orientation this situation has to change!

Understanding the QoS types is therefore an important challenge. Agility, capacity, availability, and security also have complex inter-relationships with one another. For example, measures that help increase capacity may decrease agility, and vice versa. Key challenges are involved in trading off quality attributes in line with business goals and technical capability.

In the case of traditional systems, these challenges are often addressed at a fairly high level of granularity – for example, application-level capacity – and in many cases the challenges are dealt with on a piecemeal basis. In addition, the quality requirements are one-dimensional – they do not, for example, fluctuate with time.

In the context of an SOA, this will not do. QoS levels must be established as part of the SOA, as part of a planned strategy. At the same time, the approach must be workable: it must not be so heavy with details that it overwhelms those who seek to use it. Like a ship's engine, it must be powerful enough to move the ship without sinking it through its sheer weight.

The concept of QoS criteria is designed to provide a classification scheme, which is part of the SOA, for handling these concerns.

8.3.1 Layering of QoS requirements

QoS requirements should be layered to reflect scope. Layering provides a means of managing the complexity of QoS requirements, while at the same time exercising architectural control over that policy.

At the highest layer, QoS requirements may apply *across the enterprise*; for example:

- Services must be kept confidential and maintain their integrity at all times. All the service users are known, trusted, and authorized to use the services and can be clearly identified.

At the next layer down, QoS requirements may apply to a particular *domain*. Further layers (such as sub-domains and processes) may be specified depending on the scale and complexity of QoS requirements. At the lowest layers, QoS levels may be specified for a particular service, or an operation of a service; for example:

- The information provided by the Close Contract operation shall be accessible only by users of authority level = 2.

Figure 8.3 provides a full range of examples for the above scenario. Organizing the QoS requirements in this fashion expedites the creation and control of SLAs. In particular, we avoid the trap of repeating the same policy by effectively normalizing it. At the same time, we have a way of ensuring that corporate policies ripple right through to SLAs.

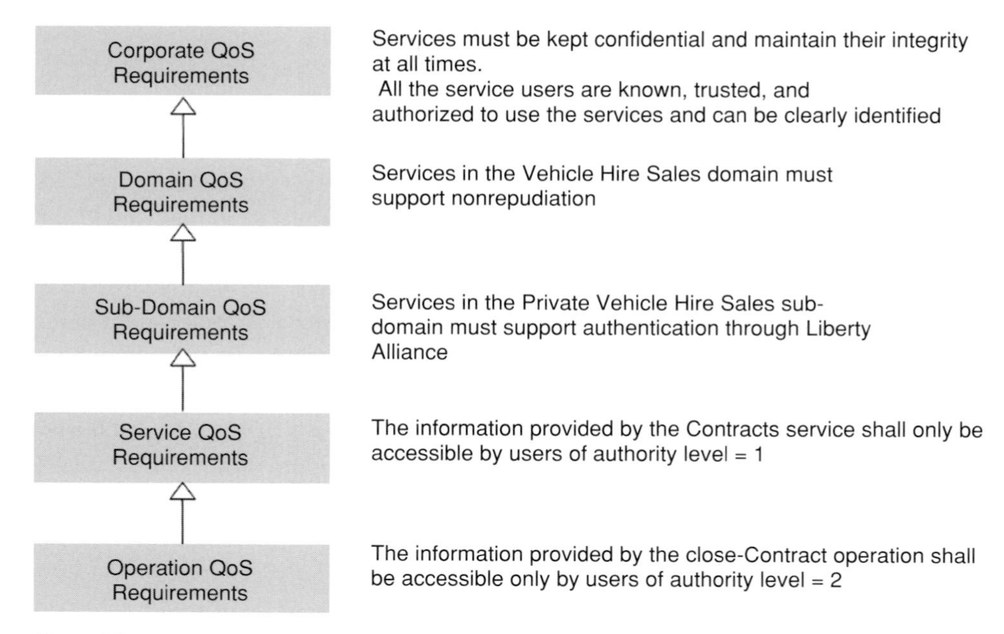

Figure 8.3 Layers of QoS requirements

The concept of layering does not apply only to QoS requirements. It can usefully be applied to all types of SOA policy, as well as other SOA artifacts. In particular, it is very useful in handling business goals and rules, as described in chapter 9.

8.3.2 Dimensions of QoS criteria

Thus far, we have discussed the scope of QoS criteria. There are several other dimensions that must be considered. Four of the most important of these dimensions, illustrated in figure 8.4, are: role, service type, time, and criticality. QoS may then be varied – qualified, over-ridden or extended – according to different dimensions and in various ways.

How much detail to actually create in the form of quality requirements for each of the dimensions is at this point an open question. A balance must be achieved between thoroughness and usefulness. However, some analysis of these different dimensions of QoS is always likely to be useful in achieving a good SOA.

8.3.3 The role dimension

Quality variation according to role reflects the entitlements of different types of service user or customer – for example, superior levels of performance for privileged customers, or upgraded levels of security against use by nonauthorized personnel. Roles are very

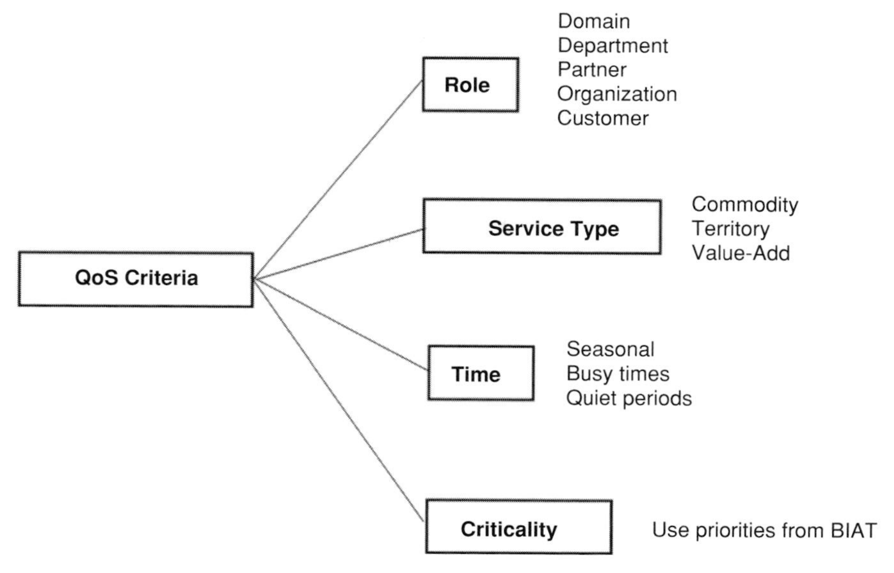

Figure 8.4 Dimensions of QoS criteria

often relative to domains. Several roles may be involved in the same domain. Layering by domains and sub-domains, as discussed in 8.3.1, can be a useful technique. Matrices in which roles (or domains) are plotted against quality attributes with cells filled in to indicate QoS levels can also be used. In chapter 9, we explore the use of service buses as a means of effectively managing QoS requirements according to domain.

8.3.4 The service type dimension

Partitioning into service types (commodity, territory, and value-add) has a significant effect on QoS. Different service types are likely to have different quality requirements and it is useful to form priorities for each type of service. Here are some overall observations.

Value-add services

Value-add services bring *innovation* and *uniqueness*. The critical success factors for value-add services may well have much less to do with QoS and much more to do with differentiation and flair. It is usually important not to compromise the innovative nature of a value-add service through overstringent quality requirements. Equally, sometimes it may actually be aspects of quality that bring differentiation – for example, a particularly rigorous security service used in the military domain. So this is not a hard and fast rule.

If an organization is particularly successful with a value-add service then it, as a service provider, might want to take it to the wider market – in which case, of course, the service may become a commodity service relative to its customers. In that case, quality criteria for commodity services start to become increasingly relevant. For example, I once worked for a large organization that developed a particularly useful scheduling system for its engineers. It was so successful that it was enhanced to cater for wider market needs and sold as a software package.

Commodity services

Regarding commodity services it is important to be able to manage these services effectively. Accurate *measurement* and *control* is a vital aspect of achieving value for money from commodity services. For example, outsourcing a commodity process may seem like a good idea but will require clear agreements that specify exactly what the provider is responsible for and the targets that must be met. The fact that a service is deemed a commodity does not mean it is relegated to unimportant, as is commonly assumed. For example, factors such as capacity, availability, and security are usually critical for commodity services. Unless these are specified clearly, the organization can quickly lose control over its ability to manage utilization of those services. Until recently, perhaps, this has not been a big deal. Now, however, it is a big deal, especially as software services play an increasingly central role in the business process. If a business process is outsourced then so too is the software part of that business process.

This is a particularly salient issue with software services invoked over the Internet with increasing geographical reach, and more generally it is an issue for utility computing. If a human gets it wrong we can somehow fudge along, but software is notoriously unforgiving. And business processes are increasingly dependent on software services.

Territory services

Territory services tend to be subject to frequent change. Agility therefore looms large. For example, I once worked as software project manager for a life insurance company. At the time, unit-linked life insurance was a relatively new concept. Products were initially offered that gave the consumer a choice of a single fund in which to invest. Then someone came up with the outrageous idea of three funds. A system was duly designed to cater for policies with three funds. There were hardly any takers for two funds, yet alone three, so surely a database design with three fixed occurrences of fund was more than adequate. Over time, however unit-linked policies grew in sophistication. Policies with multiple fund investments grew popular – up to six, seven, or eight funds and more. Because of the rigid system design, however, we found it very difficult to respond to the changing market with more flexible products.

Later down the track the company learned from its mistakes. Fund Management was seen as a territory service and a flexible parameter driven design was used for it. This was in contrast to a commodity service such as Underwriting which was eventually outsourced altogether and to value-add services such as individual Product Sales that offered lots of channel differentiation to the sales-force by using a personalized interface to guide the sales process.

8.3.5 The temporal dimension

The temporal dimension is particularly salient in the case of on-demand computing. At one extreme, quality levels might vary according to time of day (for example, morning versus afternoon versus night time). At the other extreme, quality levels might be subject to seasonal variations (for example, summer versus winter). The idea of on-demand computing requires that "just enough" IT resources are used as and when needed. For example, capacity (throughput, responsiveness, and storage capacity) may vary dynamically through time in relation to demand. Service capacity policy must set thresholds that specify acceptable tolerances for monitoring by SEM software.

8.3.6 The business criticality dimension

Not all services are created equal: some are more critical to the working of a business than others. For instance, key services that perform such functions as a credit check or execution of a market sell order are vital components of a business process where degradation in performance may bring the entire business process to a halt. Others, such as those that provide supplemental information such as product information or weather details, may be merely "nice to have."

In general, territory services may appear the most critical. However criticality is determined largely by the business context of a service rather than the commoditization level. For example, a commodity service such as vehicle cleaning (outsourced to a cleaning firm) might be absolutely critical for ensuring high levels of customer satisfaction in the car hire example. One of the best ways to gauge criticality is with reference to the BIAT, which can be used to help prioritize QoS requirements according to level of business criticality.

8.4 Design policy

The SOA should follow a consistent design approach in line with QoS requirements. The design approach consists of three main areas of design policy, which we discuss briefly in this section:

- Design rules
- Design patterns
- Design guidelines.

First and foremost, design policy should be consistent with the BIAT.

Design policy is closely related to, and influenced by, the other elements of SOA policy. For example, regarding QoS, an emphasis on agility may call for strong guidance on achieving flexible service designs. This may need to be balanced against a need for secure yet scalable services. The design policy needs to help in making these trade offs.

In addition, the sourcing and usage policy inevitably affects design policy. For example, an emphasis on using external providers over internal development may call for a minimalist approach to SOA design.

Last, but not least, technology policy has a huge influence on design policy. In particular, most organizations have existing technology portfolios. It is important to build on the body of design knowledge that will have grown up during the course of the development of these technologies within the organization. As we have emphasized throughout this book, service orientation is essentially an evolution of best practices.

In fact many of the rules, patterns, and guidelines may be little more than an extension of good overall architectural practice. However there are policies that are particularly relevant to a specifically service-oriented architecture, and we discuss some examples in this section.

8.4.1 Design rules

Design rules constrain software designs, which must be compliant with the rules. Here are four examples of typical rules.

Device independence

All services are user interface-independent. They must be usable with any type of user interface.

Service dependency

Dependencies between interfaces and software units may obtain as follows: there are no restrictions; cyclic dependencies are allowed.

Layering

All services must be layered in terms of degree of reusability (with distinguishing criteria for each).

Data management

Instances within software units must have an immutable identifier. Physical deletion of instances is allowed, though this must be logged.

8.4.2 Design patterns

In 4.2, we introduced the idea of *patterns* – approaches or solutions that have often worked in the past – in the context of business process redesign. An SOA design pattern describes a recurring design problem that arises in a specific SOA design context and presents a well-proven generic scheme for its solution. Design patterns are covered extensively in the literature (Gamma *et al.*, 1995) and generally are more relaxed than rules. So, for example, we might express the layering rule (above) in less stringent terms as follows:

Layering

Use a layered service pattern in cases where reuse and maintainability quality attributes outweigh capacity quality attributes.

8.4.3 Design guidelines

Design guidelines include suggested criteria for following different types of design rule. Here are three typical design guidelines.

Coupling and cohesion

A set of guidelines for achieving low coupling and high cohesion of services; for example, a maximum number of dependencies, consistent business purpose of operations, and so on.

Communication

A set of guidelines for dealing with when and how to employ different types of operation invocation (see 8.4.4). These guidelines should take into account technology policy on communication (see 8.6.2).

Agility

A set of guidelines for achieving agility of services; the reader is referred to chapter 9.

In addition design guidelines apply to other QoS types. The reader is referred to chapter 10 for a discussion of design guidelines for capacity (10.2), availability (10.3), and security (10.4).

8.4.4 Notes on operation invocation

Two basic kinds of operation invocation may be distinguished here.

Synchronous

A synchronous operation performs its work as quickly as possible on invocation from a calling application, which must wait for the response indicating completion of the operation. Synchronous operations are most suited to discrete tasks, often within the organization boundary, that can be completed in quick time.

Asynchronous

Rather than perform its work immediately, an asynchronous operation may acknowledge an invocation from a calling application, perform the work at a later time, and eventually return the results. The calling application does not have to wait for the response. The operation will send an event to denote that it has completed its work. This event may be broadcast to a number of recipients via a publish-subscribe mechanism. Asynchronous operations are most suited to complex tasks, such as those that span organization boundaries that cannot be completed in quick time.

For a detailed account of synchronous and asynchronous invocation the reader is referred to (Kaye, 2003, chapters 8 and 9).

Bulk transfer

A further operation invocation type, bulk transfer, may be distinguished, as in the case of Credit Suisse;[1] see also Schlamann (2004).

A bulk transfer operation carries large amounts of information from one location to another. It may be either "push" or "pull," according to which side of the connection issues the request. With a pull operation, the calling application waits for the full file. With the push operation, the transfer starts when a predefined event occurs, which triggers the bulk transfer. Bulk transfer operations are normally asynchronous, but may also be synchronous. They are typically used for batch processing and data warehouse feeding.

Note also that QoS levels may vary according to how an operation of a service is invoked. It may actually be possible to offer the same operation in synchronous and asynchronous modes, depending on these criteria. Perhaps at very quiet times a synchronous invocation may be allowed, in contrast to the remaining time when an asynchronous invocation is offered.

8.5 Sourcing and usage policy

Earlier in this book (4.6), we saw that service orientation involves major strategic *business* decisions on the sourcing and use of services. Sourcing concerns how services

[1] The Credit Suisse case study in chapter 15 provides examples of design guidelines for each of the three types of operation invocation described in this section.

should be supplied – for example, whether services should be outsourced or performed internally. Usage centers on the constituency to which services are to be offered – for example, whether to offer services only internally, to trusted partners, or outside of the organization to the wider market. Recall that these decisions form part of the company's business strategy and involve business processes and services in general, not specifically software services. The SOA must include policy that guides decisions on sourcing and usage of software services. These decisions (such as whether to outsource a software service, or whether to offer a software service to market software) are to a large extent driven by the decisions on process sourcing. For example, if a certain process is to be outsourced then it is likely that any software services that are used by that process will also be outsourced.

In addition, the BIAT should be used to guide the SOA sourcing and usage policy. At the very least, the SOA sourcing and usage policy should be consistent with the BIAT.

At the same time, consideration of the current software portfolio will often impact the process sourcing strategy. For example, if there are legacy software systems that "do a job" and it is not cost justified to replace them, then it may well be expedient to exploit those systems in support of the process sourcing and usage strategy.

In the following sections, we provide some guidelines and examples of SOA sourcing and usage policy based on the vehicle hire scenario.

8.5.1 Sourcing policy

SOA sourcing policy falls into two parts: *outsourcing* and *insourcing*.

Commodity services are usually the best candidates for outsourcing. Outsourcing should be considered where it is likely to achieve the business goals of the process better, usually through greater efficiency and cost-effectiveness. Further criteria that support the argument for outsourcing are that the service is:

- Fixed, as opposed to variable
- Generic, as opposed to specific
- Frequently used.

The service description (which forms the basis of the SLA, as described in chapter 12) should be compared with market services offered by providers using gap analysis, although, of course, the gap analysis could reveal that outsourcing of the service is not a viable strategy!

Insourcing of a service refers to the running of a service in-house. In addition, a variant of insourcing, known as backsourcing, refers to the situation where a service that is currently being outsourced is brought in-house.

Insourcing should be considered where it is likely to achieve the business goals of the process best, usually either through greater differentiation and customer value, or greater efficiency and cost-effectiveness. Territory and value-add services are usually the best candidates for insourcing. Further criteria that support the argument for insourcing are that the service is:

- Variable, as opposed to fixed
- Specific, as opposed to generic.

Backsourcing should be considered in the case of poor-performing providers. A useful account of the associated considerations with respect to the backsourcing of IT functionality is provided by Kaplan (2005). The candidate service should be examined to see where the provider is falling short of expectations using gap analysis comparing provider performance against the service description. Of course, the gap analysis could reveal that insourcing of the services is not a viable strategy – for example, where suitable in-house resources are not available. In that case, it may be that "re-sourcing" to a different provider is appropriate.

As with outsourcing, SLAs are used to mitigate risk by clearly stating the responsibilities of providers and customers, thus managing expectations and helping to avoid possible contractual disputes.

Here are nine examples of high-level sourcing policy for the vehicle hire company:

- Services shall always be executed by third parties
- External partners must be local and have a wide client base
- A partner must *bring* value without a long learning curve
- Maintenance and development of the same service can be done by different providers, in order to counter vendor lock-in
- A minimum of three candidate providers shall be considered before outsourcing execution to a chosen provider
- Only value-add services should be designed in-house
- For territory and commodity services, preference should always be given to standard off-the-shelf solutions that should be used without modifications
- The trade off between vendor-oriented (fewer partners/lower costs) and solution-oriented (many partners/best of breed) is based on availability
- The question of IT governance and conservation of enterprise critical knowledge must be independent of the sourcing model chosen.

It can be useful to consider sourcing policy further with regard to service types, and to formalize the policy using a table that provides weighting factors for each service type. For example, it may be that the policy "External partners must be local and have a wide client base" is much more applicable to commodity services (weighting: very high) than territory services (weighting: medium), and not applicable at all to value-add services (weighting: nil).

8.5.2 Usage policy

Recall that services may be offered in three ways: for internal use only, to trusted partners, or to the external market as a whole:

Internal context

Initially, it is likely that any service will be used in one context only. Only once it has been tried and tested, found useful, and evolved is it likely to be reused in different internal contexts. The more contexts in which it is used, the more it will need to be well designed, documented and implemented, and offer long-term stability. Generally speaking, therefore, the order of play for offering services internally is commodity, territory, value-add.

Trusted partner context

Services offered to trusted partners are likely to be territory services that can operate within the context of cross-organizational processes, with the objective of streamlining them. Offering services for trusted partner use is generally more challenging than for internal use. Legal implications and the consequences of changing or withdrawing a trusted partner service are a consideration, though they are likely to be covered by broader arrangements between the participating companies. The QoS requirements are also likely to be much more stringent than for internal services.

Open market context

Services offered to the external market as whole are likely to be either territory services or value-add services. In the case of value-add services the benefits of exposure to the open market must be weighed against the loss of any competitive advantage gained by keeping the service restricted internally or to selected partners or customers. Offering services for external use is generally more challenging than for internal or partner use. In particular, legal implications and the consequences of changing or withdrawing an open market service, must be very carefully considered. The QoS requirements are also likely to be much more stringent than for internal or trusted partner services.

Again, as with sourcing policy, it can be useful to consider usage policy further with regard to service types, and to formalize the policy using a table that provides weighting factors for each service type.

8.6 Technology policy

Most organizations are constrained to at least some degree by their current technology portfolio that has typically evolved over the years into a mishmash of different technologies often hooked together via proprietary interfaces. We must face the fact that the SOA technology policy must cope with this reality. It might be tempting to conclude that nothing more than traditional technology policy – for example, in terms of run-time platforms – is required for SOA.

However, that would be a misleading argument. If it is important to consider the "here and now" of technology policy it is equally important to consider the SOA vision for moving forward. The computing vision of service orientation is one in which software services collaborate on an as needed basis. In this world, organizations do not own their own platforms, they use provider platforms that are supplied on a "pay as you go" basis. The idea is that a provider can supply a hugely more reliable and cheaper platform than an individual enterprise. The notion that all resources are controlled and owned by a central organization was presumed in the design of old-generation solutions. In the world of software services, resources can be spread across different organizational domains.

The SOA must therefore provide technology policy that caters for this vision while being realistic enough to allow effective migration from the existing set of in-house technologies. If you do not know where you are, you cannot plan the journey!

Often technology architects and applications architects work in a somewhat disconnected fashion. This may be mirrored in the way in which the technology architecture and the SOA develop. It is important to guard against this trend and treat technology policy as an integral part of the SOA. For example, it may be that use of software services leads to performance bottlenecks, especially at the network level. Proactively dealing with increased XML traffic requirements means that the technology and application architects work together to understand and resolve these issues.[2]

In a nutshell, the SOA and the overall technology architecture evolve iteratively to accommodate the higher work-loads, more complex messaging requirements, and changing usage patterns that come with the successful uptake of service orientation.

Technology policy is also closely related to the other aspects of SOA policy that we have discussed. For example, both security and capacity quality requirements will impact technology policy.

In this short section we briefly consider some of the key aspects of SOA technology policy.

8.6.1 A matter of specification

SOA technology policy is closely related to the software unit architecture and plays a significant part in the *specification* of software units. As we saw in chapter 2 in the Prague metro analogy an SOA provides a logical picture, but physical pictures are also needed. The right software units (in particular, components) must be identified and implemented in relation to the SOA. The problems in implementing and deploying the underlying software units do not magically disappear, as has often been erroneously assumed in

[2] In fact, these issues have given rise to an entire marketplace of XML appliances that each addresses some combination of today's XML traffic and processing requirements.

the past. Software units are required not only to implement the services that have been identified by business process redesign; they are also required to implement technical infrastructure services such as database connectivity as well as the management of QoS requirements such as capacity, availability, and security. It is necessary to state the technology policy that governs the specification of these software units. This policy is part of the SOA.

The SOA technology policy defines the rules governing how software units offering the services are implemented in a given technology. This policy shapes the design of software units and will be referenced (and possibly elaborated) by the software unit architecture, as discussed in chapter 9.

8.6.2 Elements of SOA technology policy

The overall technology architecture documents the organization's IT technical infrastructure in terms of actual software such as operating systems, languages (including modeling notations), middleware, and databases as well as hardware platforms and networks. This documentation includes the policy that governs these items as well as software and hardware inventories[3] and associated information (such as platform or network diagrams). The overall technology architecture helps to shape the SOA. For example, if the overall policy is to always use the .NET platform this will inevitably affect SOA policy on communication.

In one sense, the SOA technology policy can be considered an outgrowth of overall technology architecture. However, as we have stressed throughout this book (particularly in chapter 2), SOA requires some significant gear changes in all areas, including technology. We consider below six of the more significant elements that are likely to require special treatment.

Standards

The standard technologies (for example, languages and protocols) that are to be followed for implementing software services.

Communication

An important aspect of the SOA technology policy is how the services communicate (see 8.4.4). This policy must cover the infrastructure software (or "middleware") that must be used to implement the different communication mechanisms; the reader is referred to Schlamann (2004) and to the case study in chapter 15 for examples.

[3] These inventories are logically a part of the asset inventory discussed previously.

Asset management

An explanation of how service definitions are to be stored in the asset inventory, how they are version managed, and how they are published and discovered.

Capacity

The infrastructure software required to manage capacity QoS requirements.

Availability

The infrastructure software required to manage the availability QoS requirements.

Security

The infrastructure software required to manage the security QoS requirements.

8.7 Where to next?

We have outlined the elements of SOA, with a particular emphasis on the policy aspect. This provides a context for discussing service design (in chapter 9) and QoS infrastructure design (in chapter 10).

9 Service design

9.1 Agility

Software applications are seldom built for the sort of general purpose uses that a computer is suitable for. Despite our ingenuity in building computer chips that can run everything from aircraft to video games, the software itself is good for only one kind of activity or another. The result is that all too often software that is anything but "soft" – it is very brittle and difficult to change.

In chapter 8, we emphasized that an SOA addresses this long-standing problem by treating software quality – and in particular agility – as an integral part of software architecture. It is important to note that this does not mean formalizing the development process, as in the "engineering" approach to software quality, exemplified by the CMMI, the established approach to formalizing the processes of creating and managing software.

The fact is that when we build software to meet specific, predefined requirements, it almost always falls short of meeting the true requirements for that software, primarily because the actual requirements for software are typically in a state of constant flux. Software must be agile enough to meet as yet undefined requirements at some point in the future.

In this chapter, we therefore focus on how to design *agile services*. This takes us to the heart of SOA: modeling techniques that help get the right set of services. But more than that, the techniques can help in structuring the interfaces that will be offered by the services, as well as in determining how best to achieve a good underlying set of software components to implement the interfaces.

Too often the term "agility" is bandied about, especially in relation to SOA, without any real meaning. In this chapter, we therefore begin by homing in on agility and providing some concrete guidelines for definition and measurement of what is a sometimes mercurial concept.

9.1.1 Measuring agility

In chapter 1, we identified agility as one of the main aspects of service orientation. Agility is the ability to act quickly and with economy of effort in accurate response to change and also to initiate change for business advantage.

Table 9.1 *Service agility template*

Quality attribute	Definition	Example measurement criteria
Reusability	Repeated use of a service.	Number of different invocations of a service.
Replaceability	The ease with which a software unit providing the service can be substituted with another having different implementation code while offering the same set of interfaces.	Cost and time to implement a software unit using a different language or design, while retaining the same interfaces.
Interoperability	The ease with which a software unit providing the service interacts with other software units at run-time.	Cost and time to ensure conformance to the software execution environment.
Flexibility	The ability of a service to perform even outside an intended context, or at least it has the feature that its functionality degrades gradually outside its context; software should be "soft!"	Cost and time to change software units providing other services when applying service in a different context.
Adaptability	The ability of a service to change efficiently in terms of its extensibility to meet new requirements and portability across implementation environments.	Cost and time to add interfaces to support service use in a different context.

Accuracy and *clarity* are preconditions for agility. If a service is not accurately defined and clearly understood then any discussion of the other quality attributes is rendered meaningless. The different types of models and specification techniques used in SOA should help in establishing the accuracy and clarity of services.

Agility is usefully expressed in terms of five constituent quality attributes, as shown in table 9.1.

Reusability, replaceability, interoperability, flexibility, and adaptability all relate to the quality of the service design. The service agility template provides metrics that can be used to assist in the design process. The models and specification techniques used in service design (and considered later in this chapter) should assist in weighing up these sometimes conflicting criteria, and in arriving at a well-designed set of services.

9.1.2 The business benefits of agility

The metrics provided by the service agility template are design metrics. It is also necessary to glean a business-driven view of agility. Agility quality attributes are usefully considered in relation to the high-level business goals of the BIAT (see 8.2.2). This allows the benefits of the agility attributes – which are often quite difficult to explain to business people – to be readily clarified. In the case of our vehicle hire company we use a table to summarize the agility benefits in relation to overall business direction; (see table 9.2). This information will be used to help guide the development of the SOA in terms of service design.

Table 9.2 *Agility benefits for vehicle hire company*

Quality attribute	SOV 7	Benefit
Reusability	Multi-channel capability	An SOA helps remove dependencies on specific technologies through the use of standardized interfaces. Services are designed that can exploit this capability and be reused in different channels and technology contexts.
Replaceability	Adaptation	Regarding associated technology changes, a service can be smoothly replaced with another having different software unit implementation code.
Interoperability	Customer fit	Exposing the line of commoditization allows the organization to focus on designing value-add services that ensure customer fit through interoperability with other service types.
Flexibility	Adaptation	Services can be designed to smoothly accommodate changes to information and rules.
Adaptability	Adaptation	Regarding business change, services can be extended in "plug and play" fashion in different business contexts.

9.2 Service design techniques

The six main structural parts of the SOA are as follows:

- Services[1] and their dependencies
- Service descriptions (which form the basis of service specifications; see 9.3.4)
- Interfaces and their dependencies
- Software units and their dependencies
- Software unit descriptions
- Service buses.

Service design brings four key questions, at increasing levels of detail, to bear on the initial set of services identified as a result of the business process redesign efforts:

- What dependencies obtain between the services?
- What interfaces are required to support the services?
- How are the interfaces realized in terms of software units?
- How are interfaces offered in terms of service buses?

In this section, we concentrate on the first of these questions, moving on to discuss the remaining three in the next section. The approach builds on component-based development (CBD) techniques pioneered in Allen (2000) and Cheesman and Daniels (2000). Figure 9.1 summarizes the overall focus of this chapter.

[1] Note that, for the sake of brevity, unless specifically stated otherwise, the term "service" is used to mean "software service" in our subsequent discussions of SOA.

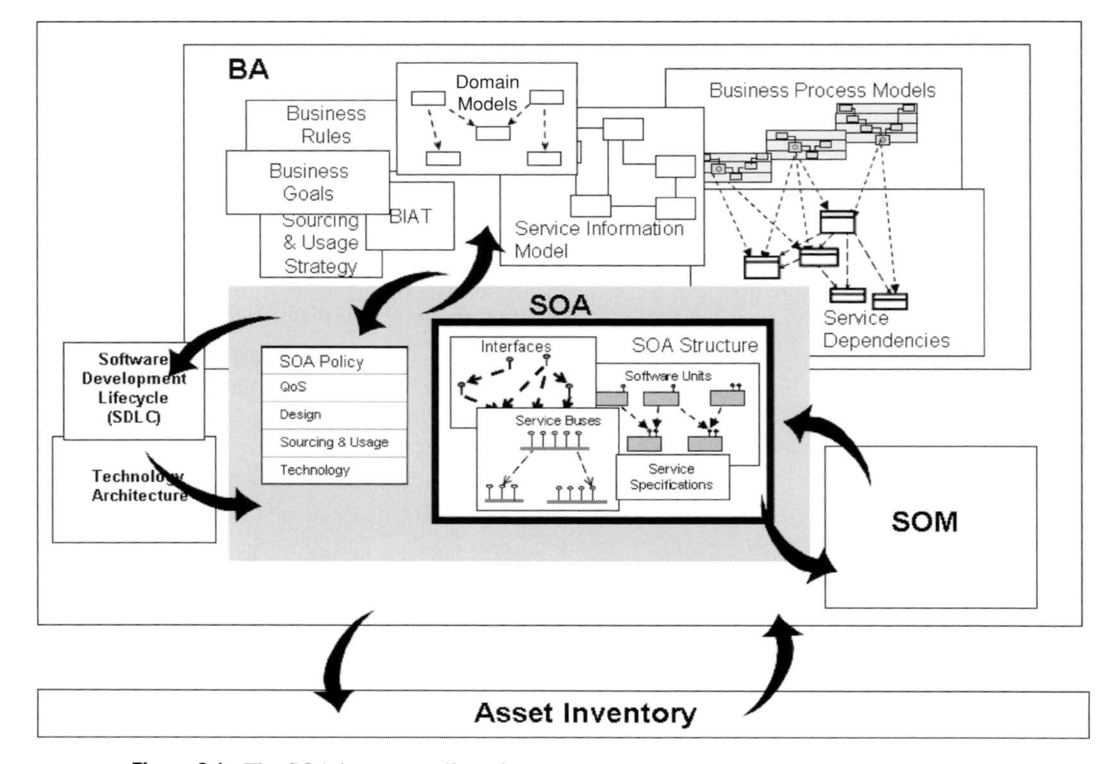

Figure 9.1 The SOA in an overall roadmap context

For the sake of explanation, we have assumed a single enterprise SOA scoped around the vehicle hire domain. The reader should note that in practice there are likely to be at least two levels of SOA (project and enterprise), as discussed earlier, in order to provide the different levels of abstraction that are necessary to manage the scale and complexity of such an undertaking in an enterprise of any size.

Unfortunately, the topic of SOA is too often clouded by ill-defined terminology. Before going further, we need to make sure we do not fall into that trap by setting out some definitions.

There are two main types of dependencies:

- *Existence dependency*: A relationship between two modeling elements in which a change to one modeling element will affect the other (the dependent element). For example, deletion of the former causes deletion of the latter, because the latter has no existence without the former.
- *Usage dependency*: A dependency in which one model element (the dependent element) requires the presence of another element for its correct functioning on implementation. Typically a usage dependency is expressed by a call from the dependent element to the other element.

We have actually already employed usage dependencies. However, in discussing interface and software unit dependencies we shall also need to be mindful of existence dependencies. We now need five more definitions:

- *Interface*: A set of semantically related operations that is used to define all or part of the external behavior of a software unit. An interface is used to offer all or part of the functionality of a service.
- *Operation*: A discrete unit of behavior specified on an interface, that a piece of consumer software can invoke at run-time.
- *Software unit*: A large-grained, cohesive collection of software that is usefully modeled as a single building block.[2]
- *Software unit architecture*: A model that identifies a proposed (or existing) set of *software units*, plus their *interfaces* and dependencies.
- *Service bus*: A grouping of deployed services that must conform to defined corporate policies and standards.

Much of SOA modeling employs forms of a Unified Modeling Language (UML) (OMG, 2004) class diagram with stereotypes to indicate the modeling element (for example, service, interface, or software unit), and lollipops to indicate interfaces. In subsequent sections, we are fairly loose in the application of UML for the sake of brevity. For example, stereotypes are omitted unless they are not obvious from the diagram name and not strictly relevant. The concepts discussed above are shown using the UML notation in figure 9.2.

9.2.1 Developing the service information model

Now we have the concepts to resume the vehicle hire scenario that we last encountered at the end of chapter 6. The service information model introduced in chapter 6 (figure 6.6) is repeated for convenience in figure 9.3 and is now modeled to a greater degree of detail to include further attributes and business types. This activity takes place iteratively with the business process redesign effort, which in turn follows the service-oriented business process redesign pattern (described in chapter 4). Process redesign and service design evolve side by side. Recall that the pattern starts from the set of existing resources and aims to deliver measurable business value and process improvements in small chunks. We have already gleaned quite a lot of this information from consideration of the different processes described in chapters 5 and 6. Further information evolves through consideration of these processes – for example, attributes of the type Delivery are added to the model.

[2] A software unit specification may be realized through several different *software unit implementations*. A software unit specification acts as a compliance document for implementers to work to. In this book, we use the shortened form "software unit" to mean software unit specification, reserving the term "software unit implementation" for the actual software used for realizing the software unit.

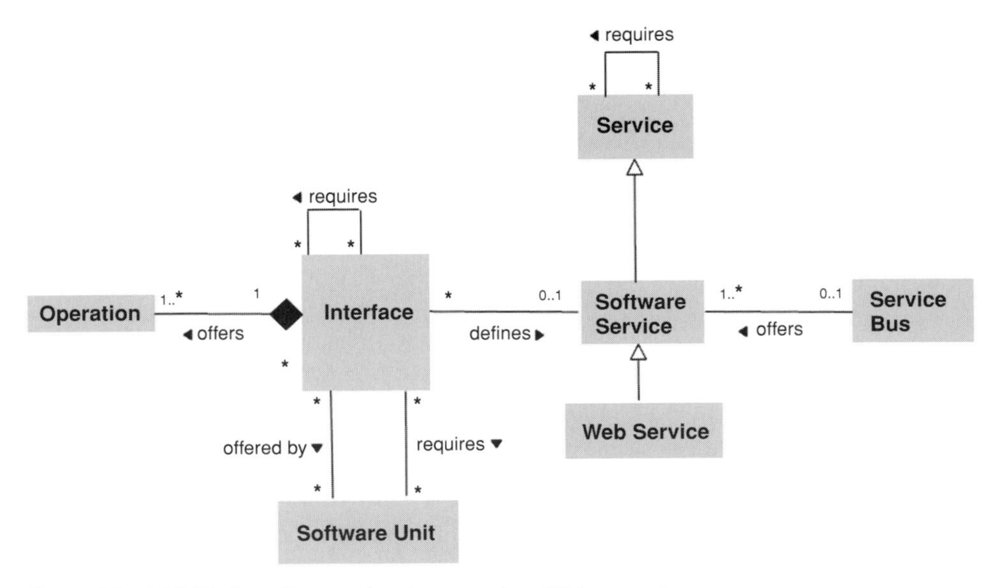

Figure 9.2 A UML class diagram showing some key SOA concepts

At the same time, the pattern seeks to evolve a set of services aimed at long-term business agility, as emphasized earlier in this chapter. The approach to developing the service information model must be consistent with this strategy. Recall that underpinning the business process redesign effort are three parallel ongoing activities that are aimed at achieving long-term business agility:

- Establish service policy in relation to business goals
- Understand domains and identify services and their dependencies
- Survey and catalog assets.

In our vehicle hire example, we can therefore expect to glean information from consideration of:

- Service policy (for example, hire types and terms)
- Domains (for example, the vehicle hire domain centers on subjects such as contracts, reservations, and deliveries)
- Existing internal assets (for example, the existing Customer and Vehicle systems)
- Existing external assets (for example, the third-party Credit Checker and Vehicle Finder interfaces).

As a result of this ongoing analysis further attributes, types, and associations are added and others are removed from the model.

Each type within the service information model is be categorized as being "core" or "detailing." *Core* business types are independent from other types (the detailing business types) in that they may exist independently; they have no mandatory association with

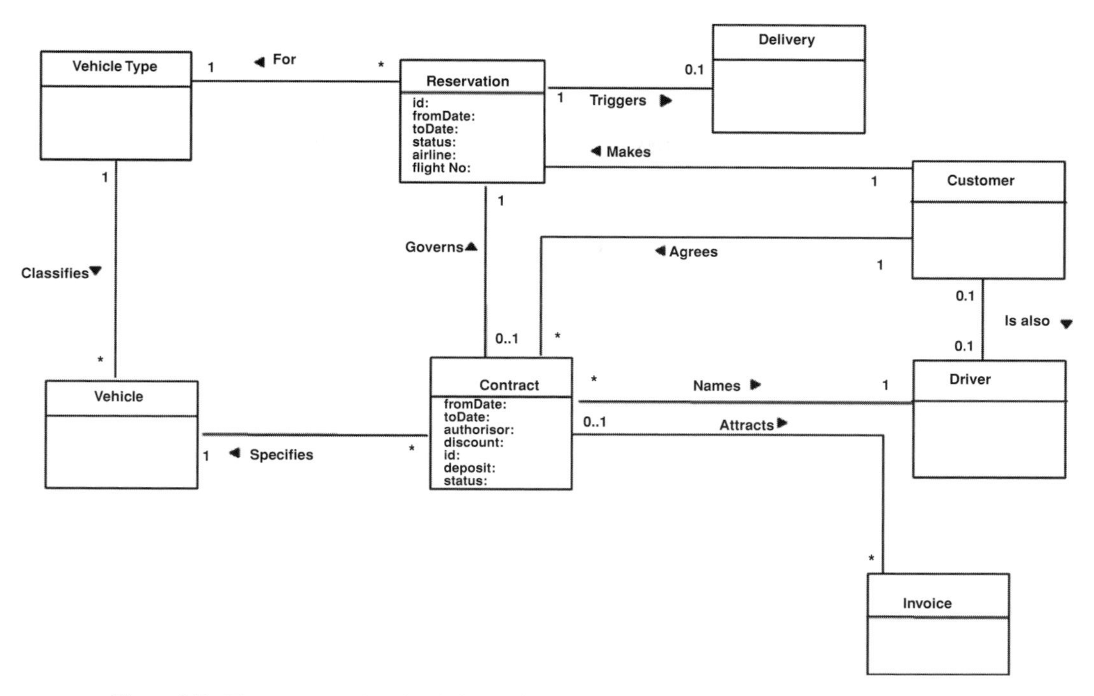

Figure 9.3 The contracts Service information model

any other type. The core types are highlighted with heavy lines in the top-left corner of their boxes, as shown in figure 9.4. Core types are used to help determine interfaces, as described later in this chapter.

9.2.2 Understanding service information needs

In order to identify the information required by each service, the services must be mapped to the detailed service information model. In the vehicle hire example the services (repeated for convenience from chapter 6 in figure 9.5) are mapped to the service information model by overlaying the representation of services as shown in figure 9.6.

9.2.3 Assigning specific goals to services

Specific goals or "targets" (as opposed to the high-level ones discussed earlier) are extrapolated from the business process models. These goals are assigned to services where appropriate to form part of the eventual SLAs for those services. For example, the Record Handback activity might have a target of "Contract must be closed within

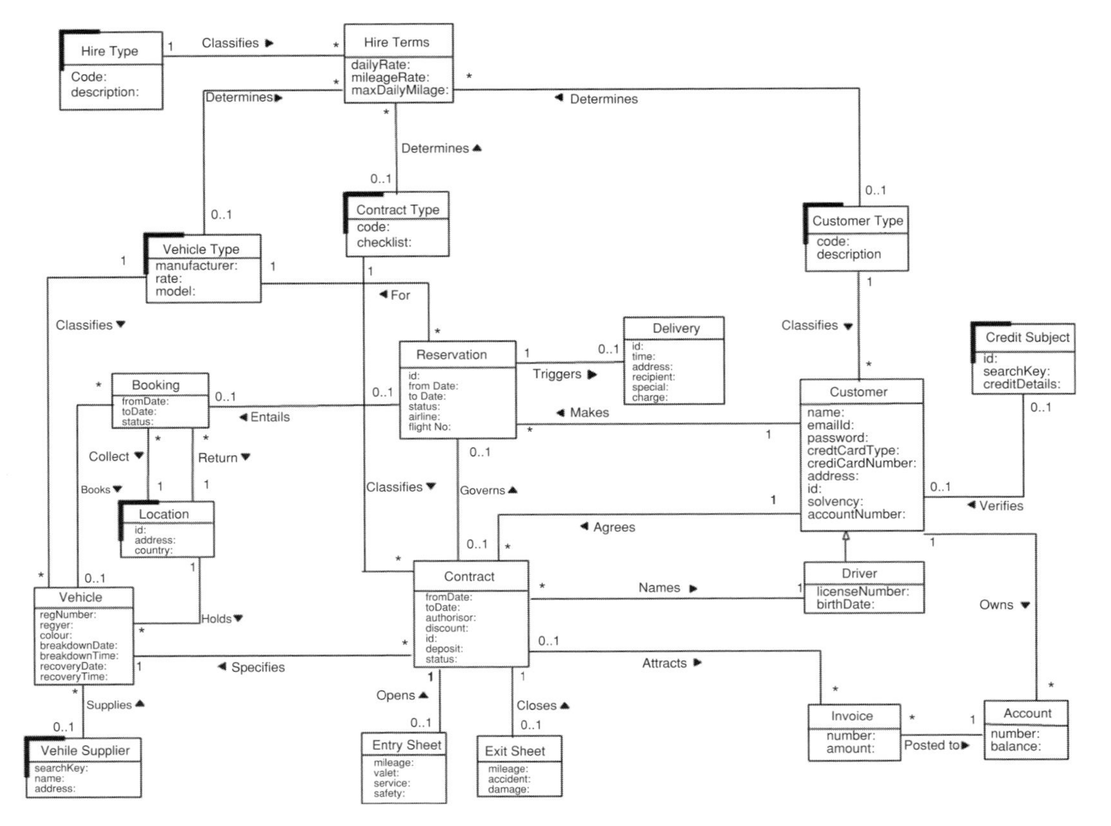

Figure 9.4 The vehicle hire detailed service information model

5 minutes." This target can be referenced by the service description of the Contracts Service. Similarly, targets such as "50% decrease in numbers of phone reservations" and "200 online reservations per day" which apply to the hire *process* can also be referenced by the service description for Contracts as Contracts clearly plays a contributing role in achieving these targets (see 9.2.7).

Further targets that apply to all operations of a service may be deduced from the process targets and assigned to the service. For example, all operations of Contract must be capable of handling up to 500 invocations per day.

Analysis of goals can become quite complex. For example, a high-level business goal may decompose into lower-level business goals, which may decompose into targets, which may decompose into further targets, and so on (Penker and Eriksson, 2000 pp. 283–8).

Further complexity can arise in that the same activity or service may have different targets depending on the role to which the service is offered. For example, the above target "Contract must be closed within 5 minutes" may apply in the case of a partner,

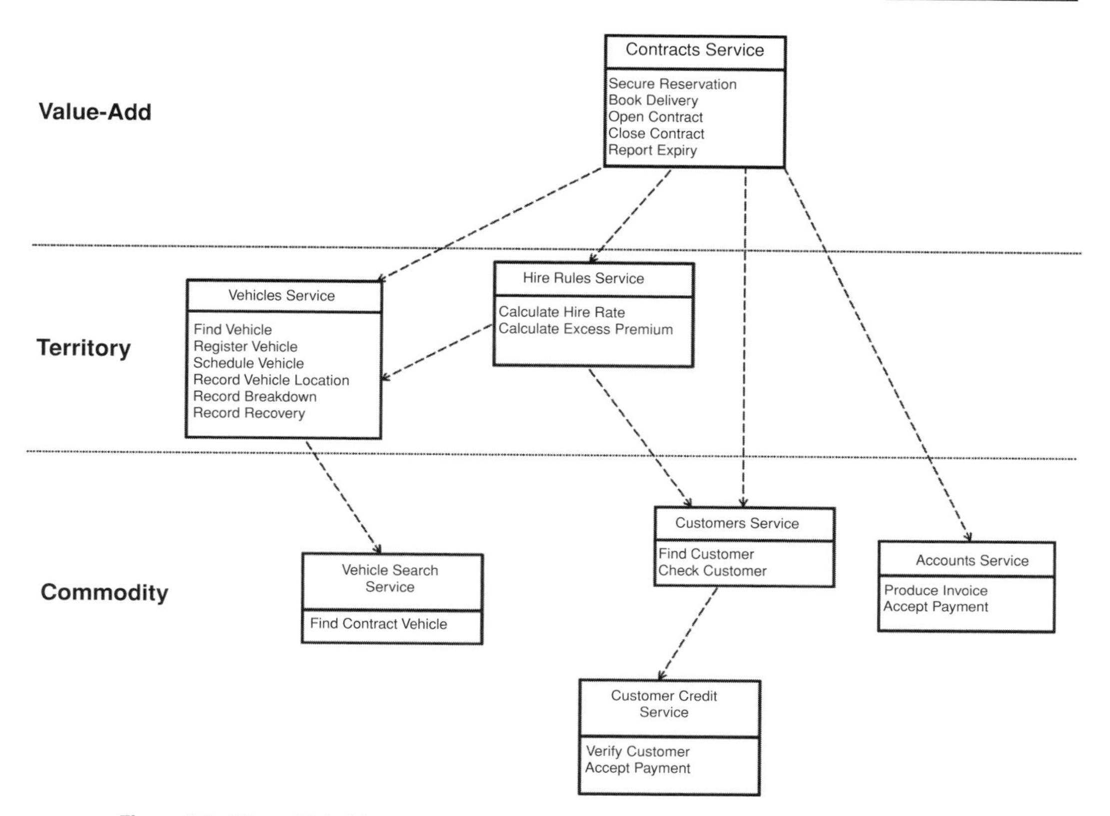

Figure 9.5 The vehicle hire service dependency diagram

whereas it is relaxed to "Contract must be closed within 15 minutes" in the case of an employee.

The temporal dimension is also a factor. The same role may be entitled to different QoS levels of the same service, depending on the time of day or week. Roles themselves may be further divided – for example, a gold star partner versus a bronze star partner.

We need to guard against too much complexity and concentrate on the high-level goals and targets that are most critical from a business viewpoint. Matrices that plot role against service, with targets indicated in the cells of the matrix, are a useful technique here.

Another way of managing complexity is to segment high-level goals and targets in the same way that we used layering of QoS policy in chapter 8. In other words goals and targets applying across the board to all processes are distinguished from those applying to specific processes, then sub-processes, then activities, and then services.

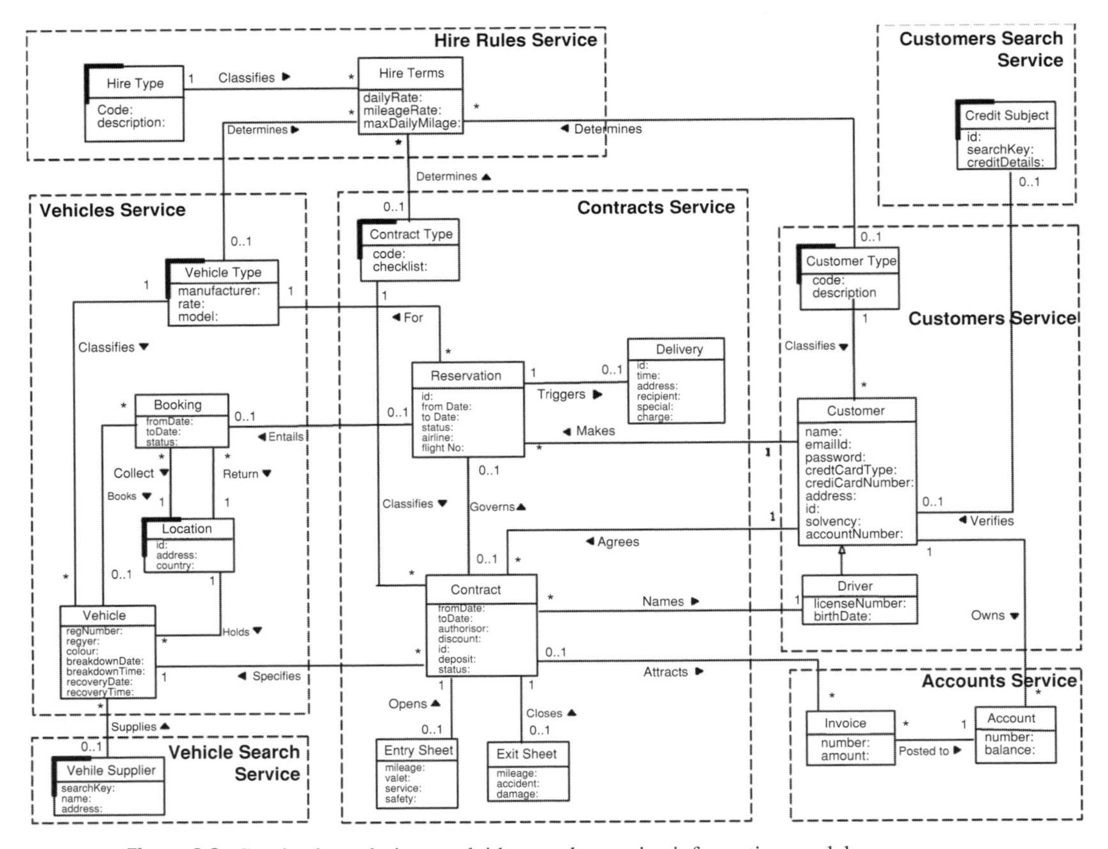

Figure 9.6 Service boundaries overlaid on to the service information model

A further guideline is that the majority of high-level goals should apply to value-add processes and services as well as to territory services that reflect core competencies. After all, the goals should be intrinsic to the value that the company is aiming to bring to market. At the same time, targets may well also be relevant to commodity processes and services, as it will be necessary to secure business measures against which suppliers can be evaluated!

9.2.4 Mapping business rules to the SOA

Because of the potential large numbers and diversity of business rules it is important (as with business goals and targets) to manage complexity and to focus on those business rules that have most business relevance. Also, many business rules are typically of a very low level of granularity, applying to specific situations that affect the flow of a business process or operation. Sometimes these are best captured as part of the process model. In other cases, where the rules are reused in different processes, the rules should be captured and maintained using a BRM, as described in 3.3.3. As far as the SOA

Table 9.3 *Business rules mapping template*

Business rule type	Model representation
Structural constraint	Service information: business types, associations and attributes
Behavioral constraint	BRM or operation of a service
Direct event constraint	Process model, BRM or operation of a service
Indirect event constraint	Process model, BRM or operation of a service
Derivation: inference	BRM, operation of a service, or attributes
Derivation: calculation	BRM or operation of a service

is concerned we are mainly interested in rules that apply to services – either to the information provided by or captured by a service, or to the external contractual behavior of the operations that are offered by a service. In cases where rules are complex and subject to change, separate rules services may be designed as exemplified by the Hire Rules Service. A rules inference engine may be used to implement such services.

A useful way of handling this situation is to use a mapping template that helps to separate out the various types of rule in relation to their SOA relevance and possible means of representation, as shown in table 9.3.

9.2.5 Assigning business rules to services

Business rules that are assigned to services may form part of the eventual SLAs for those services.

Notice that some of the structural rules are already catered for by associations shown in the service information model for the Contracts Service. For example:

- A reservation must specify a type of vehicle
- A contract must specify an actual vehicle
- A contract must name a driver.

Other structural constraints apply to attributes of Contract:

- Only the hire manager shall award customers gold status
- A deposit must never exceed 20% of the agreed reservation fee.

Regarding behavioral constraints, the following apply to the Secure Reservation operation:

- Only verified customers may reserve vehicles
- A contract can be closed only where a payment has been made.

Recall that in 6.2.5 we identified the following direct event:

- If vehicle breaks down then organize recovery and replacement of vehicle.

The business process redesign must include a guard condition to take account of this event and trigger a vehicle recovery activity. (In fact it is a real-time event sensed by a device in the vehicle and transmitted back to the hire office.)

We also previously identified the following indirect events:

- If vehicle not returned by agreed return date then issue warning notice.
- If excess vehicle mileage > 0, and the customer is not a gold customer, then an excess premium shall be payable.

The failure to return event implies that our Contracts Service needs to include an operation to detect expiry of return date and issue the warning.

The second event requires detection by the Close Contract operation of the Contracts Service. If we examine the excess vehicle mileage event a little more closely we find that it requires the following derivation rules:

- Hire mileage = recorded mileage − previous recorded mileage
- Excess vehicle mileage = hire mileage − (maximum day mileage × days hired)
- Excess premium is calculated according to a pre-stipulated formula.

The hire mileage and excess vehicle mileage derivations depend upon attributes, which are provided by the Contracts business type. "Excess payable" can then be detected by the Close Contract operation provided by the Contracts Service. The excess premium is calculated by the Calculate Excess Premium operation introduced to the Hire Rules Service that we introduced back in 6.4.4.

9.2.6 A word about BLAs

As well as playing a role in the SOA and being documented with respect to services, business rules may also have a broader, more commercial significance.

A BLA specifies the business goals and rules that govern the functionality and information described by SLAs, and the commercial criteria (such as financial penalties) attaching to these business goals and rules. The BLA provides a more business strategic view of an SLA, and can be used to supplement the SLA.

Certain of the rules described above fall into this category:

- Only the hire manager shall award customers gold status
- A deposit must never exceed 20% of the agreed reservation fee
- If excess vehicle mileage > 0 and the customer is not a gold customer, then an excess premium shall be payable.

These rules should be flagged so that they can be readily picked up when it comes to composing SLAs and BLAs.

The above analysis of the vehicle hire rules is summarized in table 9.4. Note that the BLA-relevant items are italicized.

9.2.7 Describing services

Each service has a set of operations that reflects the responsibilities of the service uncovered as a result of our work in business process redesign described in chapter 6 (see figure 9.5). In our present scope we are focusing very much on business process

Table 9.4 *Business rules mapping examples*

Business rule type	Business rule examples	Model representation
Structural constraint	A reservation must specify a type of vehicle. A contract must specify an actual vehicle. A contract must name a driver.	Contracts Service information model: business types and associations
	Only the hire manager shall award customers gold status. *A deposit must never exceed 20% of the agreed reservation fee.*	Contracts Service information model: attributes
Behavioral constraint	Only verified customers may reserve vehicles. A contract can be closed only where a payment has been made.	Secure Reservation operation of Contracts Service
Direct event constraint	If vehicle breaks down then organize recovery and replacement of vehicle.	Transition guard in Vehicle Hire process flow to trigger Recover Vehicle sub-process
Indirect event constraint	If vehicle not returned by agreed return date then issue warning notice.	Report Expiry operation of Contracts Service
	If excess vehicle mileage > 0 and the customer is not a gold customer, then an excess premium shall be payable (see below).	Close Contract operation of Contracts Service
Derivation	Hire mileage = recorded mileage − previous recorded mileage excess Excess vehicle mileage = hire mileage − (maximum day mileage × days hired)	Contracts Service information model: attributes
	The excess premium is calculated according to a pre-stipulated formula.	Calculate Penalties operation of Hire Rules service

redesign; however, it is important to appreciate that in practice other techniques, such as use cases, are employed to tease out the operations (Allen, 2000). We now consider these responsibilities in relation to the service information model and start to fill out service descriptions. The service descriptions form a basis for the service specifications (described in 9.3.4) and SLAs (described in chapter 12).

We try to ensure that the operations listed do in fact reflect the key business responsibilities of the services, and avoid listing operations (such as maintenance operations, tactics for which are described below) that are relatively insignificant.

Each service is responsible for maintaining the business type instances falling within the service boundaries depicted on the service information model. The service description simply lists these business types – it is then understood that the service will provide operations to create, read, update, delete, and list (CRUDL) instances of these types, without having to clutter the model with these basic operations; however there can sometimes be exceptions, such as where deletion is not allowed.[3] It is also sometimes the case that a CRUDL operation is sufficiently significant to warrant depiction on the model, in which case it may be shown; for example, the Find Customer operation of the Customers Service.

It may be that as a result of examining the service information model, further responsibilities are exposed. Again, other well-established techniques such as use case modeling (Allen, 2000) and CRC cards (Wilkinson, 1995) may be applied to yield further responsibilities and to tease out the operation details.

In describing the services, questions naturally arise as to whether services should be *combined* or *separated*. For example, we might consider whether the Customers and Accounts Services should be combined. However it is decided to stick with the Accounts Service on the grounds that it is a useful service that is importantly different from the Customers Service. In addition, the two services are primarily used by different business units (or roles): the Finance Department (Accounts) and the Customer Relationship and Hiring Departments (Customers). The subject matter is also importantly different. So keeping two services makes good sense.

An example service description is shown in table 9.5. Initially we just need basic information to record the high-level service requirements and allow these to be cataloged in the asset inventory. Once our understanding of service information and dependencies has stabilized we can form service specifications, as described in 9.3.4. Later we shall add QoS levels once we have examined QoS infrastructure requirements in more detail, as described in chapter 10.

9.2.8 Understanding service dependencies

Services may depend on one another in two major ways: by *usage* or by *existence*. Thus far, in the course of developing the service dependency model, the emphasis has been on usage dependencies – that is, on whether an operation of a service needs to call on an operation of another service in order to function correctly. However, our service information model also highlights existence dependencies. An existence dependency occurs when a type instance falling within one service boundary depends for its existence upon the existence of a type falling within another service boundary – for example:

- The dependencies of Contract on Vehicle and Customer
- The dependency of Account on Customer.

The original service dependency model, shown in figure 9.5, failed to show the dependency of Account on Customer. On the other hand, the original service dependency model indicated dependencies from the Hire Rules Service to the Customers and Vehicles Services that are questionable. The Hire Terms type has no mandatory associations to types falling within the latter two services.

[3] (SOSA Team, 2004) recommends the expedient of writing the codes CRUDL against each type to emphasize this, with individual codes being omitted where that operation will not be provided.

Table 9.5 *Service description for Contracts Service*

Property	Value
Service name	Contracts Service
Domain	Vehicle Hire Sales
Purpose of the service	Management of reservations, contracts, and deliveries
Responsibility by business type	Contract Type, Contract, Entry Sheet, Exit Sheet, Reservation, Delivery
Operations	CRUDL Secure Reservation, Book Delivery, Open Contract, Close Contract, Report Expiry
Dependencies	Customers, Vehicles, Hire Rules, Accounts
Commoditization level	Value Add
Sourcing policy	In-house
Usage policy	Initially restricted to hire agent. Intention is to expand usage to Customers once established.
Targets (Record Handback Activity)	"Contract must be closed within 5 minutes" in the case of a partner. "Contract must be closed within 15 minutes" in the case of an employee.
Business process support (current)	Agency Vehicle Hire
Business process support (future)	Internet Vehicle Hire
Business process targets	50% decrease in numbers of phone reservations. Target of 1600 online reservations per day. 20% increase in number of hires. 20% increase in productivity of hire agent. 50% increase in customer satisfaction rating.

This leaves the original dependencies from the Contracts Service to the Hire Rules Service and to the Accounts Service. Both are usage dependencies. In order for the former to process a contract it is necessary to call on the Hire Rules Service to determine the rate to apply. In order for the Contracts Service to close the contract it is necessary to raise an invoice by calling on the Accounts Service.

To help visualize the above issues we can show the resulting series of five dependencies in relation to the service model, through the inclusion of bold arrows from consumer to provider, as depicted in figure 9.7.

The service dependency model is accordingly modified as shown in figure 9.8.

9.3 Interface design techniques

We now address the question of how our services are going to be offered in terms of interfaces. This involves weighing up various design criteria and focusing on the *dependencies* between interfaces. In particular, operations assigned to services must now be allocated to the interfaces through which they will be offered. Once our understanding

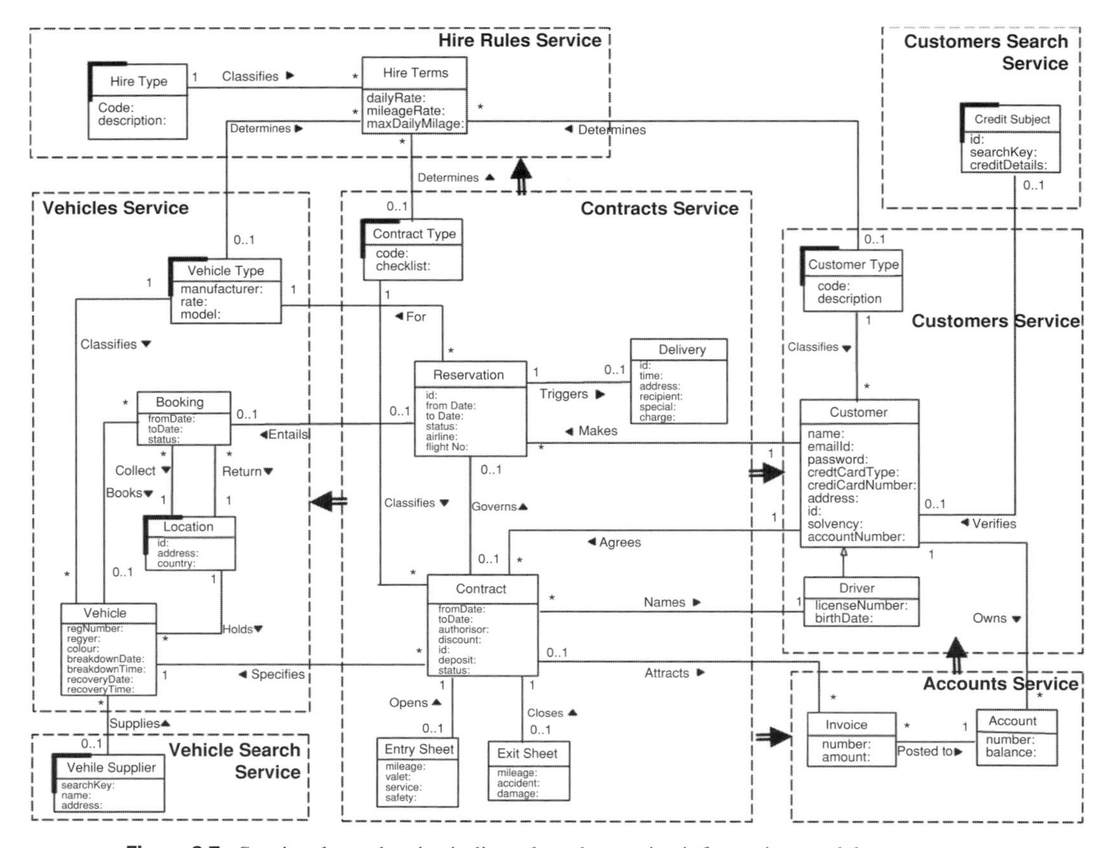

Figure 9.7 Service dependencies indicated on the service information model

of the interfaces has stabilized we can start to develop the service descriptions in more detail and begin to form service specifications.

9.3.1 Identifying interfaces

We now identify the required interfaces. Initially, we assign an interface to each core type and its detailing types. It is important to ensure that the detailing type and its core type lie within the same service boundary. In some situations, this may not be possible. For example, some detailing types may have associations to more than one core type. Where these associations cross the service boundaries questions are raised – for example:

- Is the service boundary correct?
- Should the detailing type be moved within the service boundary of an associated core type to achieve a more cohesive design?

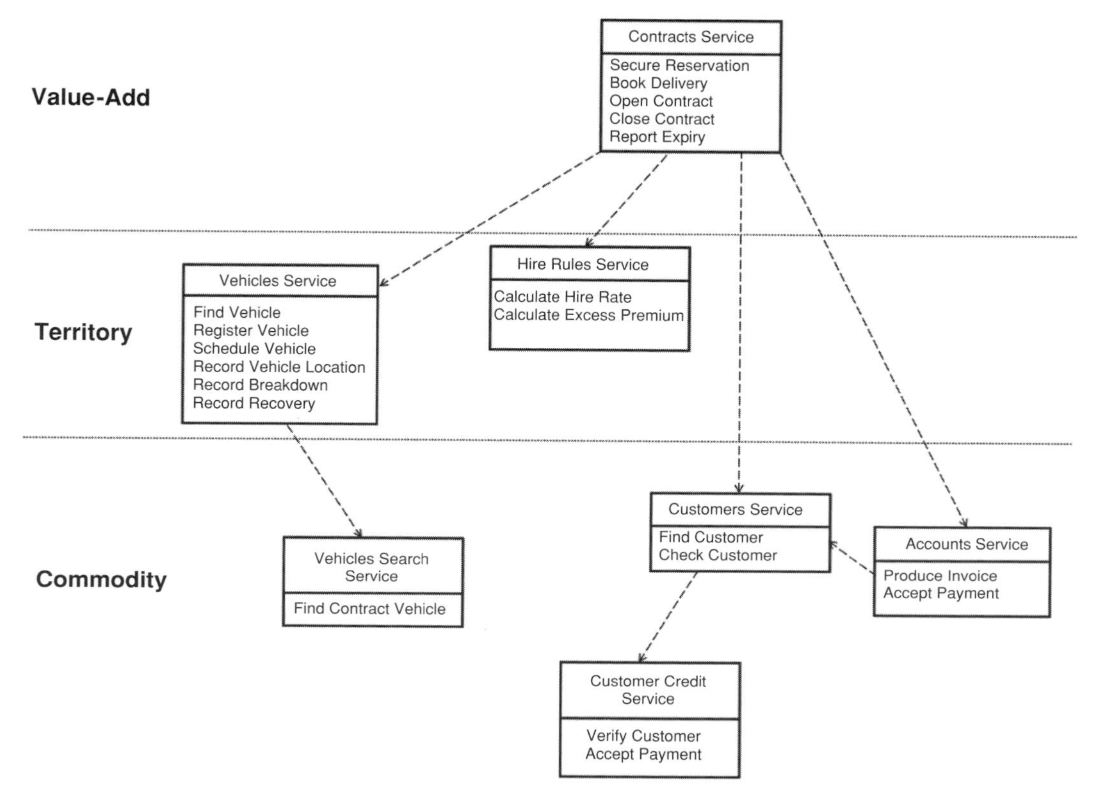

Figure 9.8 Revised service dependency model

Detailed guidelines (SOSA Team, 2004) are available that assist with these questions and in achieving a better service design.[4] For example Vehicle is a detail of both Vehicle Type and Location. It is grouped with Vehicle Type on the grounds that the subject matter is much more closely aligned with Vehicles. The manufacturer and model attributes are intrinsically descriptive of a vehicle, whereas the location address is not.

[4] In some cases, the detailing types will have multiple mandatory associations, and it is necessary to decide which core type it is "a detail of." Here are some guidelines to help in making this decision:

- If one association is identifying, then select that associated type
- If the name of the detailing type is an extended form of the name of one of the associated types, then consider selecting that associated type
- If a modification to the attribute values of this type would be considered to be a modification to one of the mandatory associated types, then it is a detail of that type.

Define one service (or interface type) per core business type but:

- Where one core type simply categorizes another (e.g. Product Group might be a core type that categorizes the core type Product), then include the categorizing core type in the same service as the categorized type – unless the categorizing type is used to categorize several types (of different core types) or has it own detailing types.
- Where there are multiple associations between two core types (counting any associations between their detailing types as well) consider including both types in the same service.
- Where a group of non-updating operations/responsibilities can be usefully separated out into another service, then consider doing so (this can be done only after defining responsibilities).

Table 9.6 *Candidate interfaces with business types*

Candidate interface	Business type
IHireRules	Hire Type
	Hire Terms
IVehicles	Vehicle Type
	Vehicle
	Booking
ILocations	Location
IVehicleSuppliers	Vehicle Supplier
IContracts	Contract Type
	Contract
	Entry Sheet
	Exit Sheet
	Reservation
	Delivery
ICustomers	Customer Type
	Customer
	Driver
ICreditSubjects	Credit Subject
IAccounts	Account
	Invoice

Any interfaces identified as a result of the sourcing and usage policy must also be added. In our example, IVehicleSuppliers and ICreditSubjects are two such interfaces.

There may be some services that do not encompass any core types – for example, the Accounts Service. In these cases, it is always important to consider whether the types might be moved to different services and dispense with the original service. We might, for example, consider moving the Account and Invoice types to the Customers Service, especially as Account is in fact a detail type of Customer. However in our analysis of services we decided to stick with the Accounts Service on the grounds that it is a useful service that is importantly different from the Customers Service – the guidelines are meant to assist in making decisions about interface identification and design, they are not hard and fast rules.

The result of our analysis is the set of candidate interfaces with associated business types shown in table 9.6.

9.3.2 Developing the interface design

We now appraise the initial set of interfaces. The main initial drivers in our choice of interfaces have been the initial partitioning into services to support business processes, cohesiveness of information, and commonality of subject matter. In actual fact, in real-life cases, there is likely to be much iteration between the service dependency and service information models. The initial set of services usually changes as a result of thinking things through at a more detailed structural level. We have kept things reasonably straightforward in our example for the sake of clear explanation.

At the same time, we must not lose sight of the main driver behind a service-oriented approach stated right at the start of this chapter: *agility*. We must ensure that the set of interfaces is sufficiently agile to meet the requirements, first described in the BIAT (see 8.2). In the case of the vehicle hire company, agility to extend the business into new areas is of paramount importance. In discussions with business leaders the need for a flexible approach to contracts turns out to be a significant requirement. We must allow for contracts that stretch beyond vehicle hire to other types of business, such as the supply of motoring information. It is also felt that the company's expertise in reservations might soon be extended into other areas such as hotel bookings that do not necessitate the legal support of a contract. This is reflected in the optional relationship between the Contract and Reservation business types.

It is decided that the Contracts Service will offer two interfaces: IContracts and IReservations, each of which can be used independently to assist with the agility drivers. We must decide which operations of the Contracts Service are to be assigned to each of these interfaces. Open Contract and Close Contract are assigned to IContracts because both operations center on information defined within the Contract type and other types within the Contracts Service boundary. Similarly Secure Reservation and Book Delivery are assigned to IReservations.

The resulting partitioning into interfaces is illustrated in figure 9.9, with interfaces represented as lighter dotted regions overlaid on the service information model.

9.3.3 Understanding interface dependencies

A significant aspect of service design is identifying dependencies between interfaces, examining the validity of these dependencies, and making adjustments to help manage them. In general, we will want to minimize dependencies on value-add services but maximize those on commodity services in the interest of reuse.

We need to ensure that all service dependencies are reflected at the level of interfaces. For example, the dependency of the Contracts Service on the Vehicles Service is reflected in dependencies of IReservations and IContracts on IVehicles. In the former case, Reservation has a mandatory association to Vehicle Type; in the latter case, Contract has a mandatory association to Vehicle.

We must also examine possible dependencies between interfaces that share the same service boundary. The existence dependency of IVehicles on ILocations is added on the grounds of a mandatory association.

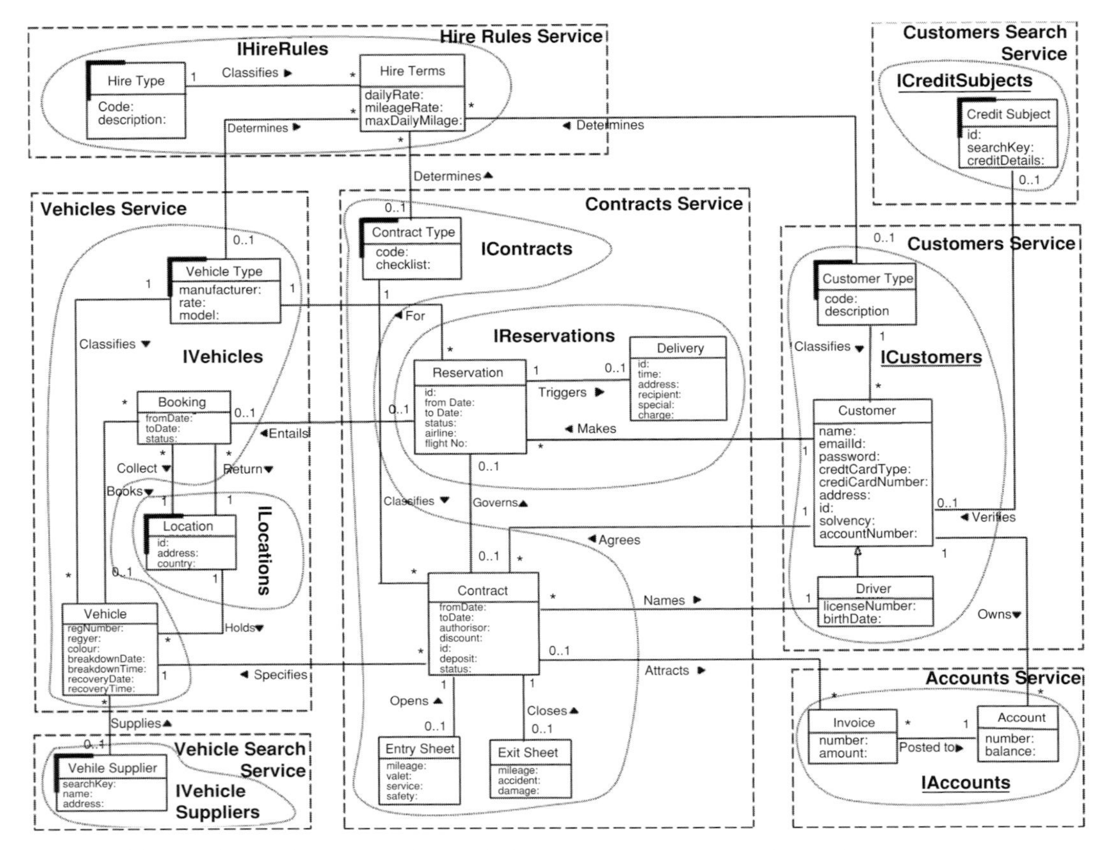

Figure 9.9 Interface boundaries overlaid on to the service information model

The Open Contract operation of IContracts requires a call to IReservations (in order for the contract to be opened, it is necessary to check reservation details). Therefore there is also a usage dependency to be added between IContracts and IReservations.

Again the service information model is used to help understand these issues. The interface dependencies are superimposed, again as arrows, from consumer to provider, as indicated in figure 9.10.

Though the service information modeling is extremely useful for helping to develop the SOA, its complexity means that it is not the best tool for documentation. Interface dependencies (such as service dependencies) are usefully shown on a separate diagram, as shown in figure 9.11.

9.3.4 Specifying services

Having established sets of interfaces and allocated these to services, we are now in a position where we can begin work on the service specifications. The service descriptions

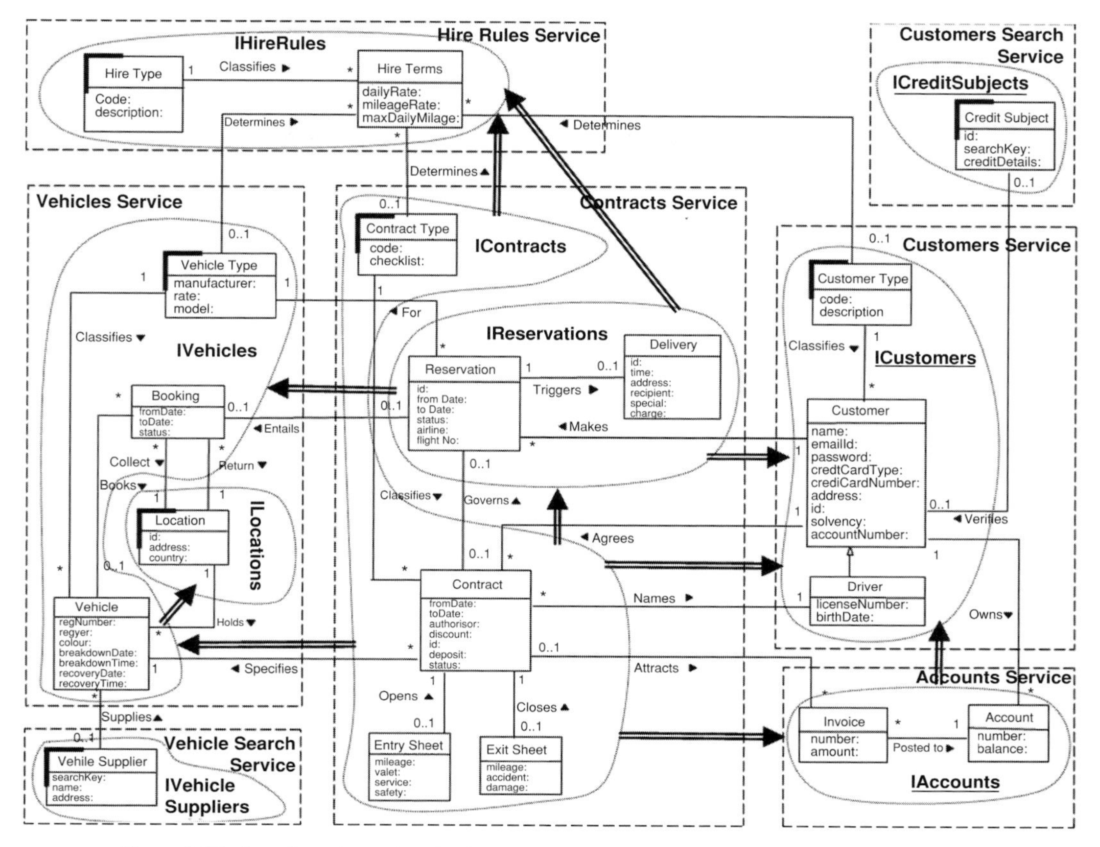

Figure 9.10 Interface dependencies indicated on the service information model

(discussed in 9.2.7) form the basis for the service specifications, which supplement the service descriptions with the following information:

- Interface type model
- Operation details
- Commercial requirements
- QoS level requirements.

Each service is responsible for maintaining the business type instances falling within the service boundaries depicted on the service information model. At this point, we specify each interface using an interface type model (a type model that defines the information that is accessible to the operations of the service at run-time). In addition, we specify each operation of the interface as follows:

- Signature (name, parameters, return type)
- Purpose (narrative)
- Revision number, date, and reviser

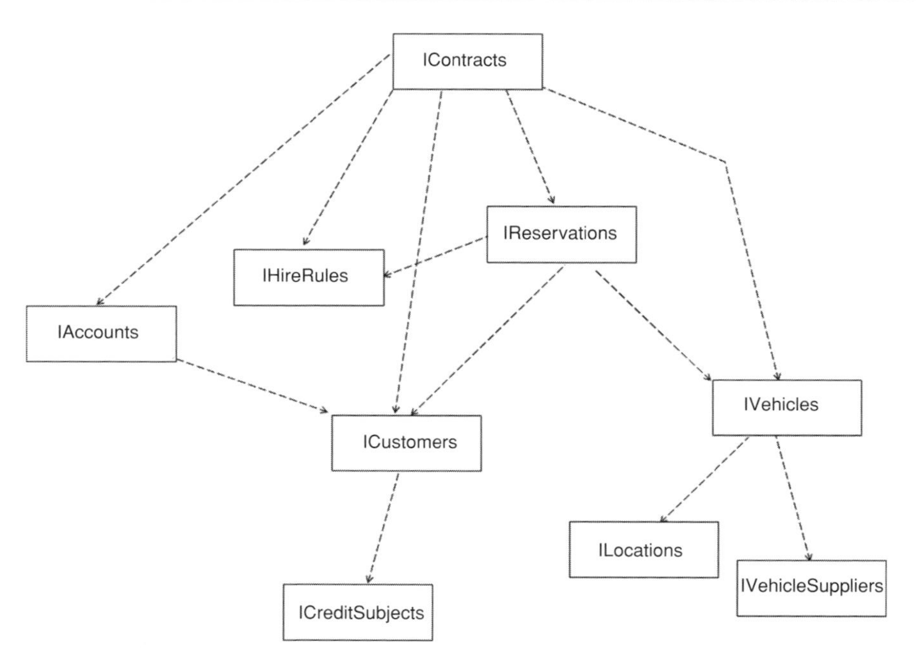

Figure 9.11 Interface dependency diagram

- Pre- and post-conditions
- Invocation type (synchronous, asynchronous, or bulk transfer).

These techniques are already well covered in the literature (Allen, 2000; Cheesman and Daniels, 2000; Dodd, 2004; Meyer, 1997), and therefore we do not go into further detail here.

Further detail in the form of both commercial requirements (for example, operation invocation cost) and QoS level requirements (for example, capacity levels) – are added as part of QoS infrastructure design described in chapter 10.

9.4 Software unit architecture techniques

We must now address the remaining main structural parts of the SOA:

- Software units and their dependencies
- Software unit descriptions
- Service buses.

This means picking up on the remaining questions, posed at increasing levels of detail at the start of 9.2:

- How are the interfaces realized in terms of software units?
- How are interfaces offered in terms of service buses?

9.4.1 Software unit architecture

Having laid out required services and interfaces, we are now in good shape to prepare the *software unit architecture*. We begin by considering how to support the interfaces in terms of software units.

Out of the nine interfaces in our example, the realization of the following six interfaces is already known:

- ICustomers
- IAccounts
- IVehicles
- ILocations
- IVehicleSuppliers
- ICreditSubjects.

The first four of these interfaces are realized in-house and therefore we have control over the implementation design of the software units employed. The ICustomers and IAccounts interfaces are realized by separate EJB component wrappers (two software units) exposed as Web services. These wrappers use functionality from two separate COBOL systems, each of which is also a software unit; note that a software unit does not have to offer an interface. The IVehicles and ILocations interfaces are realized by a single COM+ component, which also exposes the interfaces as Web services.

The two external interfaces (IVehicleSuppliers and ICreditSubjects) are also offered as Web services. However as they are supplied externally we have no control over their implementation design or over how they are actually realized. We simply need to know that they can "do a job." The descriptions of these interfaces must allow us to do that.

This leaves three interfaces: IContracts, IReservations, and IHireRules. The dependencies for the former two, both of which support the Contracts Service, are very similar. At the same time it is useful to keep the two interfaces separate as the interface users are different and also to assist in the agility requirements discussed earlier. There is therefore a strong argument for offering the IContracts and IReservations interfaces from a single software unit (Contracts Component). In addition, we decide that IHireRules is to be realized by a single software unit (Hire Rules Component), and that both the Contracts and the Hire Rules software units will be realized as EJB components.

Figure 9.12 depicts the various choices made. Note that a software unit is depicted using the UML class icon with a stereotype, ≪Name≫, to indicate the type of software unit, where known. As IContracts and IReservations are realized by the same software unit the dependency of the former on the latter is not shown on the diagram; this dependency is considered internal to the Contracts Component. For similar reasons, the dependency of IVehicles on ILocations is also not shown on the diagram.

Each software unit should be described as illustrated in table 9.7. At this point, we just need basic information to help catalog the software unit in the evolving asset inventory,

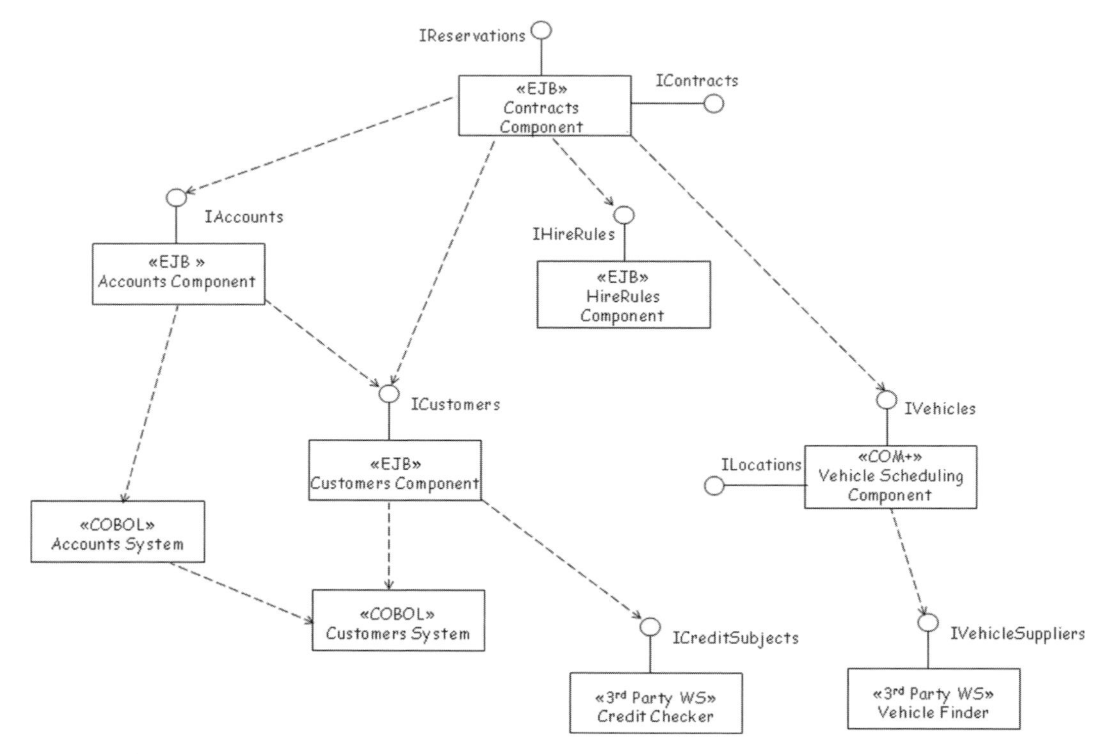

Figure 9.12 Software unit architecture diagram

introduced in 6.5. Once the software unit architecture has stabilized we can drill into further detail to form software unit specifications, though these are outside the scope of this book.

9.4.2 Use of service buses

A service bus is a grouping of deployed services that must conform to defined corporate policies and standards. These may include business goals, business rules, and SOA policy (such as QoS requirements), as well as funding and charging models.

As we have emphasized throughout this book, services must be abstracted away from the resources that implement them. This does not just mean using Web service standards and technologies to ensure platform independence, but also careful design of the SOA to ensure that the services are not tightly coupled to the way in which the current resource has been implemented – in terms of both design and technology. It is

Table 9.7 *Contracts Component software unit description*

Property	Value
Software Unit Name	Contracts Component
Purpose	To realize the Contracts Service
Services Provided	Contracts
Interfaces Provided	IContracts, IReservations
Interface Dependencies	ICustomers, IVehicles, IHireRules, IAccounts
Software Unit Dependencies	None
Sourcing Mechanism	In-house development
Run-time Platform	Must support EJBs offering WSDL, SOAP on HTTP

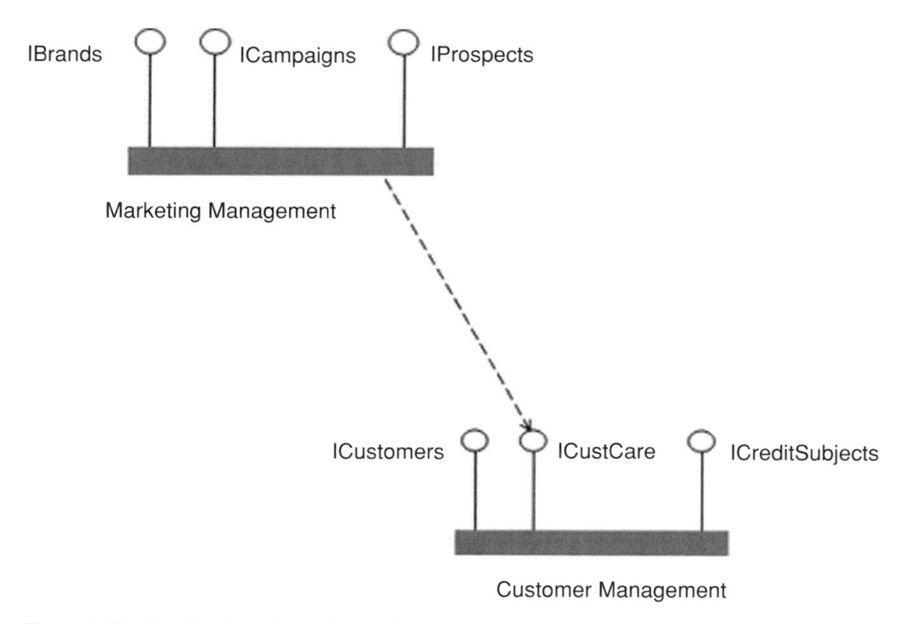

Figure 9.13 Service bus dependency diagram

important to ensure that a *bona fide* service is being used and not, for example, just the existing API to the implementation wrapped in SOAP.

Unfortunately, the term "service bus" (or some form of it) is being used very loosely by many vendors whose product implementations fail to remove the tight technology coupling typically created by existing middleware. A service bus, as understood in this book, is distinct from the idea of a service bus that some vendors have used to describe a platform with run-time support for Web services, such as support for both synchronous and asynchronous Web service calls, a UDDI compliant registry, together with publish-subscribe mechanisms that obviate the need for manual intervention by

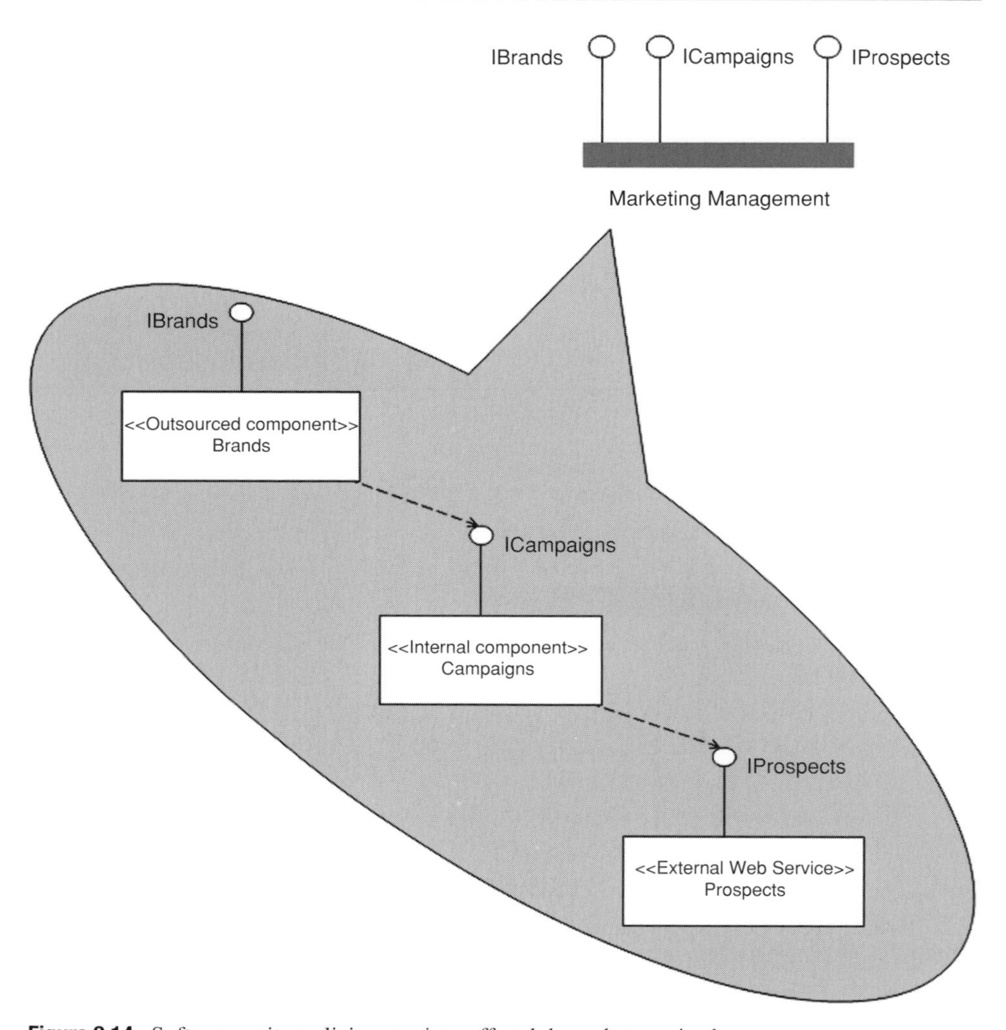

Figure 9.14 Software units realizing services offered through a service bus

the provider. The practical value of a service bus lies partly in the fact that it can be used to configure such a platform according to SOA policy.

At the same time, it is not mandatory for a service to be deployed via a service bus; applications may use other services that are not deemed to be members of the bus. Also, a service bus can use the services offered by another service bus. For example, commodity services may be grouped into service buses that are reused by "higher-level" service buses offering territory or value-add services. These kinds of dependencies can be modeled in an analogous way to software unit dependencies. Service buses are shown as horizontal bars, offering interfaces, as illustrated in figure 9.13.

A service bus also provides a practical vehicle for separating the consumer and provider aspects of services, rendering service implementations (typically components)

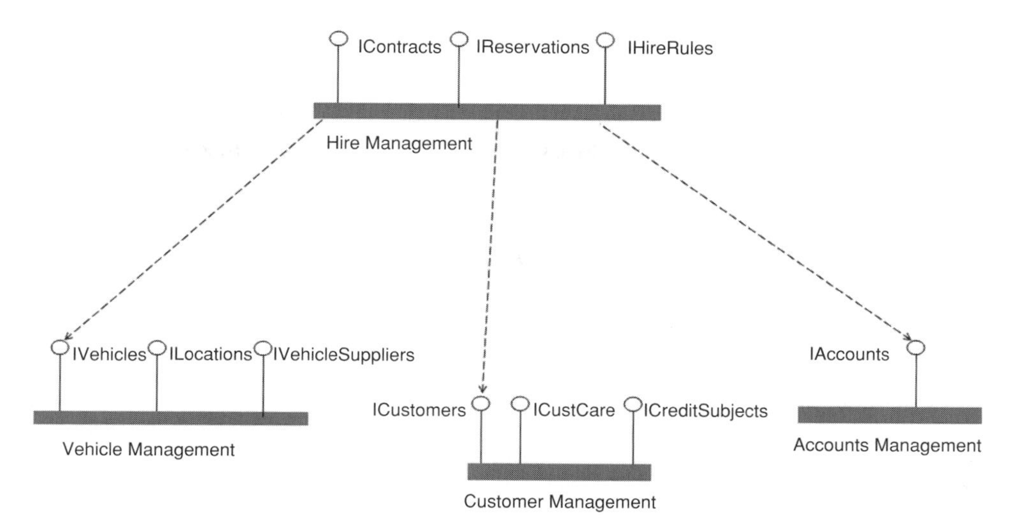

Figure 9.15 Vehicle hire service bus dependency diagram

transparent to the consumer. Recall that a service may be realized by multiple software units. For example, figure 9.14 illustrates that the Marketing Management service bus is realized by three different software units, each realized in a different way. A service bus should support the plugging in and plugging out of different software units to perform the same service, depending on different run-time constraints.

There are two basic kinds of service bus. A *business service bus* (BSB) (Wilkes, 2004b) is used to group together services related by a common business domain. BSBs are discussed in the next section. An *infrastructure service bus* (ISB) abstracts infrastructure capability (such as security, directory, messaging, management, routing, and so on) from the actual implementation by offering these capabilities in the form of services. ISBs are discussed in chapter 10.

9.4.3 BSBs

Thus far, most of the services that we have described, including those identified with business process redesign (as described in chapters 4–6) have fallen into the category of business service.

Not surprisingly, BSBs may well show a close affinity with business domains, as discussed in chapter 6. For example the Marketing Management service bus shown previously aligns with the marketing domain. This tactic of mapping domains to service buses facilitates the once-only assignment of policy that is common to the domain of a single service bus rather than for individual services; see also the layering of SOA policy in chapter 8. This is particularly useful where the different domains have different semantics.

Each BSB can be pictured as catering for the needs of different groups of business users or (more generally) roles. We can think of a BSB rather like a "virtual portal" with its own rules and representations designed for users of that portal. In the vehicle hire scenario the interfaces (from figure 9.11) are allocated to service buses as shown in figure 9.15. Notice that we have tried as far as possible to group together interfaces according to commonality of role.

9.5 Where to next?

In this chapter, we have concentrated on the first of the four quality drivers of the SOA: *agility*. This is largely a question of good service design. We now need to consider the other key quality drivers of the SOA – capacity, availability, and security. These factors are largely a matter of good SOA infrastructure design – a subject we address in chapter 10 – and that take us naturally into the realm of SOM, discussed in part 4. Another way of thinking about this is to say that we have designed the railway network, what we must now do is to figure out how to support the infrastructure of that network, and how to manage the running of it.

10 QoS infrastructure design

10.1 Preparing for service-oriented management

Some of the greatest challenges surrounding the industrial uptake of service orientation surround the abilities of the supporting technologies to scale up in the face of ever-increasing numbers of transactions, across widening geographical and commercial boundaries, and in a manner that can be trusted by all participants. While this puts the onus on effective SEM software, which we briefly discussed in chapter 2, a precondition of effective SEM software is the ability to understand and to measure what is actually to be managed. Unfortunately, the IT industry does not have a great track record when it comes to measuring software quality; "good enough" software has been the dominant trend for many companies for many years now in the face of shrinking budgets and development timescales. Software quality or nonfunctional requirements, to use the industry *lingua franca*, have always been something of a poor relation of functional requirements!

The challenges presented by software services, with their capability to cross traditional boundaries, including security zones, are an order of magnitude greater than with traditional software. The more pervasive software services become, the more critical is their role in the success or failure of the business. Effective SEM must expose and monitor executing applications so that problems can be detected and corrected as early as possible.

So things must change if service orientation is to take off successfully. The requirements for capacity, availability, and security of services must be clearly specified on the basis of *measurable quality attributes*. Effective SOM approaches can then be put in place to handle these requirements. In this chapter, we provide guidelines, in the form of quality templates, for establishing QoS requirements. We then move on to discuss the role of the the ISB in achieving a good service infrastructure design approach.

10.1.1 SOA technology infrastructure

Technology infrastructure includes hardware (processors, appliances, sensors, etc.), operating systems and middleware, and the communications infrastructure (hardware and software). In other words, the "platform" upon which the software services execute. However with services, as opposed to traditional applications, the notion of "platform" is somewhat exploded! Services may be executed in a variety of ways and on a wide range of different platforms. Some might be deployed in-house, others deployed on the processors of business partners, and others invoked on a subscription basis with no knowledge of where they are actually executed. The SOA technical infrastructure transcends the boundaries normally associated with traditional technical infrastructure. With an SOA, it is much less the technology *platform* and much more the SOA *policy* that governs the use of acceptable technologies that is of paramount interest.

At the same time, we cannot simply ignore the current technology infrastructure. A company would not normally design its SOA technology infrastructure from scratch. Rather, it would need to extend its existing infrastructure to accommodate the software services and solutions/applications that leverage the services. This may well involve a mix of different technologies provided by a range of suppliers and will evolve to accommodate higher work loads and changing usage patterns. The current technology infrastructure will also impact the SOA migration strategy.

However, it is not our aim in this book to describe this physical aspect of the SOA technical infrastructure in detail. Nor are we looking at the details and intricacies of different software technologies that may be used as part of this infrastructure. What we are primarily interested in are the SOA technology policy and the design of the infrastructure.

To be more specific, our aim in this chapter is to cover that part of the SOA infrastructure that provides the techniques required to ensure service capacity, availability, and security, as a foundation for SOM. In other words, our focus is on *QoS infrastructure design*.

Before we do that, it is useful to recap our view of QoS types, introduced in 8.1.2.

10.1.2 QoS types

Service quality divides into four types:

- *Agility* is the ability to act quickly and with economy of effort in accurate response to change, and also to initiate change for business advantage.
- *Capacity* involves three interrelated capabilities – throughput, responsiveness, and storage capacity – of services. In addition, a fourth concept – scalability – focuses on how each of the three capabilities is impacted by future demands and changes in technology.
- *Availability* is primarily concerned with keeping services free from failure within agreed limits and involves several parameters (that affect our trust in a service to do what it is intended to do) including reliability, maintainability, and resilience.

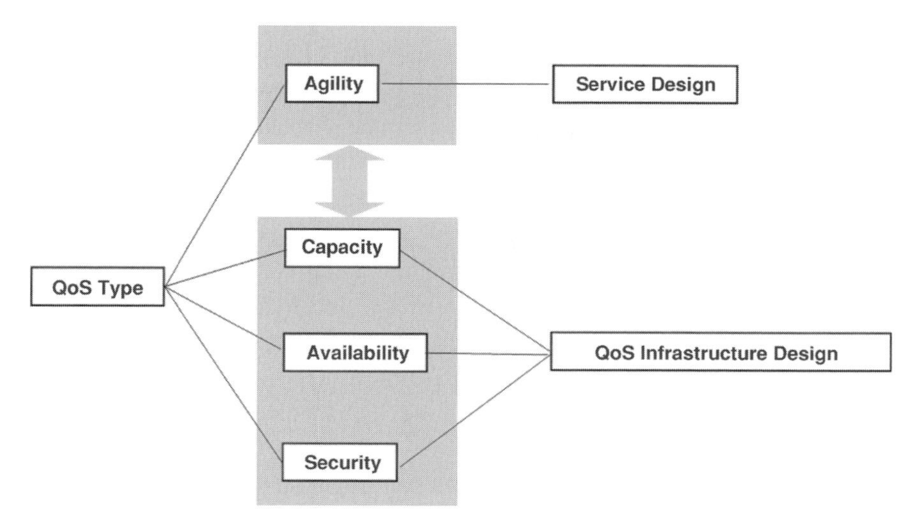

Figure 10.1 The dynamic relationship of agility to capacity, availability, and security

- *Security* focuses on the confidentiality and integrity of services, ensuring that the services are available when needed and free from denial of service attacks. In addition, security must also address three underpinning principles: authenticity, authority, and nonrepudiation.

QoS types correspond to different aspects of SOA design. In chapter 9, we focused on agility – the overarching characteristic that most strongly impacts service design in terms of the overall structure and shape of the SOA.

Capacity, availability, and security most strongly impact the infrastructure design of the SOA, along with choice of different supporting technologies. In addition, all three are intimately related to SOM and may effect the specification of individual services and operations.

Of course, as with any classification scheme, there is a certain simplification involved. These are complex interconnected concepts. Most significantly, agility is in a dynamic relationship to capacity, availability, and security (see figure 10.1), in that levels of capacity, availability, and security must be appropriate to the particular context and time. For example, agile software should be responsive in dealing with fluctuating payloads effectively, it should be easy to change when problems occur, and be able to adjust comfortably to new security requirements.

At the same time, our aim in this chapter is to provide a reasonably simple and pragmatic guide that addresses the challenges of service orientation, neglected in more traditional approaches. Capacity, availability, and security are analyzed in terms of a number of quality attributes with suggested criteria for measurement. The result is a set of *quality templates* that the reader can use as a starting point for dealing with infrastructure design and SOM.

10.1.3 Avoiding software anarchy

A further problem in many organizations is that developers make ad hoc decisions on how to design their solutions to meet quality requirements, without any regard to common definitions or guidelines. In the situation where systems are developed internally for use by a few selected internal users, the organization may have "got by" in this piecemeal fashion. As we have emphasized throughout this book, this situation no longer cuts it in the world of service orientation. There is the very real danger of software anarchy if developers are given *carte blanche*. Specification of capacity, availability, and security of services therefore forms part of the SOA policy. The guidelines are presented as part of a coherent approach to SOA.

The QoS infrastructure design part of the SOA involves weighing up what are often conflicting criteria – for example:

- Do we choose solid security measures that might adversely impact response times, or do we sacrifice elements of security in the interest of fast response times?
- What kinds of technology are available to mitigate this problem?
- How much is the company willing to invest in such technologies?
- Are the security and performance requirements actually justified?

And so on! A large part of QoS infrastructure design involves making these kinds of trade offs.

10.1.4 "You can't measure what you don't specify"

An important characteristic of agility is the ability of software to fit its requirements dynamically, to self-correct when problems occur, and to self-adjust so that costly software resources are utilized effectively. This, after all, is the whole point of on-demand computing.

However, the more fundamental point to note is that it is not meaningful to even begin to discuss these issues until the relevant QoS types – capacity, availability, and security – are clearly understood. Having defined capacity, availability, and security in the form of quality templates we have a firm foundation for providing guidelines on the design of the QoS infrastructure and for achieving effective SEM. The latter should enable the different elements comprising capacity, availability, and security to be calibrated and controlled in a flexible way that reflects run-time requirements and therefore helps to realize the promises of on-demand computing.

The good news is that, despite all the challenges of scalability, SOA-enabled applications provide a unique opportunity due to their loose coupled nature. They make it possible to add meaningful instrumentation without modifying the application code by intercepting software service requests and response messages.

10.1.5 Achieving business–IT alignment

Achieving business–IT alignment requires traceability between the artifacts developed. This is a major reason for taking the kind of model-based approach to service orientation that we describe in this book. Traceabilility of business requirements to real executing services is achieved through the models, which are underpinned by the asset inventory and enabled by a full BPM solution, as described in 3.3.3. Recall that the latter should include a BAM, a component that assesses and controls the performance of the process engine in real time. The BAM component should be enabled with connections to SEM in order to automatically monitor and control QoS levels. Again, though as discussed in the previous section, the desired QoS levels must be specified in order that they can be measured.

The service description and service specification (discussed in 9.2.7 and 9.3.4 respectively) only take us so far in respect of achieving the right level of granularity and degree of rigor for specifying QoS requirements. It is necessary to specify QoS requirements for the specific operations of a software service. Moreover, the need for traceability requires knowledge of which business activities these operations are related to, in order to understand the impact of failure or success in meeting QoS requirements. And in the wider scheme of things this is why an integrated model-based approach to BA, SOA, and SOM is so important.

In the case of our vehicle hire company, a more detailed analysis is carried out on the QoS requirements for the operations of the services designed in chapter 9. This analysis results in the identification of further requirements for operations of the Contracts Service. These requirements are of two kinds: *commercial* and *technical.* Although commercial requirements are formally covered as part of the SLA (see chapter 12), it is necessary to make some early tentative investigation into this area in order that the QoS infrastructure design and the determination of precise QoS levels can proceed on a realistic footing. For the sake of illustration, let us assume that in the case of the Contracts Service, following discussion between the service architect and other roles such as the sourcing and usage manager (see 13.3), the following requirements are established:

- *Agreed service time (AST) = all times except 02.00–03.00*
- *Unit charge per operation use: 1 cent*

Again, for the sake of illustration, let us assume that in the case of the Contracts Service, following discussion between the service infrastructure architect and other roles such as the service execution manager, the following technical requirements are established:

- *maximum of 10 concurrent users*
- *maximum outages per day = 0.02% of AST*
- *minimum/average/maximum operation invocations per day = 6400/12800/19200 with 20% increase over first six months*

These requirements are added to the service description (first developed in 9.2.7), as shown in table 10.1. Note that QoS levels are shown explicitly in the service description

Table 10.1 *Enhanced service description for Contracts Service*

Property	Value
Service name	Contracts Service
Domain	Vehicle Hire Sales
Purpose of the service	Management of reservations, contracts, and deliveries.
Responsibility by business type	Contract Type, Contract, Entry Sheet, Exit Sheet, Reservation, Delivery
Operations	CRUDL Secure Reservation, Book Delivery, Open Contract, Close Contract, Report Expiry
Dependencies	Customers, Vehicles, Hire Rules, Accounts
Commoditization level	Value-Add
Sourcing policy	In-house
Usage policy	Initially restricted to Hire Agent. Intention is to expand usage to customers once established.
Targets (Record Handback Activity)	"Contract must be closed within 5 minutes" in the case of a partner. "Contract must be closed within 15 minutes" in the case of an employee.
Business process support (current)	Agency Vehicle Hire
Business process support (future)	Internet Vehicle Hire
Business process targets	50% decrease in numbers of phone reservations. Target of 1,600 online reservations per day. 20% increase in number of hires. 20% increase in productivity of hire agent. 50% increase in customer satisfaction rating.
Cost	Unit charge per operation use: 1 cent
Capacity levels	Minimum/Average/Maximum operation invocations per day = 6,400/12,800/19,200 with 20% increase over first six months. Maximum of ten concurrent users.
Availability levels	AST = all times except 02.00–03.00 Max Outages per day = 0.02% of AST

only where these differ from the SOA policy, as in the case of both capacity levels and availability levels in our example. However, there are no differences in security levels from SOA policy and therefore security levels are omitted from the service description.

10.1.6 A word about ITIL

The Information Technology Infrastructure Library (ITIL) standards have become the dominant approach to the process of service management. Our approach to capacity, availability, and security is designed for smooth integration with the ITIL standards (which offer corresponding processes for capacity, availability, and security) and builds on those standards in a way that is tuned to the uptake of software services technologies (such as Web services). We discuss the relevance of ITIL in more detail in chapter 11.

10.2 Capacity

The industrial take up of SOA depends on the ability to cope with hugely increased volumes of service requests across rapidly expanding geographical and organizational boundaries. *Capacity management* is a discipline that cost-effectively aligns IT processing and storage capacity provision with changing business needs (Macfarlane and Rudd, 2003, p. 52).

10.2.1 Basic capacity concepts

ITIL provides guidelines for managing the scale and complexity of subject matter by outlining three principal areas of responsibility for capacity management.

Business capacity management

Business capacity management (BCM) is responsible for ensuring that the future business requirements for IT services are considered, planned, and implemented in a timely fashion. This requires a wide knowledge that includes existing service levels and future SLAs, business and capacity plans, modeling techniques (such as simulation and trend forecasting), and application sizing methods.

Service capacity management

Service capacity management (SCM) is responsible for managing the IT services that are provided to customers, and is responsible for measuring and monitoring services as detailed in SLAs, and for analyzing, tuning, and reporting on service capacity. This requires a more technical knowledge than BCM, including systems, networks, service throughput and performance, tuning, and demand management.

Resource capacity management

Resource capacity management (RCM) is responsible for managing the components of the IT infrastructure. This entails monitoring, analyzing, tuning, and reporting on the utilization of components. This requires a yet more technical knowledge than SCM that goes deeply into the use of current and future technologies and their utilization.

In our context of service orientation we are primarily interested in SCM. Clearly, for our purposes, we need to address the challenges raised in the capacity management of *services*, and in particular of Web services, which are the dominant software standard for services. This means understanding how to measure service capacity and designing the SOA to account for service capacity issues.

10.2.2 Raising the capacity bar

Notice that SCM involves addressing what ITIL refers to as "demand management," an activity that sits at interface between the business and capacity management controlling the use of resources required to satisfy demand (Macfarlane and Rudd, 2003, p. 56). Essentially what is involved here is a balancing act of cost against capacity, and of supply against demand.

This takes us in fact to the heart of the on-demand computing proposition – ensuring that service levels are maintained while paying only for those computer resources that are actually needed. The monitoring and analysis activities of the SOM life cycle (see figure 7.2, p. 133) must take account of fluctuating service capacity levels by continuously monitoring these levels against capacity thresholds agreed and documented in SLAs, which themselves may vary depending on different factors (as examined in chapter 8), such as role and time. It must do this in a way that ensures against excessive cost through overutilization of resources. So accurate measurements of capacity units and thresholds must be specified, and the SLAs must indicate acceptable cost levels.

We have already emphasized that the measurement of *service* (as opposed to application) quality demands establishing QoS policy as part of the SOA. In particular, the measurement of Web service capacity requires a finer level of granularity than the determination of application, system, or network capacity. The overall capacity of a Web service is a relatively useless metric – you need visibility into the specific operations defined in a Web service's WSDL interface description. Moreover, it is necessary to know which business activities these operations are related to in order to understand whether observed operation delays might have a business impact. In the wider scheme of things, this is in fact a major reason why this book advocates an integrated model-based approach to BA, SOA, and SOM.

Operation-level information is also necessary in order to isolate and optimize performance bottlenecks. Characterization of operation performance may also sometimes require visibility into the parameters associated with the operation calls.

The concept of service capacity involves three inter-related capabilities of services – throughput, responsiveness, and storage capacity. In addition a fourth concept – scalability – focuses on how each of the other three capabilities is impacted by future demands and changes in technology.

We have emphasized throughout this book that one of the defining characteristics of a service-oriented approach is the shift to a supply–manage–consume model. For commodity services in particular, this involves a requirement to compare capabilities of candidate service providers as part of the process of choosing the best one for a particular service. Ideally, we would like to be able to rank candidate service providers

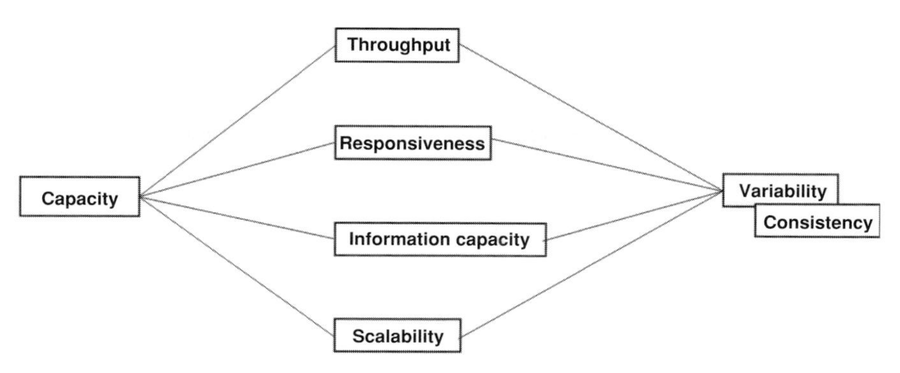

Figure 10.2 Service capacity quality attributes

based upon their scores for different quality attributes and enable detailed service provider performance comparisons of capacity trends.

Two further concepts are therefore necessary: *variability* and *consistency*. Variability centers on the degradation or improvement of capacity attributes over specified time periods. Consistency focuses on the stability of capacity attributes over specified time periods. Both of these ideas are connected with trust in suppliers to maintain or improve service levels, and may also be applied to availability (as discussed in the next section). Figure 10.2 provides an overall picture of the elements that comprise capacity.

We stated that one of the main reasons for the inclusion of variability and consistency is to provide a comparative measure. One of the difficulties we face here is the potential complexity of some the mathematics involved in related concepts such as standard deviation. A unique challenge of a comparison facility is to create meaningful capacity *indices* that are independent of absolute measures such as service response time. In other words, it is difficult to compare software services having a wide range of response times. In the world of Web services, for example, there is no standard granularity of functionality, so it is common that providers of similar Web services take very different approaches to service design. One provider may choose to offer an extensive functionality via a single Web service and another may offer the same functionality via a set of many services.

This problem can be addressed via time-neutral indices calculated using the absolute metrics. The *responsiveness consistency index* provides a way to examine the stability of the response times produced for a given operation of a software service. For example, if the set of response times produced by an operation of the software service *A* is 5ms, 10ms, 18ms, and 2ms, and the set of response times produced by a corresponding operation of the software service *B* is 50ms, 51ms, 48ms, and 50ms, then the operation of software service *B* will have a higher consistency index and will be ranked higher on that measure than the operation of software service *A*. Notice that it is *operations* (not software services) that can be specifically measured in this way. However, it is still

Table 10.2 *Responsiveness history of two providers of ICreditSubjects*

Time period and measure	Serious finance	Secure money
Week 1 Average Responsiveness	5 msec	19 msec
Week 2 Average Responsiveness	7 msec	17 msec
Week 3 Average Responsiveness	35 msec	22 msec
Weeks 1–3 Standard Deviation of Responsiveness	23.72	3.55
Weeks 1–3 Consistency Index of Responsiveness	76.28	96.45

possible to provide "average" scores for software services based on the overall results of their constituent operations.

In our vehicle hire scenario the software unit Credit Checker is a third-party Web service offering the interface ICreditSubjects. Market investigation reveals that there are two main candidate providers for this service: Serious Finance and Secure Money. The responsiveness history of these providers is analyzed in terms of standard deviation[1] and consistency index, as shown in table 10.2.

10.2.3 The service capacity template

Table 10.3 provides a starting point for establishing service capacity policy. Table 10.4 provides an example of enterprise wide QoS capacity policy for the vehicle hire company. The service description (table 10.1) is used alongside detailed knowledge of the technical infrastructure constraints to help in deciding what the various QoS levels should be. Note that these QoS levels apply to each operation of the Contracts Service. The values may be overridden and specialized at varying levels of detail, as described in 8.3.1.

10.2.4 Service capacity design

Capacity design is a discipline of great width and potential depth that involves making trade offs against other quality attributes and expressing these with complex conditions. As we have seen, SOA raises further challenges in this already complex field. In this section, we summarize some of the major aspects of capacity design that are particularly relevant from the point of view of the SOA.

Introduction of design mechanisms

Appropriate mechanisms must be put in place to try to ensure that a service complies with the threshold levels for throughput, responsiveness, and information capacity expressed in the capacity quality template.

[1] For an elegant explanation of standard deviation, the reader is referred to http://www.robertniles.com/stats/stdev.shtml.

Table 10.3 *Example service capacity template.*

Quality attribute	Definition	Example measurement units
Throughput	The ability of the operations comprising a service to process units of work in units of time.	Operation invocations per unit time (worst, average and best[a]).
Responsiveness	The time taken by the operations comprising a service to execute each invocation.	Operation response time (worst, average and best) in unit time.
Information capacity	The capacity of the messages associated with the operations of a service to hold content defined in units of storage.	Message content size (worst, average and best) in storage units.
Scalability	The ease with which a service can retain its intended capabilities (throughput, responsiveness, and information capacity) under increasingly demanding conditions.	Percentage increase in numbers of calling applications over specified time period.
Variability	The degree of acceleration or deceleration of throughput, responsiveness, or information capacity over unit time.	The rate of increase in throughput, responsiveness, or information capacity over specified time period.
Consistency	The opposite of variation – the degree of stability of throughput, responsiveness or information capacity over unit time.	Consistency index (throughput, responsiveness, or information capacity) over specified time period.

[a] Further details such as burst loads (Kaye, 2003, p. 125) may also be determined. It will be important to consider when these peaks are likely to occur and what mechanisms are to be put in place to cope with them.

Table 10.4 *Vehicle hire company enterprise capacity QoS specification*

Quality attribute	Worst	Average	Best
Throughput	0.1 /sec	0.15 /sec	0.2 /sec
Responsiveness	50 msec	20 msec	7 msec
Information capacity	2.3 mb	3.4 mb	4.0 mb
Scalability (six months)	0.5%	0.8%	1.2%
Responsiveness variability (one month)	+1%	+4%	+9%
Responsiveness Consistency Index (one month)	91	94	97

Four further issues must be considered in capacity design:

- Determining the most effective mechanisms for invoking software services in different contexts
- Deciding how to manage multiple instances of a software service concurrently, and how to share workload between them
- Determining the most effective mechanisms for closing down software services in different contexts

- Identifying mechanisms for the management of software services capacity with respect to the SOM life cycle (see figure 7.1, p. 131), including management of scalability and variation, as well as management of problem situations. For example, if a software service is responding sluggishly:
  - Where does the ultimate problem lie?
  - Is the bottleneck in the application server, the network, the database, or somewhere else?
  - Who is responsible for fixing it?
  - Can any simple action be taken to improve matters quickly?

Standards compliance

Another area of concern is the evolving standards for Web services management such as WSDM. Immaturity and incompleteness of these standards can be mitigated to some extent by the use of a new class of SEM software products that provide implementations of current standards and proprietary solutions for gaps in current standards. Most of these products are provided with an assertion that the proprietary elements will be converted to implementation of standards through controlled upgrades and releases.

Layered approaches

One way of managing the potential complexity of the above issues is to centralize the implementation of capacity management into a dedicated *management software layer*. However, care must be taken that such a layer does not itself become one of the performance bottlenecks that it seeks to eliminate. It must be ensured that the physical infrastructure can indeed support the new layer.

Increasingly, the different elements that comprise service capacity may benefit from separate, distributed, implementations. As we shall see a little later in this chapter, an ISB can provide an approach that caters for mixed capacity implementations while ensuring control of SOA QoS policy.

10.3 Availability

Successful uptake of SOA depends on the continued availability of software services in the face of ever-more demanding scenarios that stretch the capabilities of the services to the limit. Not only that, but the scope of the potential effects of service disruption or failure expands to greater and greater numbers of calling applications, often belonging to different companies. The result is yet greater demands on service availability.

10.3.1 Basic availability concepts

One of the difficulties with availability is that it is not an easy concept to define. While people usually have an innate sense of what is meant, when you ask detailed questions it becomes something of a struggle to pin things down. As we shall, see part of the reason for this is that availability has both general and specific meanings. Availability in a general sense covers a range of *quality attributes* from reliability to maintainability and resistance. In a specific sense, however, availability commonly refers to the up time of a service, or more specifically to the *percentage of AST* for which it is available.

10.3.2 Raising the availability bar

The benefits of service reuse and interoperability are in sharp contrast to the difficulties encountered in attempting to achieve significant reuse and interoperability with previous generation application-centered approaches. However, as services become employed in more and more different contexts, and used by business processes that cross organizational boundaries, so the requirements for service availability grow increasingly demanding. These challenges are hugely magnified in comparison to those faced by application-centered approaches.

As software services scale up, so they can very quickly evolve in complexity, becoming highly distributed and dependent on a mass of independent components. The result can be surprisingly many points of failure.

At the start of this section, we stated that availability has both general and specific meanings. Availability in a specific sense has a clear definition.

Availability is the percentage of AST for which the service is available

However, availability in a general sense (following ITIL) is more wide-ranging. If we look closely, it comprises availability (in a specific sense), reliability, maintainability, and resilience. It is perhaps tempting to dismiss these quality attributes as no more than we might require in a traditional software engineering approach. However, that would be a misleading assumption. With a service-oriented approach, because of the business criticality of many of these software services, their availability measurements must be increasingly precise and at a low level of detail. For example, a downtime of 1 hour in 10,000 hours might have a major detrimental effect on business in the case of a service used as part of a mission critical business process, such as order taking over the Internet. A measure in the region of 1 second out of 10,000 would probably be much more appropriate. However there are further reasons why service availability is at once both different and challenging.

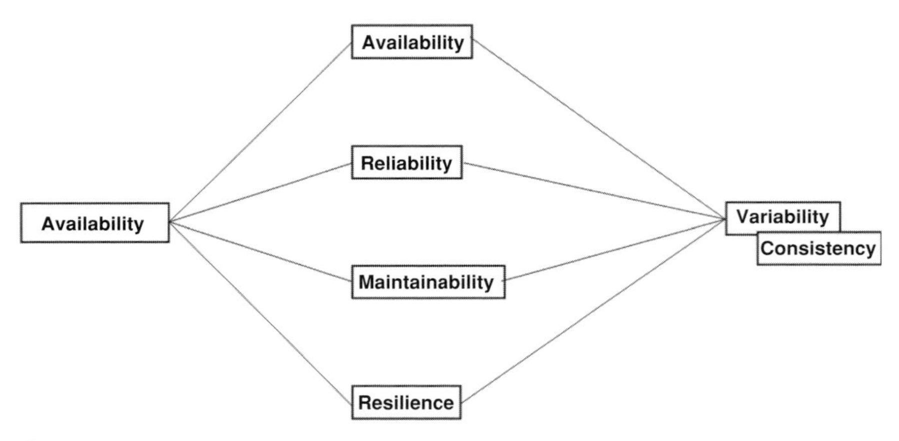

Figure 10.3 Service availability quality attributes

As with capacity, we would like to be able to rank candidate service providers based upon their scores for different availability attributes and to perform comparisons of availability trends. Variability and consistency are therefore also applied to availability. For example, an *availability consistency index* provides a straightforward indicator of the percentage of time that measured operations of a software service respond to a remote client. If a given operation of a software service responds to a higher percentage of the client transactions directed to it than others do, it will receive a higher availability consistency index score and thus rank higher.

In summary, service availability breaks down into the five component quality attributes as shown in figure 10.3.

10.3.3 The service availability template

Table 10.5 provides a starting point for establishing service availability policy.

Table 10.6 provides an example of an enterprise-wide QoS availability policy for the vehicle hire company. The service description (table 10.1) is used alongside detailed knowledge of the technical infrastructure constraints to help in deciding what the various QoS levels should be. Note that these QoS levels apply to each operation of the Contracts Service. The values may be overridden and specialized at varying levels of detail as described in 8.3.1.

10.3.4 Service availability design

SOA raises the stakes for software availability in what is already a complex discipline. In this section, we summarize some of the major aspects of availability design that are particularly relevant from the point of view of the SOA.

Table 10.5 *Service availability template*

Quality attribute	Definition	Example measurement units
Availability	The percentage of AST for which the service is available.	$\frac{\text{AST} - \text{Downtime}}{\text{AST}} \times 100\%$
Reliability	The minimization of failure and the ability to keep the service operable, ensuring continued operation of security controls.	Mean time between failures (MTBF)
Maintainability	The ease with which service errors errors can be diagnosed and corrected, without adversely impacting levels of security This should be a reflection of the ability to restore services back to normal operation.	Mean time to repair (MTTR)
Resilience	The ability of a service to continue to operate properly, in spite of the malfunctioning of one or more sub-systems – ensuring no gaps in the security provision occur when a sub-system fails or when resilience failover is triggered.	Number of additional software units needed to withstand the effect of errors.
Variability	The degree of acceleration or deceleration of any of the above attributes over a specified time period.	The rate of increase in any of the above attributes over a specified time period.
Consistency	The opposite of variation – the degree of stability of any of the above attributes over a specified time period.	Consistency index (of any of the above attributes) over a specified time period.

Table 10.6 *Vehicle hire company enterprise availability QoS specification*

Quality attribute	Worst	Average	Best
Availability	97%	98%	99%
Reliability	2,000 hours	3,000 hours	4,000 hours
Maintainability	4 hour	2 hour	1 hour
Resilience	7	5	3
Availability Variability (one month)	+3%	+5%	+7%
Availability Consistency Index (one month)	92	94	96

Service failover

A consequence of successful SOA is a greatly increased and diverse population of service users. The failover capabilities must be designed in such a way that they are transparent to these users. Service failover design may well entail a complex system of redundant and replicated services, so that in cases of failure business continuity is maintained. It is important to manage this complexity.

Standards compliance

As with service capacity, another area of concern regarding service availability is the evolving standards for Web services management. Again, immaturity of these standards can be mitigated to an extent by the use of a new class of SEM software products that provide implementations of current standards and proprietary solutions for gaps in current standards. Most of these products are provided with an assertion that the proprietary elements will be converted to implementation of standards through controlled upgrades and releases.

Layered approaches

One way of managing the potential complexity of these issues is to centralize the implementation of availability policy, along with capacity management into a dedicated management layer, as described in 10.2.4. However, care must be taken that such a layer does not represent a single point of failure that can shut down an enterprise. Since so many dependencies will be placed on this layer it must be ensured that the physical infrastructure can indeed support the new layer.

The different elements that comprise service availability may benefit from separate, distributed, implementations. An ISB, discussed later in this chapter, can help to provide the necessary flexibility while ensuring control of SOA policy.

10.4 Security

The SOA vision is one in which globally distributed services, many of which are publicly available shared services, can be readily incorporated into numerous applications. Knowing who is using these services, what these users are authorized to do, and how to protect all information at the various nodes, is a major challenge to the successful take up of SOA.

10.4.1 Basic security concepts

Software security is traditionally associated with confidentiality, integrity, and availability. *Confidentiality* is concerned with protecting of sensitive information from unauthorized disclosure or interception. *Integrity* centers on safeguarding the accuracy and completeness of information and software. *Availability* focuses on ensuring that information and vital IT services are available when needed.

Confidentiality and integrity are often confused with each other. To clarify, integrity is concerned only with whether or not the service has been altered in some way and, unlike confidentiality, does not guarantee that the service has not been used by an unauthorized consumer. Furthermore, it is possible to guarantee the integrity of a service without encrypting it; for example, a digitally signed document that is sent as unencrypted text, which is open to any consumer to read but that cannot be altered without detection.

Availability concerns the freedom of services from disruption as a result of security incidents. It therefore has a narrower meaning than overall availability (described in 10.3) – in effect, availability in the context of security is a sub-set of overall availability.

10.4.2 Raising the security bar

Software services, and Web services in particular, raise some significant further challenges. For example, as we saw in chapter 2, transport-based security like SSL provides one type of protection for messages in transit – server-to-server non-persistent confidentiality – but offers little or no control at the application level, which is the place where the messages have meaning. Messages need to remain persistently secure not just while in transit, but also when residing in permanent storage. Instead of a focus on perimeter or network security, what is needed instead is persistent message-level security that addresses both confidentiality and integrity.

While the evolving and maturing Web services security standards provide XML-based abstraction layers for established security technologies, these standards must be applied in a way which meets the security requirements and policy of the particular organization. Piecemeal approaches will not do!

What, then, are the special challenges of software services security and how can we address these challenges effectively?

Confidentiality

Individual messages may need to be encrypted, or signed, or both, in order to protect the information that participants are exchanging. The keys for encryption and decryption, and the identity of the owner of the keys, are exceedingly complex when the software service end points are not all within a single organization that can authenticate the parties to transactions. In addition, intrusion detection and attack protection may increasingly be required.

Integrity

We must guarantee that message content is consistent. The information received must be in an identical state to the information transmitted through a software service. Constant monitoring of message integrity is another absolute requirement for keeping a software service in compliance with business and regulatory security policies.

Availability

We want to insure freedom of services from disruption as a result of threats such as denial of service attacks. Protection from such attacks must be provided at the level of messages, not only the end points of those messages.

We must also address three underpinning security principles that further characterize a distributed message-based architecture:

- *Authenticity*[2] We must be able to identify accurately a message sender or recipient. Constant monitoring of an individual's authenticated identity is an absolute requirement for keeping a software service in compliance with business and regulatory security policies. In addition, single sign-on (SSO) management (see 10.4.4) is an increasingly important feature.
- *Authority* Consumers must have permission to use a message or access stipulated message content. An individual's authority to access stipulated information and functionality must be monitored for compliance with business and regulatory security policies. In the case of messages of any size this might involve quite complex interpretation of message content in order to map user identity to access permissions on individual content items.
- *Nonrepudiation* We must be able to prove participant identities in a specific instance of message usage. In certain circumstances – for legal or business reasons – it must be possible to prove that events occurred, such that (for example) buyers cannot deny placing orders and sellers cannot deny receiving them.

The above discussion is summarized in figure 10.4, which depicts the constituent quality attributes of service security.

10.4.3 The service security template

Table 10.7 provides overall guidelines for services security.

Table 10.8 provides an example of enterprise-wide QoS security policy for the vehicle hire company. The values may be overridden and specialized at varying levels of detail, as described in 8.3.1. In theory, it is also possible to add columns for lower and higher acceptable QoS levels, and rows for variability and consistency for each quality attribute. In practice, however, security is a much more absolute concept than either capacity or availability, and it is the overall QoS policy itself, rather than specific QoS levels, that is likely to be more relevant, as suggested below.

Note that although we have added means of measurement for the security attributes we are limited here to *known* breaches of security. We might, for example, measure the numbers of these breaches and their MTTR. However, in a significant sense this rather

[2] Authenticity applies at three levels (or combinations of these levels):
- Something you *know* – for example, a password
- Something you *have* – for example, a token
- Something you *are* – for example, biometric data such as fingerprint.

The reader is referred to Rosenberg and Remy (2004) for an excellent discussion.

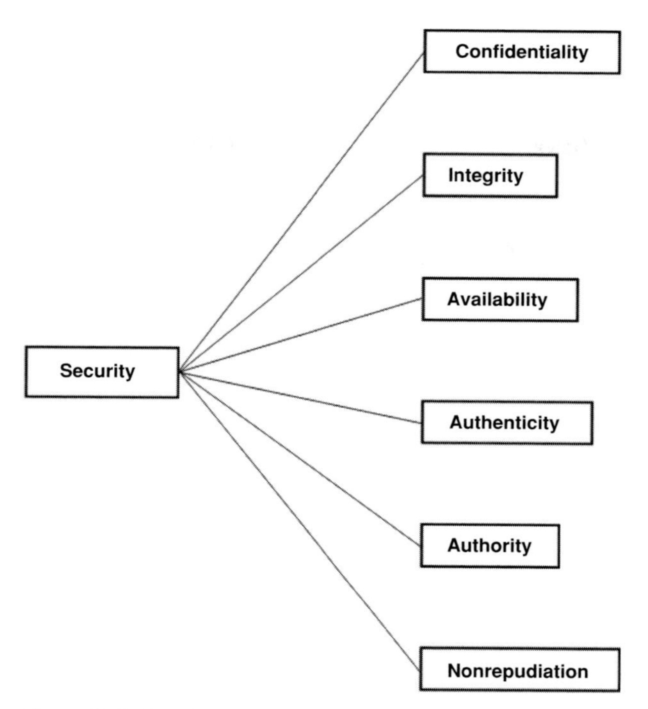

Figure 10.4 Service security quality attributes

begs the question of how secure the services really are. The security quality attributes are much more binary than the availability quality attributes, which can be assigned more meaningful QoS levels. It will be the mechanisms used to ensure confidentiality and integrity, as described in the technology policy and the design policy, that are of prime concern and debate rather than the QoS policy itself, which is likely to take the form:

All services must be kept confidential and maintain their integrity at all times.

Similarly, it is the mechanisms used to ensure authenticity, authority, and nonrepudiation (as described in the technology policy and the design policy) that are likely to take primary significance. Again, the QoS policy itself may well take the form of a quite general statement:

All the service users are known and authorized to use the services and can be clearly identified.

Finally, it is worth reminding ourselves of the scope of this book. Of necessity, we have concentrated throughout this chapter on software services, and in particular on the messaging aspects of those services. All this attention on achieving a good secure set of software services will be for naught if the remaining aspects of the software portfolio – such as databases and application code – are insecure.

Table 10.7 *Service security template*

Quality attribute	Definition	Example measurement units
Confidentiality	Protection of the message from unauthorized access.	The percentage of messages that are not penetrated by unauthorized consumers.
Integrity	Consistency of message content in terms of information received being in an identical state to the information transmitted.	The percentage of messages that are untouched by any sort of corruption, removal, change or addition to the service[a] between inception and completion of a service execution.
Availability	A sub-set of overall availability: the percentage of AST for which the service is free from security outage time (SOT), such as time lost from denial of service attacks.	$\frac{\text{AST} - \text{SOT}}{\text{AST}} \times 100\%$
Authenticity	Accurate identification of a message user.	The percentage of messages, requiring authentication that are successfully authenticated.
Authority	Permission of an authenticated user to use a message or access service elements[b].	The percentage of messages requiring authority that are successfully authorized.
Nonrepudiation	Proof of participant identities in a specific instance of message usage.	The percentage of messages requiring nonrepudiation that satisfy this requirement.

[a] Integrity applies to the information, logic, and rules, that comprise the service.
[b] Service elements include the operations, rules, and information (including specific attributes) provided by a service.

Table 10.8 *Vehicle hire company enterprise security QoS specification*

Quality attribute	QoS level (%)
Confidentiality	99.95
Integrity	99.8
Availability	99.7
Authenticity	99.95
Authority	99.6
Nonrepudiation	99.1

10.4.4 Service security design

Service security design is a complex subject in its own right. The reader is referred to Rosenberg and Remy (2004) for a detailed approach to Web services security design and to Erl (2004) for consideration of how to integrate Web services security with legacy applications. In this section, we summarize some of the major aspects that are particularly relevant from the point of view of the SOA.

Designing for single sign-on

SSO has become something of a necessity for most SOA applications. SSO allows a user to log in once with a recognized security authority and use the returned login credentials to access multiple resources. This approach involves mapping security credentials transmitted by services to various systems, so that the authentication process is streamlined. The complexities inherent with an in-house design approach to SSO render a third-party approach attractive (see below). Both Microsoft's .NET Passport and the Liberty Alliance were inspired by the need for SSO, especially with Web services.[3]

Gauging the impact on capacity

It is important to gauge the increased levels of latency incurred by introduction of security capabilities, and ensure a good understanding of the capacity requirements that will be determined by your chosen security approach.

Layered approaches

One way of managing the potential complexity of these issues is to centralize the implementation of security policy into a dedicated security layer. Despite the overheads of another layer added to the SOA, especially in terms of initial investment, this can appear to be an attractive strategy. There is, for example, the promise of increased overall control and ability to enforce security quality policy, resulting in long-term cost savings.

At the same time, a security layer requires careful design if it is not to become a performance bottleneck and represent a single point of failure that can shut down an enterprise. Since so many dependencies will be placed on this layer, the importance of scalability and availability loom large. It must be ensured that the physical infrastructure can indeed support the new layer.

Increasingly, it is also the case that the different elements that comprise service security may benefit from separate, distributed, implementations. For example, it may be appropriate to employ a third-party authentication service alongside an in-house integrity service, with management software to ensure availability outsourced to a partner. As we shall see in 10.5, an ISB can provide an approach that caters for mixed security implementations while ensuring control of SOA service quality policy. An ISB can also be used to furnish a similar approach to management of availability and capacity.

Remote security services

Third-party security services may appear to be an attractive way around some of these problems. Certainly third-party authentication services based upon standards such as

[3] The Web services standard for SSO is Security Assertion Markup Language (SAML) (see www.oasis-open.org).

Liberty Alliance and .NET passport are gaining in momentum. However, we need to tread carefully here. Even if an organization impresses with its policies and guarantees, there is no way of knowing if a change in management or ownership down the line will endanger the quality of service initially agreed. Therefore at this point in time most organizations choose to keep at least some of their security services in-house.

Security context

Service security design depends upon the security context of the services (Kaye, 2003, p. 172). The "security context" refers to the duration and physical spread (across different nodes) of the service. Clearly a synchronous software service restricted to a single node raises fewer security problems than an asynchronous software service (that might take days or weeks) involving several different nodes owned by different organizations, each with different security policies that must be reconciled.

10.5 Infrastructure service buses

A constant theme in our discussion of capacity, availability, and security has been the tension that exists between the need for standardized approaches to gaining control of these infrastructure services (as part of the SOA) and optimizing the management of these services. Employing an ISB can provide an integrated approach to the management of capacity, availability, and security in line with SOA policy, while at the same time allowing the freedom to use distributed approaches to the implementation of infrastructure services.

10.5.1 The concept of ISB

An ISB (Wilkes, 2004b)[4] abstracts infrastructure capability – such as management (capacity and availability), security, directory, messaging, routing, and so on – from the actual implementation by offering these capabilities in the form of services: see figure 10.5. This is analogous to the way in which a BSB abstracts business capability from the actual implementation by offering these capabilities in the form of services.

10.5.2 Designing an ISB

The advantage of using an ISB is that the infrastructure capabilities are offered in a fashion that is location- and platform-independent. This allows externally provided

[4] The CBDi Forum also introduce the enveloping term "enterprise service bus" (ESB), which they define as a uniform service integration architecture of infrastructure services that provides consistent support to business services across a defined ecosystem. The ESB is implemented as an SOA using Web service interfaces.

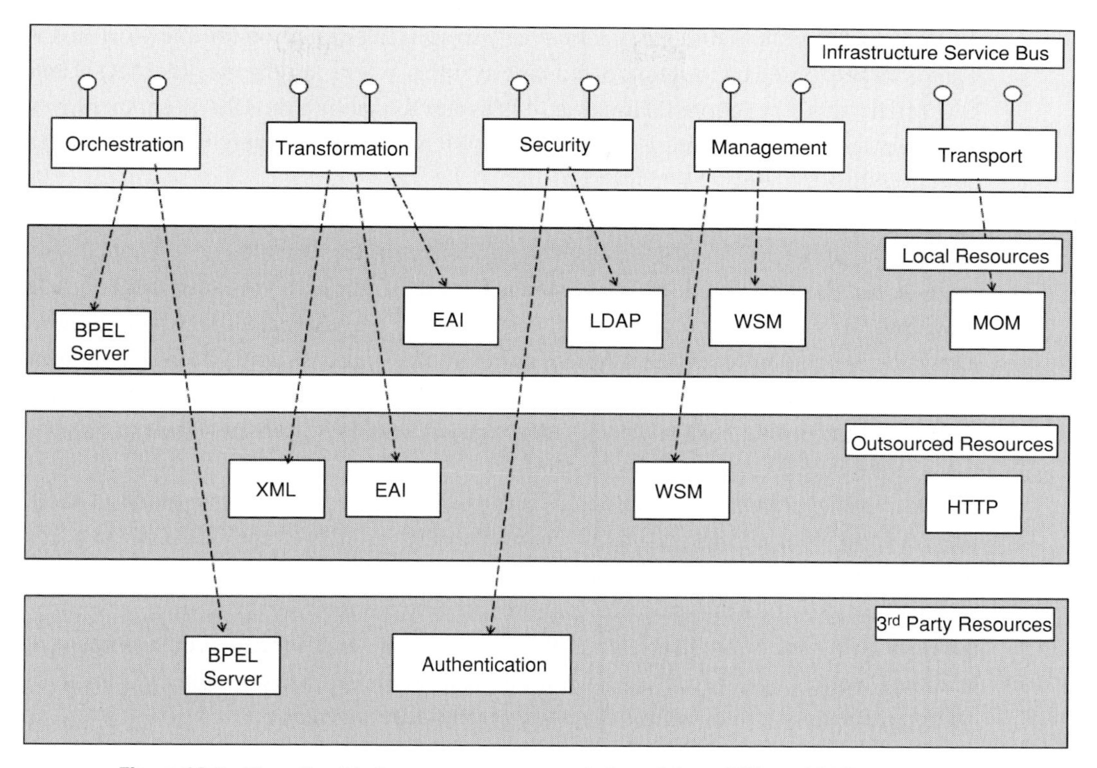

Figure 10.5 Virtualized infrastructure resources (adapted from Wilkes, 2004b)

services (such as identity services) to be readily included as part of a holistic infrastructure portfolio.

The ISB should be designed to reflect SOA technology policy and service quality policy. Technology decisions (such as which routing mechanism to call) can be made according to SOA policy applied to run-time situations. The services offered to business processes can be cleanly separated from the infrastructure services. Infrastructure decisions can be made on the basis of results from business services. For example, routing decisions could be made on the basis of real-time business information.

This is in contrast to the situation where infrastructure services are invoked on a piecemeal basis according to decisions made by individual developers within the code used to implement services. For example, it is common to find infrastructure services called via proprietary APIs from within the code that is used to implement business logic.

The ISB should reflect SOA policy held as meta-data, and ideally defined by an open standard. Web service standards and protocols are well placed to support this vision. Not only do we have emerging standard protocols for security, reliability, and transaction message headers and service definitions, we also have emerging standard infrastructure services, as will be defined by initiatives such as WSDM for service management, or Liberty Alliance for identity and authentication (Wilkes, 2004b).

As we have emphasized, the practical way to service orientation for many organizations is largely based on the reuse of existing systems, where middleware like MOM and EAI often cannot be ignored. However, it is useful to position the ISB as another layer that leverages the capabilities offered by the MOM and EAI layers without creating tight coupling to them.

By acting as a layer that sits above technology resources, the ISB facilitates an evolutionary approach to developing an SOA, in which services are offered to applications in graduated fashion. While an ISB extends the use of infrastructure resources outside the platform or organizational boundary, it allows for coexistence of new technologies alongside existing middleware. Choices as to which resources should be used to realize the services can be rationally based upon the SOA policy. Although working against agility, tight coupling may be deemed necessary in certain situations – for example, to achieve the desired capacity requirements.

While loosely coupled infrastructure services help to create a more agile SOA there are yet more compelling reasons to include an ISB as part of the SOA. As services move increasingly toward the on-demand vision (described in chapter 2), relocation of resources may require that the infrastructure resources supporting the business resources are relocated, too, or are available at the new location. In a federated environment, it becomes increasingly necessary to communicate with the infrastructure resources of other participants, providing third-party infrastructure services.

10.6 Where to next?

We now need to organize both the service design and the SOA policy reflected in the QoS infrastructure design in such a way that it can be clearly communicated to consumers and providers of services. These considerations take us naturally to the topic of SOM.

As the SOA becomes more useful and successful, so the need to scale up our software solutions grows and the need for effective SOM increases. The SOA must integrate effectively with SOM.

Look at this a slightly different way. Focusing on policy, particularly service quality policy, will be all for naught if the policies cannot be successfully implemented, if they do not achieve tangible results. The SOA policy must be executed! Not only that, but it must be managed! In this regard, the need for service security perhaps takes pride of place. Close behind it, though less obviously indispensable, come capacity and availability: the need to monitor, audit, and regulate the workings of dozens, hundreds, or thousands of software services.

An additional keynote of a service-oriented approach is that services are executed in a way that balances consumer demand, resource availability, and financial constraints. In short, services must be executed *dynamically*. Our approach to SLM must recognize this fact. It must encompass SEM software that is both flexible and in compliance with the SOA.

Until recently, this kind of approach has not been possible, partly because the technologies and standards have been emergent but not proven. Partly it is also because our traditional environments have divorced the responsibilities of software architects and operations specialists. This situation is now changing fast. In chapters 11 and 12, we move on to address these issues.

Part 4 Service-oriented management

Management of services is commonly treated as an entirely separate subject from topics such as business improvement, business process redesign, software architecture, and software development. Worse still, both the management of run-time software and the management of SLAs, are traditionally dealt with as separate *IT* concerns that have little to do with each other.

In part 4 we examine the imperative for this situation to change and introduce a new discipline, SOM, to bring together the areas of SEM, SLAs, and SLM in a cohesive fashion. We show how this approach integrates with both SOA and BA to round off an effective execution strategy for service orientation.

Chapter 11 paints the "big picture" of SOM. We explore the major gear shifts that are required to move from traditional management of software operations to execution management of software services. In particular, we explore the connection points between the various disciplines and examine their relationship to ITIL.

Chapter 12 describes a streamlined yet rigorous approach to the key topic of SLAs that is in tune with the ITIL guidelines on service support and delivery. We set out an approach that is integrated with both BA and SOA, and that is designed to cater as easily as possible for the migration of business processes toward composition from services.

Chapter 13 discusses some of the cultural issues involved in transitioning to service orientation. The required changes are, above all, about getting the best out of the organization's people and about understanding the market in which the organization chooses to operate. We provide two key takeaways: a role catalog and a set of market contexts. This information is augmented with guidance on the ownership and financial aspects of service orientation.

11 The "big picture"

11.1 A cohesive approach

The IT industry is full of niche comfort zones. One of the unfortunate consequences of the niche mentality is that the execution management of software is treated as an entirely separate subject from topics such as analysis, design, and software development. While separation of concerns is necessary for specialization and discipline, it becomes unhealthy when important interconnections and relationships between domains of interest are ignored – when separation leads to divorce!

In our present context, this trend surfaces in the separation of SOA from both overall operational management of IT services (in a general sense) and the specific execution management of software services. The consequence of divorcing these subjects is extra painful in the case of services. In particular, if services are to adapt in an on-demand fashion, execution management of the services must ensure that SLAs are satisfied, in compliance with SOA policy, with respect to ever-changing run-time conditions. As we argued in chapter 2, these new challenges call for gear shifts in our approaches to service management, toward SOM: a discipline aimed at ensuring that our architectures (BA and SOA) are not just "theory," but actually reflected in real production running software services.

SOM is dependent upon good specification of the services that it seeks to regulate. To underline a constant theme of this book: "You can't manage what you don't measure," but more than that "You can't measure what you don't specify." Incidentally, for the more technically minded, if you are still unconvinced about that, another way of thinking about it is to consider that much of the evolving WSDM standard depends on WSDL.

This chapter paints the "big picture" of SOM, focusing on both SEM and SLM. SLAs are explored in some detail in chapter 12. We also consider SOM with respect to current thinking on IT service management, as exemplified in the ITIL guidelines.

11.1.1 The elements of SOM

Recall that SOM refers to the combination of SEM with approaches to SLM enhanced to handle the challenges of service orientation. SOM is also dependent upon

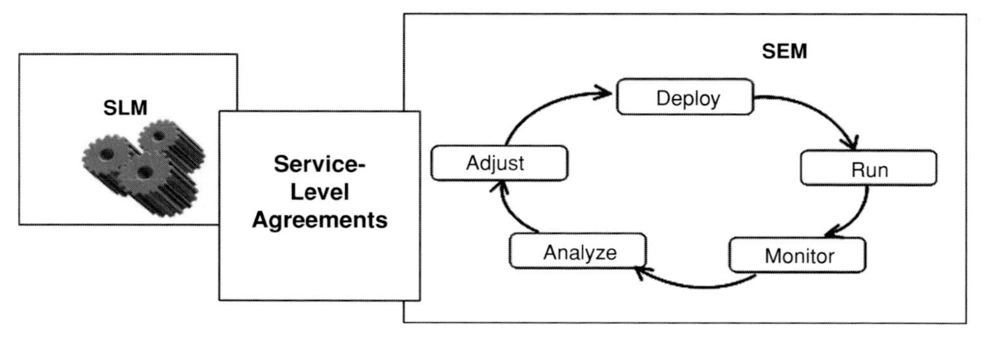

Figure 11.1 The elements of SOM

good consistent SLAs, as illustrated in figure 11.1. (repeated for convenience from figure 7.2).

11.1.2 SOM in context

Up until recently, it may have been possible for SOA and SOM specialists to work in their own isolated comfort zones. In chapter 7, we emphasized the need for a holistic approach to BA, SOA, and SOM, which is highlighted on our roadmap illustrated in figure 11.2.

As we have emphasized throughout, the SOA evolves in harmony with the various business models shown at the top of figure 11.2, and with the technical architecture and asset inventory shown at the bottom of the figure. We have also stressed the importance of an integrated solution development life cycle that works in parallel with these activities.

Clearly the relationship between the SOA and SOM is a *dynamic* one. In order to control this dynamic it is important that software services are catalogued and documented in a UDDI compliant registry. This registry should be a part of the wider asset inventory that we described in chapter 6 and that forms the underlying repository for change management of services and software units.

11.2 Service execution management

The services described in the SOA must be implemented efficiently and executed in line with QoS requirements. In particular, measures must be put in place to ensure that wherever possible the software implementing the service "self-corrects" with the minimum of disruption to users or to other services. Hence the need for SEM. This section builds on the SEM concepts, introduced in 7.3.1, and examines the relationship of SEM and SOA within the context of SOM.

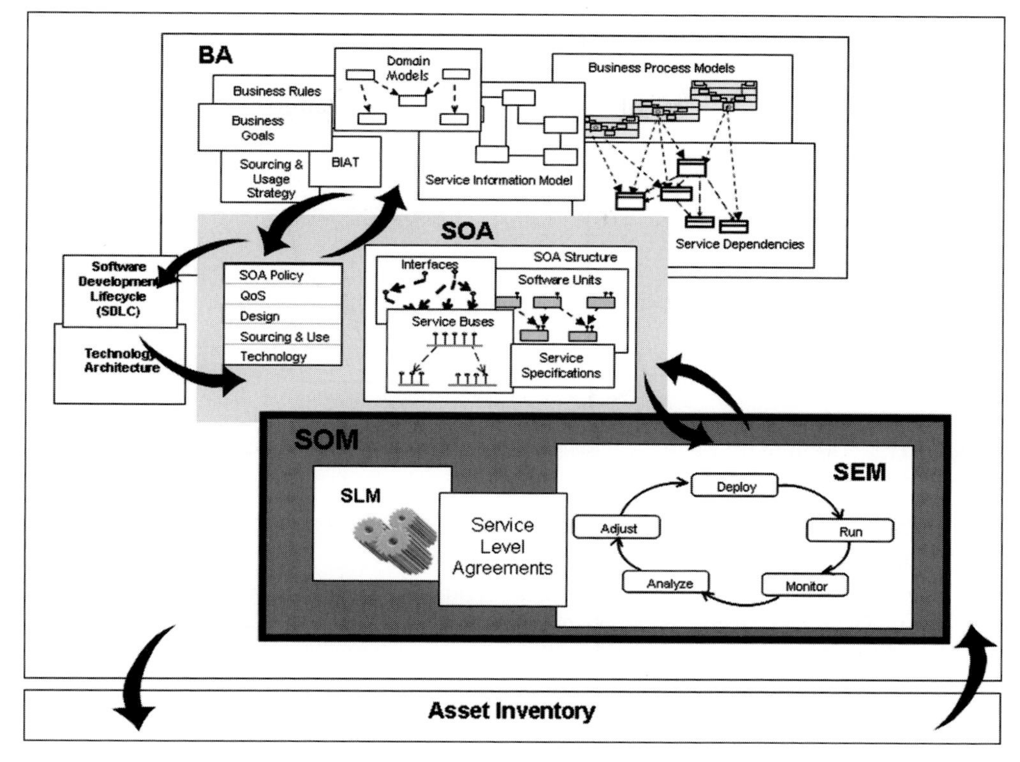

Figure 11.2 SOM in an overall roadmap context

11.2.1 The SEM life cycle

SEM concerns the run-time execution of software services and consists of five main cyclic activities, depicted in figure 11.3.

Let us examine each of these five SEM activities in a little more detail:

- *Deploy* This activity moves a service from its acceptance-tested state into production. While this may sometimes just be a matter of copying or re-installing software modules to another location, it could also involve establishing monitoring arrangements, operating procedures, and a phased rollout where similar software is installed on many processors or for many consumer communities. When replacing existing software, the previous version should be securely stored, so that operations can fall-back to this earlier version if serious problems are encountered with the new release.
- *Run* This activity covers the day-to-day operation of the deployed services, so that they remain available for users and consuming software to execute. It also involves the scheduling and execution of batch jobs. It includes the regular running of backups and other "housekeeping" activities. While we naturally think of this as an automated activity, it must also cover the operations support capability to correct service software

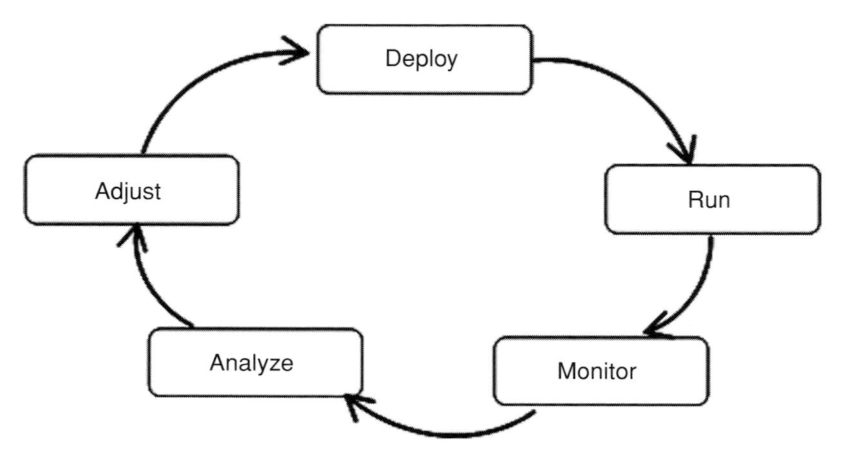

Figure 11.3 The SEM life cycle

problems in a temporary manner where emergency fixes are needed, and to fallback to an earlier release of the software where necessary.

- *Monitor* This activity defines the tasks required to monitor the health of deployed production services. This includes setting thresholds (that reflect QoS requirements), observing system performance, monitoring the observations, and reporting on findings, in the form of alerts, online queries, dashboard displays, and periodic reports. The monitoring is aimed at ensuring that SLAs and BLAs are adhered to. Issues are passed to the Analyze activity in the form of monitoring data. These issues range from shortfalls in meeting QoS targets such as capacity, availability, or security requirements to failures in meeting business requirements.
- *Analyze* This activity examines the monitoring data and attempts to diagnose faults and their possible remedies. Further, more detailed, reports may be produced showing diagnostics and possible courses of corrective action. Wherever possible the problem is passed to the Adjust activity for automated platform "tuning." In more difficult cases, the problem may require attention to the design and implementation of the SOA as well as to the design of specific services.
- *Adjust* This activity covers the actions taken to tune the technical infrastructure – hardware, network, and software platform for the services – to alleviate problems identified by analysis – or, better still, to prevent the future problems that can be foreseen by the monitoring. It also covers the discovery of services that are better able to meet service requirements, ready for redeployment at the start of the cycle.

11.2.2 A closer look

Looking at the SEM cycle in a little more detail, from the point of view of the SOA, the Monitor–Analyze–Adjust part is particularly interesting. The monitor activity senses for run-time situations (such as exceeding alert thresholds) that require analysis. Specifically, it is certain elements of the service specifications such as agreed capacity, availability, and security levels against which executing services are scrutinized by the monitor activity.

In some situations, analysis may call for some simple automated tuning by the Adjust activity. In other situations, analysis may call for a more complex response. A complex response may be automated by the Adjust activity. That, after all, is the whole point of self-correcting software that heals itself! However, to be realistic this might also require manual intervention in the Adjust activity that requires infrastructure specialists to tune the platform.

More far-reaching action may sometimes be required to solve the problem. It may be more than a relatively simple matter of platform tuning, in which case the symptoms may call for changes to the SOA. These changes often impact the design of capacity, availability, or security management and therefore involve the infrastructure design (chapter 10). For example, it may be necessary to redesign availability measures (such as failover) in order to correct an unwelcome trend spotted in the SEM analyses or to correct an immediate urgent problem.

In more extreme cases, however, it is possible that the SOA design itself is impacted (chapter 9). For example, a fault might involve changing the service's implementation, which might require a different software unit to be identified and installed. Again, one of the attractions of software services is that it should be possible to achieve an automatic rerouting to a different software unit. At the same time, and to be realistic, we should be prepared to handle this as part of the SOA design work.

It may even be that changes to the service specification itself are required, though this should be avoided as it would often impact all consuming software. Nevertheless, it may be required on occasion – for example, a poorly performing operation may need to be broken into smaller or more specialized operations.

On the other hand, if affected users of the poorly performing operation are very localized – for example, only those sending messages with a big payload – then a new version of the service, with an amended operation, could be brought into production specifically for these users.

In summary, we ignore the relationship between SOA and SEM at our peril. The SOA provides the specifications against which services are monitored. Conversely, SEM provides input to the SOA design in the form of regular analyses of service behavior as well as requests to correct problems. It is usually the SOA infrastructure design that is impacted. However, in extreme cases the service design itself may also be impacted. Up until recently, it may have been possible for SOA and SEM specialists to work in their own isolated comfort zones. Right now, though, the need is for a holistic approach to SOA and SEM. This need can only get greater with increased uptake of software services.

11.2.3 The role of SLAs

SLAs sit at the heart of the SOA–SOM dynamic. They are created for the services described in the SOA. These SLAs drive the SEM process. The SOA is in turn driven by requirements to adjust services from SEM, in circular fashion.

The *asset inventory* supports this process with consistent definitions, including facilities for effective reuse and version management. New and amended services are designed as part of the SOA and registered in the asset inventory. Service execution data is posted by SEM to the asset inventory.

We must not beg the question here: guidelines are needed for definition of SLAs that are both straightforward and workable, on the one hand, and based on integrated SOA and SOM concepts, on the other. In chapter 12, we therefore turn our attention to this neglected area.

SLAs reflect a trade off between what customers want and what providers can provide at the best value. How SLAs shape up depends to a large extent on their commercial and cultural context. There is a need to manage SLAs in relation to the different roles that play a part in their formation and maintenance: hence the need for SLM, our next topic.

11.3 Service-level management

SLM centers on the SLA. One of the main differences between SOM and traditional approaches is that deliverables, not procedures or processes, sit center stage. The most finely tuned process in the world is no substitute for clear deliverables. SLAs are the deliverables that are used to manage expectations between customers and providers of services. We focus on the critical topic of SLAs in chapter 12. Right now, we look at the overall requirements, the "big picture," for SLM.

11.3.1 Challenges and opportunities

SLM is the setting up, agreeing, monitoring, and controlling of SLAs. It is a discipline that involves comparing actual performance of services with pre-defined expectations, analyzing any gaps between the two, and determining appropriate actions. As we have seen, SLM sits at the meeting point of SOA and SEM. The roles involved may be quite complex depending on the particular context, but at a minimum will involve a service-level manager, a provider, and a customer. These roles are discussed in more detail in chapter 13.

Traditional approaches to SLM are more tuned to the control of the operational aspects of applications in an internal IT context rather than the increased demands represented by distributed software services across different business boundaries. For example, in chapter 10 we examined in some detail how software services raise the bar for management of capacity, availability, and security.

The challenges of SLM are multiplied further by the utility computing dimension of time, especially with respect to monitoring and control of software services, with the ability to change implementations dynamically in response to QoS levels reaching various thresholds. These challenges also raise the bar for IT service management in general, which has traditionally been positioned as an operational support function; we discuss these issues in 11.4.

At the same time, software services bring new potential opportunities. More generally, there is the opportunity to fine tune software support for business processes so that the software is utilized in the optimum way.

11.3.2 Raising the SLA bar

In the traditional world of application development we can be reasonably sure that the software evaluated before purchase will continue to operate as it did when it was tested because it will be used in an environment that is under the control of the organization that purchased it. The world of service orientation takes that comfort blanket away because the software used by an organization is no longer products but services. The quality of those services can change from day to day, hour to hour, and even second to second. While pre-purchase evaluations and due diligence remain on the agenda, the requirements for execution management far outstrip anything we have become accustomed to until now.

In particular, managing the quality of software services requires new types of management software designed to cope with these new challenges, alongside a gear shift in processes and skills. Just as testing and warranties are critical to managing packaged software, so SLAs are vital in managing software services.

However, one of the difficulties with the traditional idea of an SLA is that it is seen very much as an operational agreement concerning the outsourcing of the running of an application. It is not very well suited to the challenges of software services. A second difficulty is the complexity that surrounds the topic. Just as we need to raise the bar in other areas – such as capacity management, availability management, and security management – so we must also raise the bar in our approach to SLAs. In chapter 12, we examine these issues in some detail and describe a streamlined approach to SLAs that integrates with the specification concepts that we have applied in our treatment of SOA.

Right now, we simply want to highlight the centrality of the SLA to integrating SOA and SOM. However, we should be mindful that this is all a part of the much more significant cultural transition that involves breaking down the traditional barriers between operations and development in IT.

11.3.3 A word about software services

SLM centers on SLAs in the general sense of the term "service." At the same time, our approach to SLM must be integrated with effective SEM: the planning, monitoring, and control of the workings of (potentially thousands of) *software* services. The technologies and standards required to underpin SEM are complex and emerging.

In particular, in chapter 10, we emphasized that a production application comprising software services is potentially complex, distributed, and dependent on a mass of independent components. These factors require a raising of the execution management bar to cater for:

- Capacity: coping with the many factors that can degrade performance
- Availability: ensuring reliability against many points of failure
- Security: guarding against vulnerability to many types of abuses of security.

Inter-enterprise systems of this kind need more, not less, execution management than traditional in-house applications. A consequence of platform neutrality of software services is that applications can now cross different platforms. Our approach must be strong enough to cope with these greatly increased demands.

A good SEM tool should really look after capacity, availability, and security, along with related issues such as version management. It should also help to insure users against the disruptive effects of changes in the underlying and emerging Web standards by providing support for these standards "out of the box."

However all is for naught if the SEM tool works in isolation from SLM. Effective SOM requires that our approaches to SEM and SLM integrate through clear SLAs.

11.4 The role of ITIL

There are aspects of IT service management, such as problem and incident management, change management, and financial management, that are absolutely vital. There is a fairly comprehensive literature that covers these aspects. Perhaps the most noteworthy example is the ITIL approach, which continues to grow in influence at the time of writing.

In this section, we provide a very brief overview of ITIL (references to the details are provided) before moving on to consider its place in relation to SOM.

11.4.1 ITIL in a nutshell

Developed in the late 1980s, ITIL has become an increasingly influential process framework for IT service management. Starting as a guide for UK government, the ITIL framework is now known and used worldwide, and has emerged as a *de facto* standard for IT service management.

ITIL focuses on the activities that are involved with the deployment, operation, support, and optimization of applications. Its main purpose is to ensure that deployed applications can meet their defined service levels (OGC, 2002).

ITIL does not cast in stone every action required on a day-to-day basis because that is something which differs from organization to organization. Instead it focuses on *best practice* that can be utilized in different ways according to need. It has been very successful at improving tactical decision-making and aligning influence on the tasks of IT service management. Lack of communication and cooperation between various IT functions is minimized by emphasizing the relationships between the processes of IT service management and encouraging the use of a common language.

Another key facet of ITIL is that it is a customer-oriented approach that emphasizes managing IT as a business, in contrast to previous approaches to IT operational management that are internally focused and concentrate on technical issues.

11.4.2 The ITIL library

In 2004, the IT Service Management Forum (*it*SMF) announced a library series of books that will comprise five principal elements, each of which has interfaces and overlaps with each of the other four:

- The business perspective
- Managing applications
- Delivery of IT services
- Support of IT services
- Managing the infrastructure.

The most mature and well established of these are the books on service delivery (OGC, 2001) and service support (OGC, 2000); it should also be noted that guidelines have also been more recently developed on IT Security (Cambray and Hodgkiss, 2003).

Service delivery is concerned at a tactical level with what IT service the business requires of the provider in order to provide adequate support to customers in the following five processes:

- Capacity management
- Financial management for IT services
- Availability management
- SLM
- IT service continuity management.

Service support is concerned at an operational level with ensuring that the user has access to the appropriate IT services to support his or her business functions in the following six processes:

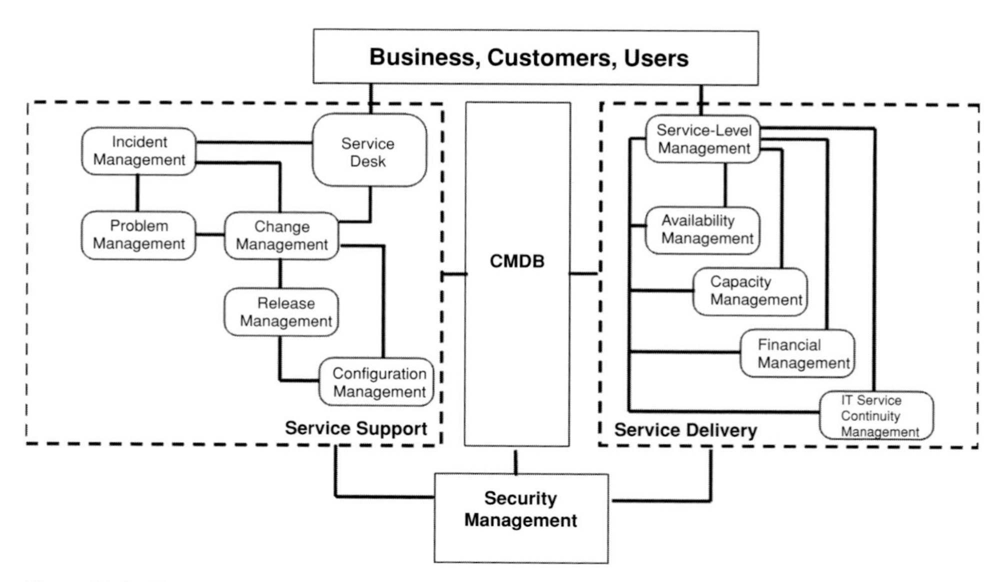

Figure 11.4 The ITIL process framework

- Service desk
- Incident management
- Problem management
- Configuration management
- Change management
- Release management.

Security is dealt with by ITIL as a separate discipline, but with links to both Service Delivery and Service Support.

11.4.3 The ITIL process framework

The various ITIL processes form a framework. Tactical processes are centered on the relationships between the IT organization and its customers. Service delivery is partially concerned with setting up agreements and monitoring the targets within these agreements. Meanwhile, on the operational level, the service support processes can be viewed as responding to the changes needed in, and any failures in, the services laid down in these agreements. On both levels, there is a strong relationship with quality systems such as ISO 9000 and quality frameworks such as the European Foundation for Quality Management (EFQM).

Figure 11.4 provides a picture of the ITIL process framework. Note how the configuration management database (CMDB) underpins the whole process, providing a central source for managing all supporting information.

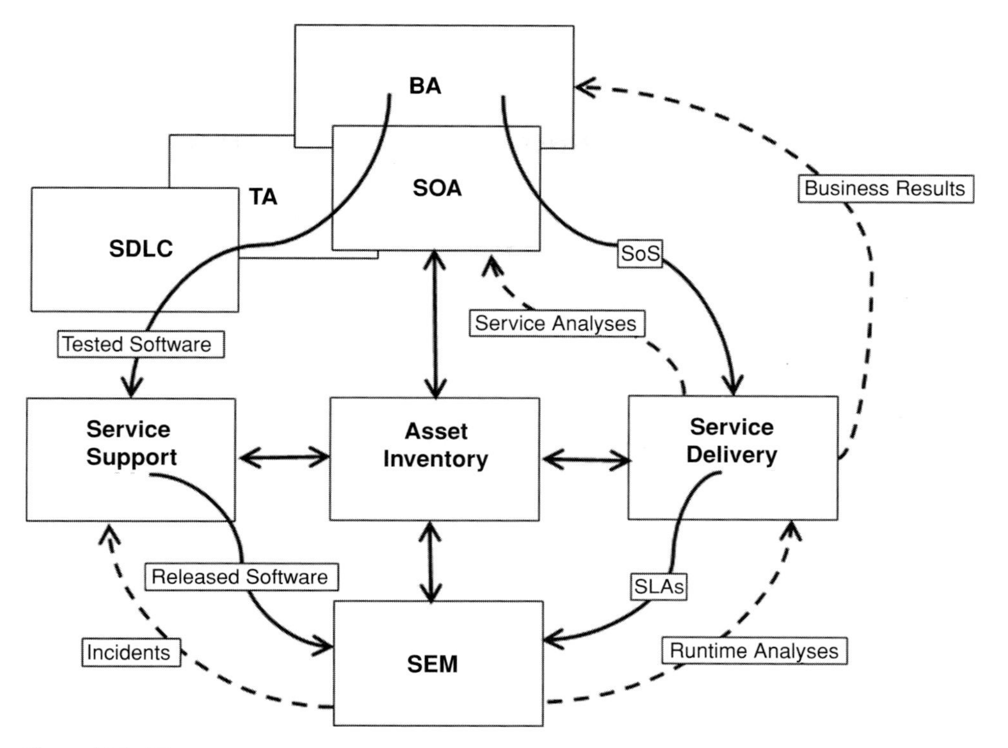

Figure 11.5 The soup to nuts process

11.5 Bringing it all together

We can now turn back to the “big picture” that we began to sketch out earlier in this chapter and start to develop a clearer understanding of where service delivery and service support fit into this picture.

11.5.1 Soup to nuts

Service support and service delivery both sit at the interface between SOA and SEM, as depicted in figure 11.5. In line with ITIL, SLM should be pictured as part of *service delivery*. Looking at figure 11.5, we see there are forward loops (solid arrowed lines) and backward loops (dashed arrowed lines) through the overall soup to nuts process. These loops are annotated with the names of their associated deliverables.

There are two inter-related cycles of activity, to the left and right sides of figure 11.5. The left-side activity has to do with development, deployment, release, and subsequent incident identification and resolution of software solutions, including services architected within the SOA; it is important to note that not all software solutions will employ services. The right-side activity is concerned with designing services in alignment with business needs, agreeing SLAs, and ensuring the executing services meet their goals.

The solution development and assembly route emerges as a result of business requirements, goes through the SDLC, resulting in tested software, which is input to Service Support for deployment and eventual release. The released software is managed by SEM. Incidents detected by SEM may be dealt with automatically or returned to Service Support for treatment.

In parallel to the above route, SOA raises specification of service (SoS)[1] and QoS requirements for software services comprising the solution. SOA works in dynamic collaboration with the SLM piece of service delivery to furnish SLAs that reflect the agreed needs of customer and provider. SLAs are pushed forward to SEM to provide a basis for execution management of the services. SEM produces real-time analyses of service execution for evaluation by SLM, which in turn reports results back to either SOA (service analyses impacting SOA issues) or BA activities (business results).

11.5.2 The SLM gearbox

SLM is an integral part of the right-side activity of figure 11.5 that controls the interplay between SOA and SEM. The aim is twofold: to achieve SLAs that reflect the agreed needs of customer and provider, and (once the services are in production) to evaluate and control live executing services against these SLAs. SLM is, in effect, the gear box that connects the worlds of SOA and SEM, as illustrated in figure 11.6. Note the position of SOM, which overlaps service delivery and embraces a number of disciplines, as already discussed.

11.5.3 Raising the SLM bar

The ITIL approach has been instrumental in organizations gaining much better control of the operational aspects of IT infrastructure. It is important that we build on this hard-won knowledge. Indeed, we have already seen that our approach to QoS infrastructure design (described in chapter 10) is structured so as to fit smoothly with the ITIL guidelines on management of capacity, availability, and security.

At the same time, to continue the earlier gearbox analogy, the SLM gearbox is under increasing pressure to work with newer, more powerful engines, on the one hand, and with more demanding drivers, on the other. While Web services standards are still maturing, they are opening up increasing opportunities to share business processes

[1] SoS is a term used to cover specification artifacts from the SOA; the reader is referred to 12.2.1 for more details.

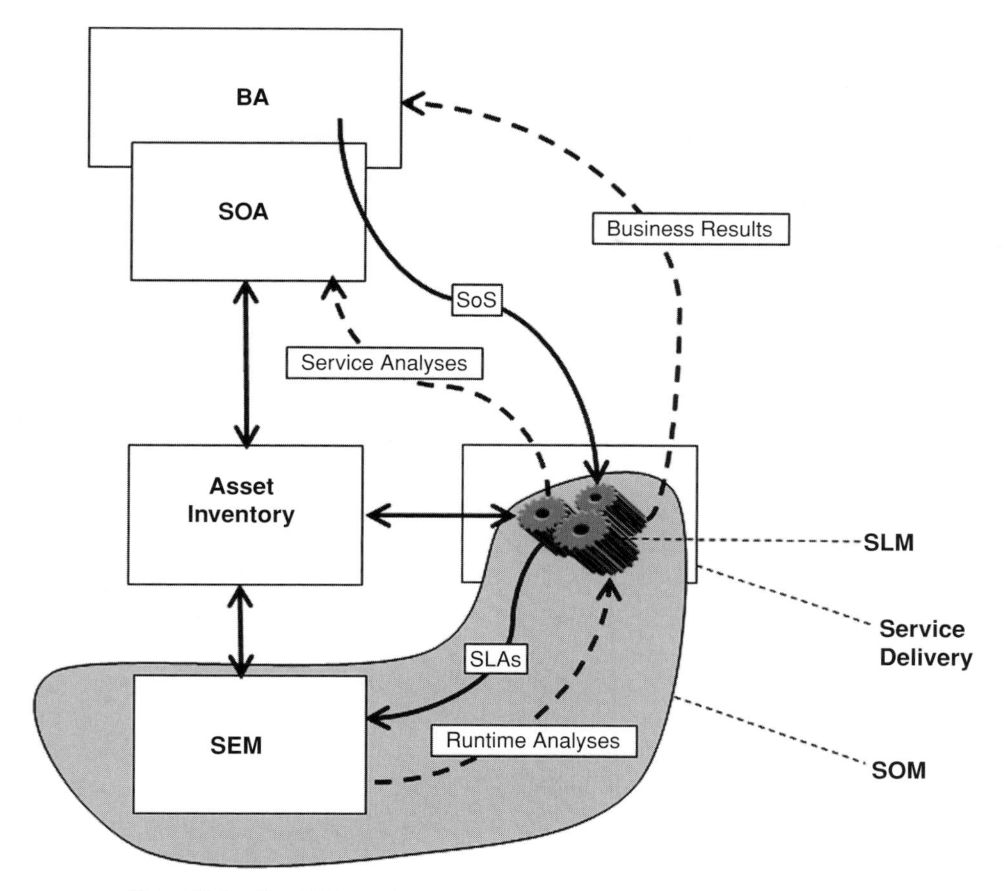

Figure 11.6 The SLM gearbox

across both organization and technology boundaries. In this context, it is much less applications, much more *services* that need to be managed. This calls for a much more rigorous and streamlined approach to the *specification* of SLAs that is integrated with the specification of services. Our approach to SLM must integrate with both SOA and SEM.

There is also the fact that business itself is fast becoming service-oriented and demands both SOA and SEM to be closely aligned to business needs. Until now, SLM has been commonly understood as very much a specialized operational management discipline limited to IT as a business, rather than being understood as part of the business itself. SOA is too often viewed as just another IT application architecture, rather than a way of helping to identify and structure sets of agile services. The application mind-set is deeply ingrained in the IT psyche that has dominated much of the thinking around SLM. Again, there is a need to focus on services in the true sense.

Another factor is that the software tools required to support SEM have not been available until very recently. The result is that SLM tends to be seen as a reactive support function rather than a strategic necessity. In contrast, the world of service orientation mandates SOA as very much a strategic necessity that is supported by SOM.

Yet another factor is that traditional approaches to change management do not cut it in the world of service orientation. As soon as an organization moves beyond early tactical use of software services to overcome the limitations of technical dependence, change management starts to raise its head. Change management of services, interfaces, and software units is an order of magnitude more challenging than change management of applications and modules. As services become an integral part of business so more comprehensive change management becomes a key requirement.

Change management of services should be based upon the underlying asset inventory. It is also worth emphasizing the role of *repository technology* in supporting the asset inventory that sits center stage of the "big picture" sketched earlier. Our approach to definition of the asset inventory meta-data must encompass clear definition of SLAs and related art facts such as services, interfaces, and software units. For an excellent discussion of the issues of service change management the reader is referred to Veryard (2004b).

At this point, the reader may be feeling quite daunted at these challenges! It is a feeling that the author has also often experienced. At the same time, if we are realistic enough to face the challenges and admit that there are problems to solve then we can start to do something about them. The place to start is the root. And the root of the problem lies in the need for *clear specification*.

Above all else, it is the SLA that requires attention if our approach to SLM is to cope with the challenges of service orientation. If these artifacts can be defined and agreed we have a solid basis upon which to build procedures for SLM. Without clear SLAs any process will inevitably be based upon shifting sands. The SLA must integrate with the SOA through service specifications and QoS criteria. SEM can then operate on the basis of agreed measures and goals, which should tie up with business requirements as defined through the SOA.

11.6 Where to next?

There are a huge variety of procedures involved in SLM. There is a rich and plentiful literature, best exemplified by the ITIL guidelines, that provides process guidance. At the same time, there is a definite need for SOA and SEM to be integrated not just with SLM but with service management in general. It is anticipated that this guidance will evolve gradually as organizations move more and more toward service orientation. It is not our aim to reinvent this process. What we will do in chapter 12, however, is to set out the ground for integration of SOA, SEM, and SLM through a clear approach to SLAs. It is important to restate that our approach is *deliverable-based* – once you have worked out what you want to do then you can consider the various ways of how you are going to do it! A good integrated approach to SLA definition should, after all be flexible enough to work with any number of processes.

12 Service-level agreements

12.1 Managing expectations

One of the main differences between service orientation and traditional approaches to both doing business and developing software to support business is that *deliverables*, not procedures or processes, sit center stage. The most finely tuned process in the world is no substitute for clear deliverables. In chapter 11, we emphasized that SLAs are the deliverables that are used to manage expectations between consumers and suppliers of services. We make no apology therefore for repeating the dictum:

> **"You cannot manage what you don't measure."**
>
> **Not only that but pre-defined expectations must be agreed between provider and customer.**
>
> **"You cannot measure what you don't specify."**

And this dictum does not just apply to nonfunctional requirements. What is commonly overlooked is that the business requirements must be identified and the software specified in such a way that it meets those requirements and can be accurately measured against them. It will be important to ensure that SLAs are clearly specified in such a way that business metrics can be applied to software services where appropriate.

Unfortunately, the topic of SLAs has traditionally been fraught with complexity and inconsistency. There is a need for a streamlined yet rigorous approach that is in tune with current guidance on IT service support and delivery, as described in chapter 11. In this chapter, we set out an approach to SLAs that is integrated with SOA and service specification and that is designed to cater as easily as possible for the migration of business processes toward composition from services.

12.1.1 Managing risk

Although SLAs are contractual and binding deliverables, they are much more a means of managing risk than formally setting out legal criteria (in fact, as we shall see, SLAs are not themselves legal documents, though they may well reference such documents). SLAs help to mitigate risk by clearly stating the responsibilities of the various roles involved in the consumption and provision of services, thus managing expectations and helping to avoid possible contractual disputes.

An SLA cannot guarantee that you will get the service it describes, any more than a warranty can guarantee that your car will never break down. The SLA is a communication tool that specifies what will happen if something goes wrong. It is a communication tool for exposing assumptions and managing expectations. And, again, it is a tool for helping to trade off the sometimes conflicting criteria of high quality and low price. In particular, an SLA cannot make a good service out of a bad one. At the same time, an SLA can mitigate the risk of choosing a bad service.

Unfortunately, SLAs are commonly seen as a one-way street in which the onus is solely on the provider to stick to the terms of the SLA. However, SLAs are very much a two-way street. On the one hand, the customer ideally wants the very best service. More typically, however, he or she wants achievable and acceptable service levels at a fair price. Equally, on the other hand, the provider wants to ensure that the service is used "correctly" by its users without degrading the service or causing disruption. Consumers, as well as providers, must comply with the conditions specified in the SLA. Consumer accreditation schemes (as described in 14.5.2) are a useful strategy for minimizing the risk of abuse by consumers.

12.1.2 The SLA as a trade off

The SLA acts as a bridge between the SOA and the commercial world. On the one hand, it is driven by the services specified as part of the SOA, along with the SOA policy. On the other hand, it is heavily influenced by the pragmatics of the commercial marketplace. SLAs are essentially a trade off between what the organization wants to achieve (the SOA) and what it can achieve via its internal and external resources (the marketplace). It is important to understand that SLAs are not always appropriate, and to choose the appropriate level for each market context, as described in chapter 13.

12.2 Terminology

This section introduces a meta-model for specification of services, together with terminology that provides a foundation for integrating SLAs with the concept of SOA.

12.2.1 Specification of service

The SoS must very clearly communicate the capabilities of a service to its consumers and providers. For example, customers must be able to readily comprehend these capabilities

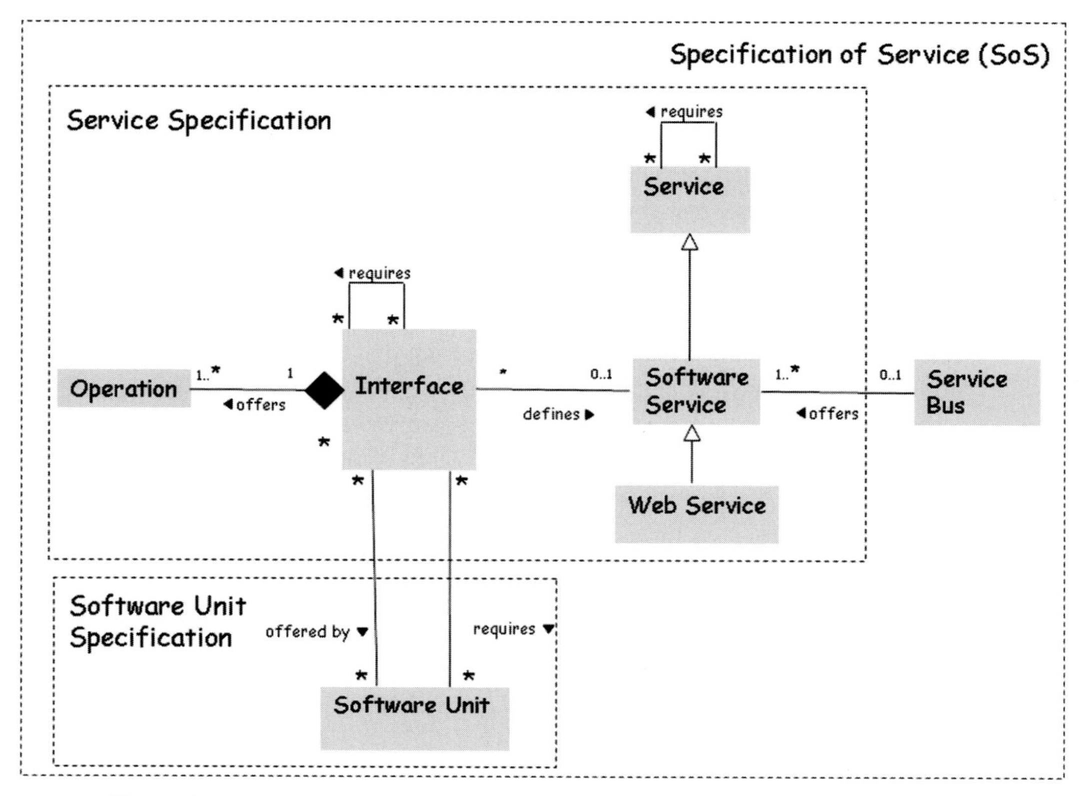

Figure 12.1 Meta-model of Specification of Service

and to benchmark providers offering a service against the specification. At the same time, providers must be able to ensure that the implementation of a service, by a supplier, complies with the specification.

As we saw in chapter 9, two kinds of specification are therefore required:

- The service specification addresses the consumer-oriented aspect of a service in terms of functionality, cost, and QoS requirements. It will include a separate specification for each interface offered by the service and for each operation of each interface.
- The software unit specification addresses the supplier-oriented aspect of a service in terms of implementation constraints that govern a software unit realizing the service. A software unit may realize a number of interfaces.

Looking at figure 12.1 (which builds upon figure 9.2) we can see that the total picture is in fact a good deal more complex than might be implied from the simple two-way split. Note, in particular, the scope of the term "service specification," and that in addition a service bus may offer a set of services, which typically are subject to QoS policy that is common to a set of roles. The term SoS is introduced to refer to the totality of the concepts shown in figure 12.1. The SoS will provide a very useful bridge to the SLA.

The SoS and its constituents may be a hard copy or electronic report, or an online "model" accessible through a Web browser or other graphical user interface. Once published, the SoS must be versioned and subject to change control procedures as part of the asset inventory.

Note that the SoS is independent of particular consumers, suppliers, and providers. It is a standalone entity. Keeping the SoS independent of particular participants in the process is a key to effective sourcing and usage of services. An independent approach provides customers with optimal choice – for example, in submitting offers to tender a service. Conversely, providers are able to reach as wide a potential consumer base as possible.

12.2.2 Service-level agreements

An **SLA** is an agreement between a customer and provider about what services are to be offered by the provider and to be used by the customer. It must specify the measurable levels of those services that the provider must achieve, and the terms of use that the customer must comply with.

Again, notice that the SLA is a back-to-back agreement that covers obligations on *both* the provider and the customer. A common mistake is to think of the SLA in terms of the customer getting the service promised by the provider. For example, a provider agrees to make a service available 99.9% of each day. Equally, however, it is intended as insurance for the provider that customers do not breach the terms of use. For example, a software service that is used by three times as many of the maximum number of permitted users might cause the software service to suffer unacceptable response times.

Recall that an SoS is independent of particular customers and providers. It is a standalone entity. An SLA applies the SoS to a particular customer – provider pair (or set of pairs) in terms of a commercial and binding contractual agreement that governs the provision and the use of a service (or set of services). The SoS acts as a base role-neutral specification and may be the subject of different SLAs involving different role participants. In this way, the SoS supports reuse of specifications in different contexts, helping to address the problem of duplicated and reinvented information that is common in many organizations today. The SoS is, after all, a part of the asset inventory!

In referencing a service, the SLA by default also references all elements that are associated to that service within the dotted boundary of the SoS as depicted in figure 12.2. At the same time, we might override this default in certain circumstances. For example, it might be that not all of the software units associated with a service apply to

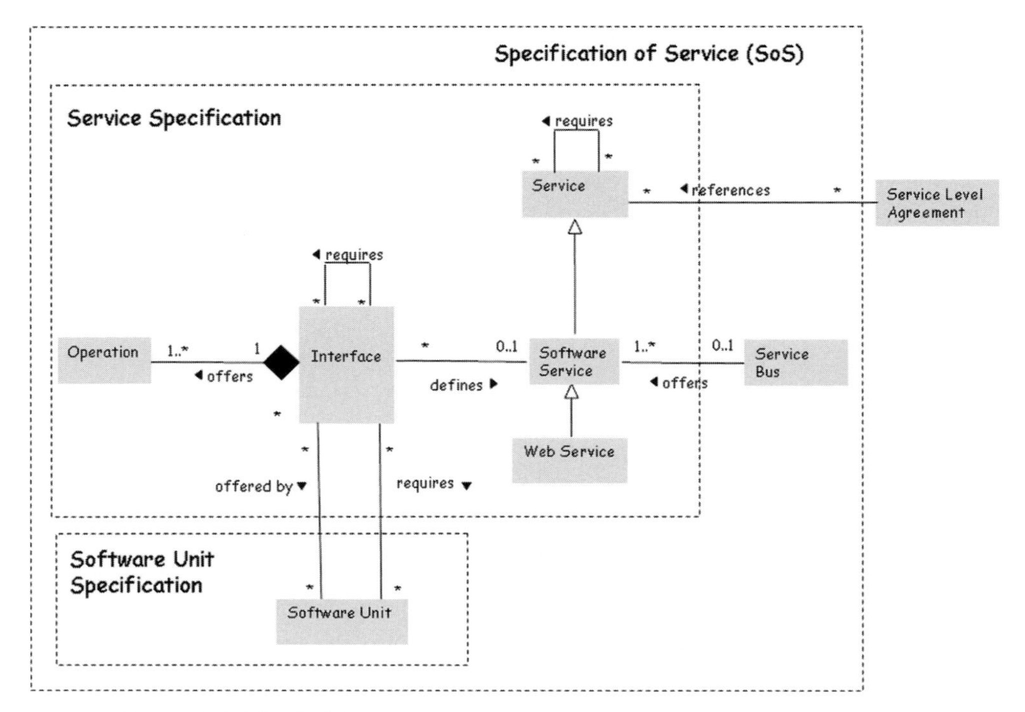

Figure 12.2 The SLA in context

an SLA. There could be different SLAs for the same service, each of which stipulates that a particular software unit is (or is not) to be employed.

If the customer and the provider are separate legal entities, the SLA normally references legal documentation. However the SLA itself is not a legal document that contains contractual information on matters such as support obligations. ITIL recommends the use of separate agreements for internal and external purposes:

- *Operational-level agreements* (OLAs) define support requirements internally – for example, between the service desk and internal support groups
- *Underpinning contracts* ensure that all targets contained within both SLAs and OLAs that rely on external suppliers are underpinned by the appropriate level of maintenance and support contracts.

12.3 Structuring the SLA

While various definitions of SLAs have been around for some while now, a truly service-oriented definition has been lacking. This section therefore provides a template for defining SLAs based on *services*. We also describe some approaches to the layering of SLAs that are designed to help manage their potential scale and complexity.

Throughout this book, we have emphasized the importance of integration of SOA and SOM, not only from a design and technology viewpoint but, more importantly, from a business viewpoint. The concept of BLA is central to this principle. We therefore explain how to define BLAs in consistency with our approach to SLAs.

12.3.1 An SLA template

Table 12.1 provides a template for structuring an SLA.

12.3.2 Layering of SLAs

To assist in managing complexity SLAs may be layered in analogous fashion to the layering of QoS requirements, described in 8.3.1. ITIL (Macfarlane and Rudd, 2003) recommends the following multi-layer structure for SLAs:

- *Corporate level*: covering all generic SLM issues appropriate to every customer throughout the organization
- *Customer level*: covering all SLM issues relevant to the particular customer group, regardless of the service being used
- *Service level*: covering all SLM issues relevant to the specific service, in relation to a customer group.

In our present context, we are focusing very much on the third of these three types of SLA. More generally, SLAs should also inherit characteristics from SOA QoS policy applying to groups of services of which the services described by the SLA are members, as discussed throughout chapter 8. This aids manageability and helps enforce SOA QoS policy.

12.3.3 BLAs

A BLA specifies the business goals and rules that govern the *functionality* and *information* described by an SLA, and the *commercial criteria* (such as financial penalties) attached to these business goals and rules.

The BLA is covered by the business-level entry in the SLA. However, because of its strategic importance we may well choose to elevate the status of this section to a separate document.

The underlying idea here is that a BLA should be monitored by SEM software so that customers know when (and *when not*) a service is meeting its business goals and so that users know when (and *when not*) the service is performing according to business rules.

Table 12.1 *SLA template*

Purpose	• A concise description of the scope and objectives of the SLA • Any unique or defining features such as restrictions on other customers taking this service[a] • Reference to legal documents – for example, "this agreement shall be interpreted according to the laws of the State of New York" • Definitions of terms to be used in the body of the document • References to SOA QoS policy covering this SLA.
Version and release number	Up-to-date identification criteria.
Start and end dates	Start and end dates that cover the period for which the SLA is valid.
Roles	A description of roles involved, the following is a list of *possible* roles: • Customer • Provider • Supplier(s) • Users (may include limits on numbers of minimum or maximum users) • User software. Note that in the case of a software service it is a software application that invokes the software service, therefore we adopt the convention "User via software application" in describing the user. For a list of all possible roles the reader is referred to 13.3.
Owners	Business owner and technical owner.
Services	References to the services covered, including version numbers.
Software units	References to the software units covered, including version numbers.
Business levels	*Relevant* targets and rules.
Capacity levels	As per service specification, including qualification of SOA capacity QoS levels (where necessary) and references to capacity design policy including implementation mechanisms.
Availability levels	As per service specification, including qualification of SOA availability QoS levels (where necessary) and references to availability design policy, including implementation mechanisms.
Security levels	As per service specification, including qualification of SOA security QoS levels (where necessary) and references to security design policy, including implementation mechanisms.
Monitoring mechanisms	• Service levels to be monitored • Frequency of monitoring • Responsible role • Software used to carry out monitoring • Details of monitoring outputs.
Billing details	• Unit of acquisition (for example, subscription to use service anytime or pay per use) • Ownership and usage rights (for example, is it only rights to execute the service that are purchased, or is ownership transferred to you?)

Table 12.1 *(cont.)*

	• Unit charge • Time variations on charge • Payment request method • Payment making method • Provider penalties associated with not meeting agreed service levels • Customer penalties associated with late payments or breaching usage terms.
Rules of engagement	• Problem reporting procedures • Problem escalation procedures • Regular review meetings • Help facilities • Change or improvement request procedures (for example, how are changes in requirements, or updates implemented?).
Special responsibilities (by role)	This section should include Customer responsibilities. Examples for Customer role include security of passwords, required accreditation levels (see 14.5.2), supplying evidence of reported problems, and achieving agreed usage criteria, such as minimum usage rates. (In complex cases, a role-responsibility matrix may prove useful).

[a] For example, it may have been designed for the exclusive use of this customer, and its specification and design must not be made known to other parties.

Traditionally, systems management software has focused on the monitoring of non-functional requirements. Now, however, in the world of service orientation, the bar is raised for management software, in that business features must also be monitored. This is a particularly salient point in relation to the *Optimization* viewpoint of SOV7, which refers to the ability to offer services in real time at high performance levels.

For example, the organization may have been caught out in the past by offering products at far lower discounts than intended because of clerical error. I am in fact reminded of one organization that offered a digital camera at a special knockdown price for customers that bought another product in its range. Unfortunately, someone forgot to program the link that made this offer available only to purchasers of the product – it was made generally available without the seller's knowledge. The problem was that the offer was posted late on Friday. By Monday morning, countless knockdown transactions had taken place, as the chat rooms buzzed with the news, with no legal come-back. A real-time alert to managers' cell-phones that reported crazy prices or abnormal numbers of transactions would have prevented the huge loss that was incurred. Better still, by including an automated check implemented by SEM software that, say, "that the ratio of knockdown products over total amount of products sold cannot be > 1:1" the problem could have been nipped in the bud much earlier by having the SEM software disable all further knockdown transactions.

It is worth noting that these kinds of events occurring in complex combination are of growing importance as organizations become more service-oriented. Complex event processing (CEP) is a set of tools and techniques described by David Luckham (2002) for analyzing and controlling such events. This emerging technology helps to quickly identify and solve problems and more effectively utilize events for enhanced operation, performance, and security.

In particular, Web services are the enabling technology that makes BLAs practical. This is because Web services can reflect meaningful business activities better than traditional messaging technologies. For example, a Web services request might raise a purchase requisition, whereas the implementation comprises much finer-grained modules and components that have no direct relevance to the business user. There is thus a better opportunity to manage Web services at the business level, monitoring compliance against business metrics. For example, all purchase requisitions must be completed within a certain time frame.

A BLA is essentially a business-centered view of an SLA. It should not be confused with business process description or agreement. A BLA covers a *service*, not a business process. At the same time, other business goals and rules that apply to a business process, rather than a service, can be referenced by the SLA where the service is used to support the business process.

As we saw in 9.2, part of the work involved in designing services involves deciding which business goals and rules are to form part of BLAs. There will be potentially thousands of business goals and rules in organizations of any size. Only the most commercially significant business goals and rules, those that must be measured as part of the SLA, will qualify for inclusion in BLAs.

In 12.4, we provide examples of SLAs and BLAs in relation to the vehicle hire case study.

12.4 A step-by-step guide

Creating SLAs is a key part of SLM. In chapter 11, we emphasized the role of SLM as the gearbox that connects SOA with SEM. It is a discipline that entails many different relationships with the different roles of service orientation. Some of these roles and relationships are illustrated in figure 12.3. We include this figure simply as a reminder of the complexity of the process of SLM, and shall return to the subject of roles in much more detail in chapter 13. Notice that SLM takes SoS as an input and produces SLAs as an output. However, it is far from a simple linear process. Above all, it is an evolutionary and iterative process that involves negotiating with the various roles and making trade offs between often conflicting forces.

In this section, we cut to the chase in providing a relatively simple step-by-step guide to creating SLAs, building on the vehicle hire examples from chapters 9 and 10

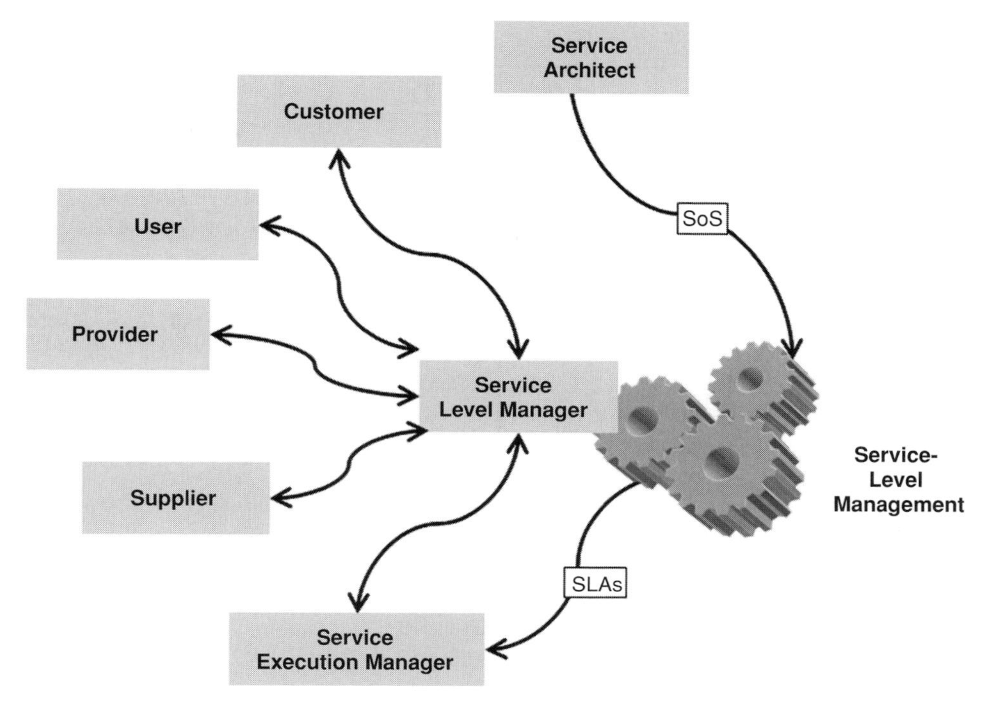

Figure 12.3 SLM-related role interactions

to illustrate the evolutionary and iterative process of creating SLAs. Our attention is largely confined to the interactions between SOA and SLM. However, it is important to realize that this is one (albeit important) aspect of a much wider commercial picture that we address in our discussion of marketplace pragmatics in 13.5.

12.4.1 A simple SLA process pattern

SLAs evolve in top-down fashion and in iteration between the disciplines of SOA and SLM. We use a process pattern (as in our approach to business process redesign described in 4.2) to structure our approach.

Macro level: establish viability of the SLA

- Evolve the SoS and SLA within the context of the SOA in line with business process redesign efforts, paying attention to business goals that will guide the creation of BLAs
- Assess and work with suitable providers, in line with SOA sourcing and usage policy
- Produce service descriptions and outline SLA in line with SOA QoS policy.

Once we are confident the SLA is viable, we can proceed to a more detailed analysis.

Table 12.2 *Enhanced service description for the Contracts Service*

Property	Value
Service name	Contracts Service
Domain	Vehicle Hire Sales
Purpose of the service	Management of reservations, contracts, and deliveries.
Responsibility by business type	Contract Type, Contract, Entry Sheet, Exit Sheet, Reservation, Delivery
Operations	CRUDL Secure Reservation, Book Delivery, Open Contract, Close Contract, Report Expiry
Dependencies	Customers, Vehicles, Hire Rules, Accounts
Commoditization level	Value-Add
Sourcing policy	In-house
Usage policy	Initially restricted to Hire Agent. Intention is to expand usage to customers once established.
Targets (Record Handback Activity)	"Contract must be closed within 5 minutes" in the case of a partner. "Contract must be closed within 15 minutes" in the case of an employee.
Business Process Support (current)	Agency Vehicle Hire
Business Process Support (future)	Internet Vehicle Hire
Business Process Target	50% decrease in numbers of phone reservations. 1,600 online reservations per day. 20% increase in number of hires. 20% increase in productivity of hire agent. 50% increase in customer satisfaction rating.
Cost	Unit charge per operation use: 1 cent
Capacity levels	Minimum/Average/Maximum operation invocations per day = 6,400/12,800/19,200 with 20% increase over first 6 months. Maximum of 10 concurrent users.
Availability levels	AST = all times except 02.00–03.00 Max Outages per day = 0.02% of AST

Micro level: detail SLA

- Establish QoS levels and consider how QoS policy may need to be qualified for SLAs
- Detail out the SoS; for example, refine service descriptions into service specifications
- In parallel, detail out the SLA, paying particular attention to detailed BLA aspects.

12.4.2 Developing SLAs at the macro level

Much of our work in the vehicle hire scenario has been carried out at the macro level as far as the SOA is concerned. Regarding the Contracts Service, an enhanced service description was developed in chapter 10; table 10.1 is repeated for convenience in table 12.2.

Table 12.3 *Outline SLA for the Contracts Service*

Purpose	The purpose of this SLA is to establish expectations for the Contracts Service.
Version and release number	V1.0 This version applies internally, to hire agent employees. It is anticipated that a further version will be developed that broadens scope of usage to include partners and customers.
Start and end dates	1 January 2006 to 31 March 2006:
Roles	• Customer: vehicle hire sales director • User: hire agent via vehicle hire application • Provider: vehicle hire service delivery department • Asset inventory manager: vehicle hire senior architect • Supplier: vehicle hire IT department
Owner	• Business owner: customer • Technical owner: supplier.
Services	Contracts
Software units	Contracts Component

We now need to consider the service description in the light of SLM. The first step is to evolve an outline SLA, making sure this is consistent with the business process redesign goals that will be used to form BLAs. An important part of this step is ensuring that the SLA is in line with sourcing and usage policy, to make sure that all roles are covered and create a firm commercial footing. The top-level rows of the SLA template are completed, as illustrated in table 12.3.

Note that for the sake of explanation we are keeping things reasonably straightforward here. For example, it may sometimes be that several services are encompassed by the same SLA. Another factor to consider is that the SLA may raise issues that impact the SOA.

12.4.3 Developing SLAs at the micro level

We now turn our attention to the Contracts Service description, which must be expanded to include further details, including QoS policy. Where such details are part of overall SOA, the required QoS levels are simply inherited by default by the service. However, where QoS levels are specific to a particular service or one of its constituent interfaces or operations, QoS levels must be recorded as part of the service description. Note that this same principle also applies to other related artifacts, such as the interfaces and operations that comprise a service.

Recall that quality templates are used for the purposes of capturing quality attributes for agility (9.1.1), capacity (10.2.3), availability (10.3.3) and security (10.4.4).

We decide that the SOA QoS policy for the vehicle hire enterprise is to apply to the Contracts Service with one exception: capacity QoS levels. This is because of the increased amount of data expected to be exchanged through the Contracts operations.

Table 12.4 *Vehicle hire company enterprise capacity QoS specification*

Quality attribute	Worst	Average	Best
Throughput	0.1 /sec	0.15 /sec	0.2 /sec
Responsiveness	50 msec	20 msec	7 msec
Information capacity	2.3 mb	3.4 mb	4.0 mb
Scalability (6 months)	0.5%	0.8%	1.2%
Responsiveness Variability (1 month)	+1%	+4%	+9%
Responsiveness Consistency Index (1 month)	91	94	97

Table 12.5 *Contracts capacity QoS specification*

Quality attribute	Worst	Average	Best
Throughput	0.1 /sec	0.13 /sec	0.16 /sec
Responsiveness	90 msec	50 msec	15 msec
Information capacity	3.3 mb	4.4 mb	5.0 mb
Scalability (6 months)	0.3%	0.6%	1.0%
Responsiveness Variability (1 month)	+0.75%	+3%	+7%
Responsiveness Consistency Index (1 month)	88	91	94

In chapter 10, we established the required enterprise service capacity QoS levels in the form of an enterprise service capacity specification as shown in table 12.4 (repeated from table 10.2).

The capacity QoS specification for the Contracts Service is adjusted to reflect the expected higher volumes of data with the Contracts Service, as illustrated in table 12.5. These details are added to the service description, which is in effect refined to form a service specification.

We now refine the service specifications, by including a reference to table 12.5. Recall that the service specification includes further details, as described in chapter 9. The Contracts Service employs two interfaces. Each of these interfaces is specified in terms of an interface type model and operation specifications. Further information such as invocation methods is also added to the resulting service specifications as described in 8.4.4 and 9.3.4.

The outline SLA is now expanded with particular attention to commercial requirements, as shown in table 12.6.

12.4.4 Creating BLAs

Recall that the BLA is covered by the business levels entry in the SLA. Because of its strategic importance, we may well choose to elevate the status of this section to a separate document.

Table 12.6 *SLA for the Contracts Service*

Purpose	The purpose of this SLA is to establish expectations for the Contracts Service.
Version and release number	V1.0 This version applies internally, to hire agent employees It is anticipated that a further version will be developed that broaden scope of usage to include partners and customers.
Start and end dates	1 January 2006 to 31 March 2006:
Roles	• Customer: vehicle hire sales director • User: hire agent via vehicle hire application • Provider: vehicle hire service delivery department • Asset inventory manager: vehicle hire senior architect • Supplier: vehicle hire IT department
Owner	• Business owner: customer • Technical owner: supplier.
Services	Contracts
Software units	Contracts Component
Business levels	To be decided
Capacity levels	As per service specification
Availability levels	As per service specification
Security levels	As per service specification
Monitoring mechanisms	• Service levels to be monitored: all • Frequency of monitoring: real-time • Responsible role: provider • Software used to carry out monitoring: XYZ management software
Billing details	• Unit of acquisition: pay per use • Unit charge per use: 1 cent • Ownership and usage rights: only rights to execute the service; provider owns service • Payment request method: weekly to customer account via Vehicle Hire Billing system • Payment making method: weekly from customer account via Vehicle Hire Payments system; the provider shall compute and credit the customer's account for all penalties due.
Rules of engagement	The provider shall log all service levels and submit logs to the customer daily in format X.
Special responsibilities (by role)	Provider shall ensure that Vehicle Hire Billing and Payments systems are updated with payment details on a weekly basis.

Specific goals or "targets" (see 9.2.3) are extrapolated from the business process models. These goals may be referenced in service descriptions where appropriate. For example, the Record Handback activity might have a target of "Contract must be closed within 5 minutes." This target is referenced by the service description of the Contracts Service. Other targets such as "50% decrease in numbers of phone reservations" and "200 online reservations per day" that apply to the hire *process* can also be referenced by the service description of the Contracts service as this service clearly plays a contributing role in achieving these targets.

Further complexity can arise in that the same service may have different targets depending on the role to which the service is offered. For example, the above target "Contract must be closed within 5 minutes" may apply in the case of a partner, whereas it is relaxed to "Contract must be closed within 15 minutes" in the case of an employee.

The temporal dimension is also a factor. The same role may be entitled to different "grades" of the same service, depending on the time of day or week. Roles themselves may be further divided – for example, a gold star partner versus a bronze star partner.

We identify the business targets and rules that have a commercial significance for the business processes and activities. The following target has already been identified for the Record Handback activity:

- *Contract must be closed within 15 minutes.*

We now need to establish the precise detail and agree on any penalty and incentive clauses that are going to be attached to this target. The challenge is to design a system of penalties and incentives that will create a continuous motivation for the provider to maintain as high a level of service as possible. The provider should be motivated to improve on acceptable performance and, where performance is unacceptable, to rectify it, no matter how much the provider has already lost in penalties.

A useful approach to constructing BLAs is to break the process target down into sub-targets that are relevant to the operations of a service. Each target should have a small individual penalty, rather than having a single large penalty that relates to a single threshold value. Good provider motivation to maintain as high a level of service as possible is achieved through "the large cumulative effect of these nagging little penalties" (Overton, 2001).

For example, it might be agreed that the Close Contract operation must complete within 6 seconds, otherwise the following penalties are payable by the provider: If completion time exceeds 6 seconds a penalty of $1 is payable; for every 2 further seconds, a penalty of $2 is payable.

Provider incentives can be introduced on similar grounds. In this case the following incentive is introduced: if completion time is less than 0.5 seconds an incentive of $1 is credited to the provider.

Other targets might apply to customers (rather than providers). For example, it might be a requirement that the customer's staff place at least 50 hires per day or a penalty is payable to the provider. This might be the case where a customer has contracted to subscribe to a partner's service at a certain monetary rate per service call. There are no examples of this kind in the vehicle hire scenario but it is nevertheless important to be aware of this possibility.

We also identify the business rules that we wish to include in the BLA. Recall from 9.2.6 that the following rules were flagged as falling into this category:

Table 12.7 *BLA for Contracts Service*

Business targets	**Penalties or incentives**
The Close Contract operation must complete within 6 seconds.	If completion time exceeds 6 seconds a penalty of $1 is payable. For every further 2 seconds, a penalty of $2 is payable. If completion time is less than 0.5 seconds an incentive of $1 is credited to the provider.
Business rules	**Action for breaking rule**
Only the hire manager shall award customers gold status.	Supervisor automatically informed and staff warning issued.
A deposit must never exceed 20% of the agreed reservation fee.	Legal requirement: automatic refund issue if breached.
If excess vehicle mileage > 0 and the customer is not a gold customer, then an excess premium shall be payable.	Supervisor automatically informed and staff warning issued.

- Only the hire manager shall award customers gold status
- A deposit must never exceed 20% of the agreed reservation fee
- If excess vehicle mileage occurs and the customer is not a gold customer, then an excess premium shall be payable.

These rules are binary in the sense that they are either obeyed (which ideally should be the case) or they are not. We need to decide, as part of the BLA, the action that shall be taken if they are not obeyed. It should be noted that in a system of any size there are potentially thousands of rules. We are picking the ones here that demand special treatment in the context of the BLA. For example, awarding customers gold status and avoidance of excess mileage payments are both open to abuse by staff. So management needs to know about these misdemeanors as soon as they occur. Supervisors must be automatically notified and the staff concerned warned if these rules are breached.

Again, the rule concerning the deposit threshold might be a legal requirement that needs an additional check as part of service monitoring. If the rule is breached then an automatic refund must be issued. By stating these rules in the BLA we are raising the need for SEM management to isolate and report rule breakages.

A BLA is created, based upon the above discussion, as illustrated in table 12.7.

12.5 Where to next?

Our focus in this chapter has been on laying out a structure for SLAs that dovetails smoothly with our approach to SOA, and that can be readily addressed by SEM software.

This is very important for clear communication between the various roles involved in SLM, and for consistency and measurement of services.

In chapter 11, we touched on process issues and sketched out a big "soup to nuts" process for service orientation. Given a clear definition of SLAs it is in a sense not so fruitful to see how this "big picture" breaks down into processes and procedures, especially since much work has already been done and continues to evolve in the area of process guidance (as exemplified by the ITIL approach). At the same time, we have been at pains to emphasize that service orientation is different, does involve new challenges, and can take various forms.

Above all, there are significant cultural gearshifts that we must address. Thus far, we have said very little about the roles involved or the types of market situation in which they are increasingly being asked to collaborate. It is therefore to those subjects that we turn our attention in chapter 13.

13 Cultural factors

13.1 Specification before process

There are a host of cultural factors affecting the uptake of service orientation. Many detailed processes are possible. In this book, we have emphasized a lightweight approach to process, introducing process patterns to assist where useful. Our main emphasis, however, has been firmly on deliverables. The view is that if you get the deliverables right then processes will usually follow.

At the same time, cultural change is a lot more than instituting the right processes. Above all, it is about getting the best out of the organization's people and about the market in which the organization chooses to operate. In this chapter, therefore, we resist the temptation to add to the groundswell of procedural guidance – the fact is that organizations evolve and adapt their processes to fit their people and their market. Our focus is on offering guidance on the roles that are required for individuals to play in relation to service orientation, and on sketching some scenarios that depict different market contexts that organizations may find themselves in. Armed with clear guidance on specification, along with role definitions and market guidance, the organization can adapt their processes to maximum advantage.

13.1.1 Supply, manage, and consume

What I first referred to as a "two-tier" process in relation to CBD (Allen and Frost, pp. 12–14, 1998) continues to be relevant to service orientation. This is a model that I further enhanced in relation to e-business in the form of a "track-based pattern" (Allen 2000, pp. 50–2). It is a major departure from the traditional models of software development to which many organizations still cling. A more recent, but essentially similar, example of what is referred to as the "service life cycle" is provided by CBDiForum (Wilkes, 2004a).

It is not our purpose here to reinvent the service life cycle. This ground is well covered in the literature, as discussed above. We have also seen that IT service management of applications is well covered by the ITIL framework. The service life cycle sits on the left-hand route of our "big picture" forming an interface between the activities of software development and service support. These high-level processes do a job. The issues that we must address, however, are much less to do with process and much more to do with clear understanding of service specification and SLAs, on the one hand, and the cultural factors influencing SLM, on the other.

13.1.2 Integrating SOM

As we saw in chapter 11, effective service orientation depends not only on the supply–manage–consume model of software development that has been around for some while now, but also on clear linkages between SOA and SOM. These links can be achieved only through a connected approach to SLM.

Our deliverables-based approach to SLM focuses on clear definition of SLAs, as discussed in chapter 12. Equally, it depends on clear role definition of the participants in the process. One can use the most thorough and detailed process imaginable, but if the people involved do not understand their roles then all will be for naught. It may be tempting to try and apply traditional software development processes and roles to the challenges posed by service orientation. However, service orientation involves roles not found in traditional software development. In addition, many of the roles used in traditional software development either no longer apply or have drastically changed. In the next section, we therefore turn our attention to the roles of service orientation.

13.2 The roles of service orientation

Good professionals know their roles. A role is a *related set of responsibilities and skills*. Roles bring objectivity to any process, in that one person (or organizational unit) may fulfill more than one role and one role can be supplied by many individuals (or organizational units). Different roles are applied as appropriate at different times.

Ask an airline pilot or air steward, and there is no question about what they have to do in flying the aircraft or keeping the passengers happy. In most industries, the situation is pretty much the same. On a film-set, you'll find a director, a gaffer, a make-up artist, and the various actors and actresses with clearly defined parts to play. In software, the situation is not so clear-cut: many of us have become generalists, epitomized in titles like analyst/designer that disguise a multitude of different roles. The larger the job, the more the lack of role definition starts to hamper the project. Service orientation adds further dimensions of scale and complexity, which demand clear definition of roles.

In this section, we provide an overview of the roles of service orientation, with particular attention to SLM, before moving on in 13.3 to provide a catalog of roles.

13.2.1 An overview of roles

Figure 13.1 depicts the various roles of service orientation in relation to the "big picture" introduced in chapter 11. Notice that some of the roles, shown overlapping on to process boxes, have specific responsibilities that relate to high-level processes. Other roles have

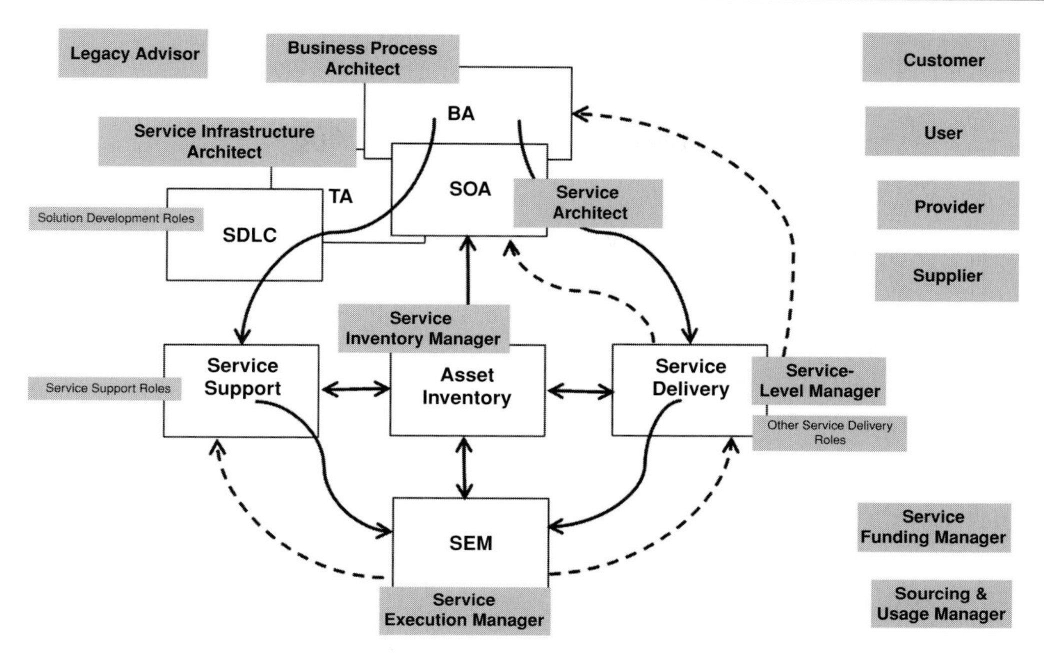

Figure 13.1 The roles of service orientation

responsibilities, shown disconnected from any processes that have wider commercial or advisory significance.

13.2.2 Dynamism of roles

One of the most important features of roles is that they interconnect and work together in dynamic fashion. Whereas figure 13.1 is essentially a static picture that maps roles to the overall process, figure 13.2 attempts to show some of the most important collaborative relationships between roles. One of the most significant features of this picture is the central place of the *service-level manager*.

Note also the position of the *asset inventory manager*. Because the asset inventory underpins the entire process, the asset inventory manager is depicted in the bottom right corner joined to the entire box, indicating the criticality of this role.

Clearly there are a great many other roles that space and context preclude us from showing. In particular, solution development should include roles from agile development methods such as Dynamic Systems Development Methodology (DSDM) (Stapleton, 1997) and from business process analysis and CBD (Allen, 2000).

13.2.3 Key SLM roles

We can take a closer look at the roles directly related to SLM, as shown in figure 13.3.

The SLA sits at the intersection point of the roles depicted above. Each of these roles has a specific responsibility with respect to the SLA, as illustrated in figure 13.4.

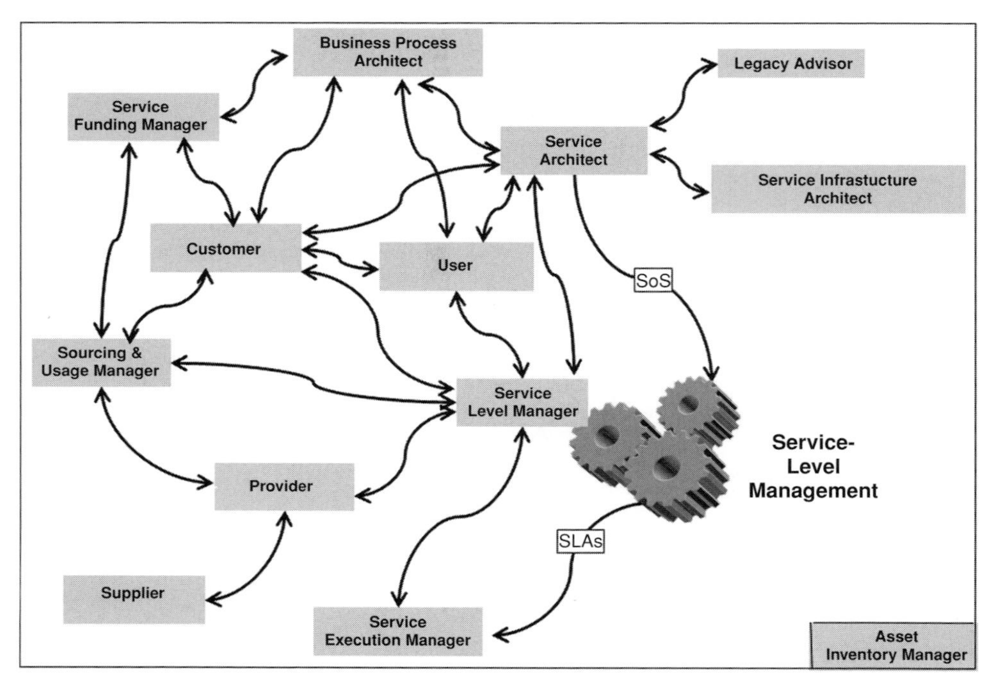

Figure 13.2 Dynamic relationships between the roles of service orientation

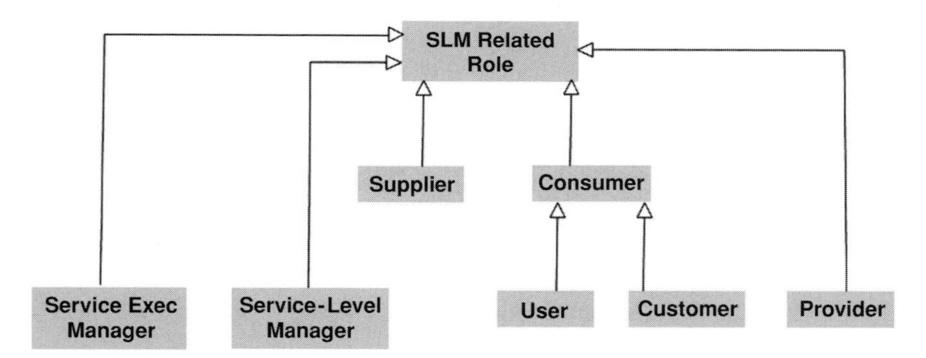

Figure 13.3 Meta-model of SLM roles

13.3 Catalog of roles

In this section, we provide a catalog of the thirteen key roles involved in service orientation:

- Business process architect
- Service architect

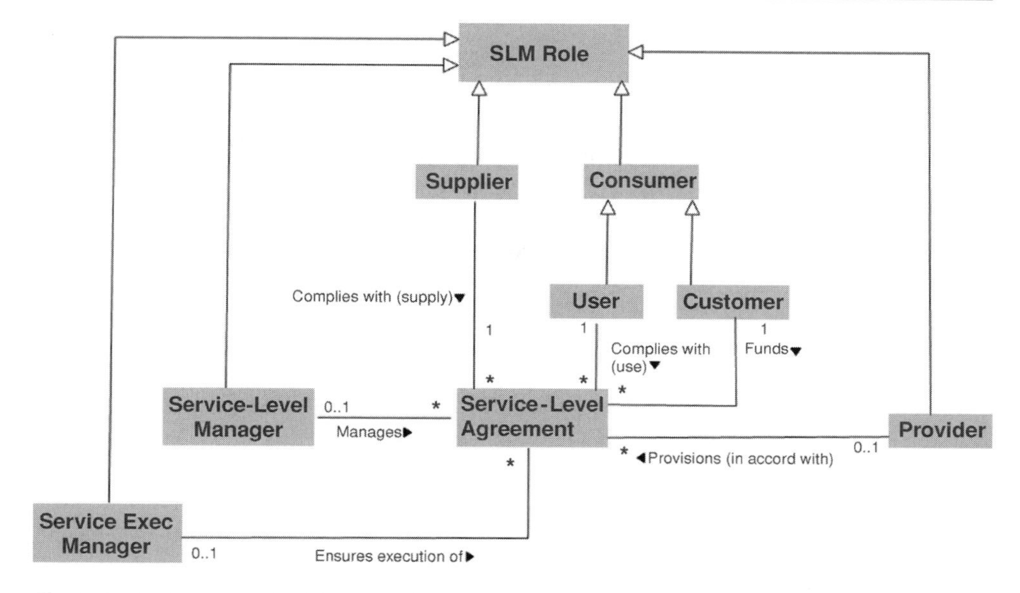

Figure 13.4 SLA usage by different roles

- Service infrastructure architect
- Legacy advisor
- Sourcing and usage manager
- Service funding and charging manager
- Customer
- User
- Service-level manager
- Provider
- Supplier
- Asset inventory manager
- Service execution manager

Readers should note that this is not meant as an exhaustive list, and you will no doubt wish to add roles of your own and qualify some of our definitions.

Again, it is important to remind ourselves that we are talking of roles, and not individuals or organizations, in this context. *It is possible for the same organization or individual to play different roles with respect to the same service.* The role catalog described in this section is not intended as a standard set that must always be implemented in its entirety. Instead, readers should think of it as checklist that organizations can adapt to their own specific needs. Much will also depend on the *market context* – supplier-driven, consumer-driven, or collaborative – as discussed in 13.5.

Table 13.1 *Business process architect*

Responsibilities	• Planning of services (in the general sense of the term), • Ensuring that services provide the business capabilities required by business processes so that each business process as a whole will operate with the services • Clarifying and prioritizing business objectives and resolving the tensions between them • Promoting the value of service orientation to the business leaders • Understanding and communicating BA industry trends and best practices.
Skills	The role must possess broad business knowledge and be an excellent communicator able to align interests of business and IT. Good business modeling skills and understanding of BPM techniques are important. The role should also possess proficiency in techniques such as storyboarding and scenario playing, to enable identification and resolution of business process integration issues. There is a need for overall vision and good awareness of strategic business needs.
Inter-working	The role must collaborate closely with business people such as the customer and with the services architect.
Sub-roles	In large organizations, the role may further sub-divide into business process designer, business process analyst, strategist, and evangelist roles.

Table 13.2 *Service architect*

Responsibilities	• Planning of software services • Understanding the business requirement for services in balance with both technology limitations and opportunities • Identifying and structuring services, interfaces, software units, and service buses • Specifying the QoS policy and design policy aspects of the SOA • Promoting the value of SOA throughout the organization • Understanding and communicating SOA industry trends and best practices
Skills	The role has highly developed software design skills, but with further business-facing and political awareness skills. The role must encourage a collaborative climate and be aware of the practical consequences of decisions. Again, there is a need for overall vision and good awareness of strategic business needs.
Inter-working	The role works closely with the business process architect in proactive fashion to identify areas for business improvement and to advise on the feasibility of these improvements. The service architect helps to develop and verify SLAs in harness with the service level manager. The role also works in tandem with the services infrastructure architect.
Sub-roles	The role may sub-divide into further specialist roles, depending on the size of the organization. These include roles that are responsible for particular aspects of the SOA: for example, service specification manager and service bus manager. However, in very advanced organizations it may be that the same person or group plays both service architect and business process architect roles, as the disciplines start to merge with increased uptake of software services.

Table 13.3 *Service infrastructure architect*

Responsibilities	• Planning of service infrastructure design • Identifying and specifying technical software units[a] • Ensuring consistency of usage of technical software units across projects • Specifying the technology policy aspect of the SOA • Tuning the service infrastructure technology in line with service execution needs • Understanding and communicating developments in service infrastructure technologies, including the impact of Web services standards.
Skills	The role has highly developed technical software design skills, with further political awareness skills. There is a need for overall technology vision, including a good knowledge of infrastructure technologies, particularly Web services standards.
Inter-working	The role typically works in tandem with the service architect, helping ensure that plans are technically feasible. The role also works with the sourcing and usage manager, who may acquire technical software units identified by the role, and with the service execution manager, to ensure that the service infrastructure is correctly tuned to actual working services.
Sub-roles	The role may sub-divide into further specialist roles, depending on the size of the organization. These include roles that are responsible for management of QoS policy: for example, availability architect, security architect, and capacity architect.

[a]Technical software units include a broad range of supporting technologies: user interface, communications, device handling, and data storage management.

Table 13.4 *Legacy advisor*

Responsibilities	• Advising on the suitability of legacy assets (applications, databases, or software packages) for servicization • Re-engineering and mining of legacy assets to enable servicization • Assessing the impact of proposed new services on legacy assets • Understanding and communicating developments in legacy enabling technologies.
Skills	A thorough knowledge of legacy assets is essential. The role should also have a keen awareness of latest technologies and techniques for legacy system re-engineering and mining.
Inter-working	The role works closely with both the service architect and the service infrastructure architect.
Sub-roles	In a large organization, the role may split into separate roles for legacy system code, legacy databases, and software packages.

Table 13.5 *Sourcing and usage manager*

Responsibilities	• Specifying the sourcing and usage policy aspect of the SOA • Identifying service users and managing the organization's relationship with these users • Identifying service providers and managing the organization's relationship with these providers • Negotiating the sourcing and usage aspects of SLAs • Understanding and communicating sourcing and usage industry trends and best practices, as well as the use of IT by competitors.
Skills	A good understanding of the business context in which the organization operates is essential, including financial and legal aspects. The role should possess a combination of relationship management skills and a high degree of authority. This is likely to require a senior executive with excellent IT industry knowledge. The role must be a good communicator, politician, and generalist.
Inter-working	There is a very close relationship between this role and the service-level manager role. Whereas the latter focuses on SLAs in themselves this role focuses on the implications of SLAs from a customer or a provider relationship viewpoint: for example, "Is this SLA in harmony with our partnership with this customer?" or "Have we dealt with this provider before and if so what lessons were learned that we need to address in the SLA?" These considerations mean that the role works closely with provider, customer, and service funding manager. However the role also has interfaces with most other roles: for example, the role works with the service infrastructure architect to explore the commercial feasibility of plans to acquire technical components.
Sub-roles	The role may sub-divide into sourcing manager, usage manager, and partner relationship manager roles (separate roles for customer relationship management and provider relationship management).

Table 13.6 *Service funding and charging manager*

Responsibilities	• Ensuring that funds are available for service usage and that customers will provide the funding • Planning how to measure and bill for service provision • Investigating and recommending funding and charging models • Assessing the economic impact of service usage and sourcing on business needs[a] • Understanding and communicating developments in service funding and charging strategies and technologies.
Skills	The role is likely to have a combination of IT management and accountancy skills. Some organizations have a dedicated IT Finance Manager, of which this role is a specialization and an extension. However, the skills required for this role are essentially emerging ones that call for an individual that is a good communicator who is willing to learn quickly and to question old assumptions and prejudices.
Inter-working	The role must work with representatives of the finance department, to ensure that approaches to service charging and funding are aligned with corporate policies of budgeting and accounting. The role also works in a consultative capacity with most of the other roles, especially the customer, the service level manager, the provider, and the sourcing and usage manager. Good communication with software and business metrics experts is necessary to keep up to date with the latest thinking and practice on measurement.
Sub-roles	It may be that separate roles are required for funding and charging.

[a] For example, comparing the strengths and weaknesses of the different approaches and articulating the benefits, against the costs.

Table 13.7 *Customer*

Responsibilities	• Ensuring that services meet business needs • Agreeing SLAs with providers, including customer responsibilities, described in SLAs • Providing funds for using services • Holding executive responsibility for a set of users to access services • Holding executive responsibility for provider selection.
Skills	The skills required for a good customer are perhaps too numerous to mention! In a nutshell, they traditionally involve a combination of business clout and commercial knowledge. In our present context, a good understanding of how service orientation can work to the organization's benefit is essential.
Inter-working	The role works most closely with the business process architect and the service architect (to identify services), with the service funding and charging manager and the sourcing and usage manager (to ensure financial viability), and with the service-level manager (to agree SLAs).
Sub-roles	The customer role may divide into executive sponsor and visionary, as defined in DSDM (Stapleton, 1997).

Table 13.8 *User*

Responsibilities	• Providing detailed operational business knowledge required to define services • Invoking services[a] in compliance with SLA terms of usage • Reviewing overall experience with service usage • Reporting immediate faults and incidents.
Skills	The role must possess thorough business knowledge of how the services will actually be used in a working environment. Good communication skills are essential.
Inter-working	The user works closely with the customer, and may report to the customer in clarifying requirements and reporting working experience of using the services. The role also works with the business process architect and the service architect in helping to identify detailed business needs. The user also interacts with the service-level manager to help verify requirements and feedback review information. Other channels of communication to do with feedback in using the service should be via the service desk (as defined in ITIL).
Sub-roles	The user role may divide into ambassador user and adviser user as defined in DSDM (Stapleton, 1997).

[a] In the case of a software service, it is a software application that invokes the service. Therefore we adopt the convention "User via software application" in describing the user.

Table 13.9 *Service-level manager*

Responsibilities	• Negotiating SLAs and acting as a coordination point between customers and providers • Reviewing SLAs regularly with customers and providers • Ensuring that SLAs are communicated clearly to SEM • Initiating monitoring, analysis, or adjustments to be carried out by SEM • Understanding and communicating best practices in SLM.
Skills	The role must possess the business clout to effectively negotiate with customers and providers on behalf of the organization, and to initiate and follow through actions required to improve or maintain agreed service levels. It is essential for the role to weigh up the viability of an SLA. This entails a good understanding of the provider's services, and the customer's business and technical context in which the services are to be used. An innovative and open mind is important. The role needs to be a good communicator and politician, to be aware of the tension that may exist between other roles and maintain a healthy balance[a].
Inter-working	The role must work closely with customers and providers to achieve balanced SLAs. However, the communication channels of the role stretch far wider than that to most of the other roles, in that SLAs sit at the very center of service orientation, and it is the service-level manager that is responsible for them.
Sub-roles	While it is possible to split the role (for example, separate roles for the commercial and technical aspects of SLAs) we would caution strongly against this. The very essence of the roles is to manage SLAs in a holistic fashion.

[a] For example, the goals of users and customers may sometimes conflict: users may demand high availability whereas customers look for value for money at different levels of availability. A certain balance is needed.

Table 13.10 *Provider*

Responsibilities[a]	• Agreeing SLAs with customers, including provider responsibilities, described in SLAs • Making services available to users in compliance with SLAs • Managing the operations that surround the offering of services during use • Controlling customer access to services (may be via service buses) • Billing customers for service usage • Monitoring supplier service levels so that quality can be assured against SLAs (by the service-level manager).
Skills	The provider must possess the ability to offer run-time services in accordance with SLAs, acting as a broker between suppliers and users. The role must be able to align the needs of customers with the capabilities of suppliers to meet those needs. Reliability and trustworthiness are key attributes of the role
Inter-working	In terms of planning and agreeing SLAs, the role works closely with the service-level manager and with suppliers. In terms of the actual execution of services, the provider is the conduit between suppliers and users of services, that makes these services available to users. Note it may be that the provider and supplier are instantiated by the same physical organizational unit, in which case the provider also executes services.
Sub-roles	The role itself is unlikely to be split, although there could be several instantiated providers for different kinds of service.

[a]The responsibilities of the provider are often referred to collectively as "provisioning" (Polan, 2002).

Table 13.11 *Supplier*

Responsibilities	• Complying with SLA terms of supply • Designing the implementation of services • Supplying the implementation of services • Executing services.
Skills	Proficiency in the techniques used to design an implementation of a service, and the technologies used to implement and run a particular service.
Inter-working	The supplier works with the provider both to verify SLAs and to supply services that the provider makes available to users.
Sub-roles	The role may sub-divide into implementation designer, implementation supplier, and execution supplier: for example, a particular implementation of a service could be designed by one supplier, then coded by another, and finally executed by a third supplier. At the same time, it is likely that the same physical organization will perform all three roles.

Table 13.12 *Asset inventory manager*

Responsibilities	• Maintaining and controlling the asset inventory of services • Cataloging services • Publicizing capabilities of services • Controlling configuration management and release management of services • Understanding and communicating developments in best practices for asset inventory management.
Skills	The role must possess a thorough understanding of the repository technology that is used to implement the asset inventory, as well as having highly developed technology administrative skills (for example, cataloging structures and retrieval mechanisms). However, the role must equally possess good communication skills.
Inter-working	The role works closely with the service architect and the service infrastructure architect. However, the asset inventory manager is responsible for the control of asset information that underpins the entire process of service orientation.
Sub-roles	The role may well split into separate roles for librarian, configuration manager, and release manager.

Table 13.13 *Service execution manager*

Responsibilities	• Handling the technical challenges concerned with managing run-time services dynamically, in real time • Identifying users of services and controlling access in line with security policy • Monitoring, analyzing, and adjusting of service execution in line with SLAs • Discovering alternative services that comply with SLA and determining the most appropriate service for use in a specific context • Metering service usage and ensuring that costs are apportioned correctly • Understanding and communicating developments in SEM technologies.
Skills	In traditional environments, the skills required are those of an IT operations manager. However, as we move further toward on-demand computing so the "skills" required are increasingly outside the bounds of human capability. More and more sophisticated SEM software is required to carry out the responsibilities in real time on larger and more complex sets of dynamically changing information.
Inter-working	The service execution manager works with the service-level manager.
Sub-roles	The role may split into separate roles for execution management of security, business, and "technology" (availability and capacity). It is important to note that the role may actually be performed by a combination of one or more software units and one or more humans.

13.4 Ownership and finance

Ownership of services and financial models for funding and charging of services are two of the most challenging cultural aspects of service orientation. In this section, we provide a brief guide to tackling some of the issues concerned with these topics.

13.4.1 Ownership

Business owners should be distinguished from *technical owners*.

The business owner owns the functionality described in the SLAs. Often this is the customer of the service. After all, if the customer funds the service it seems only right that the customer should own it! However much will depend on the market approach to service orientation (see 13.5). In a provider-driven approach it is much more likely to be the case that the provider (or the supplier on whose behalf the provider is acting) actually owns the service. Customers buy the right to use the service, but not the service itself.

The technical owner is the supplier who owns a specific implementation of the service. A service may have a single business owner but different technical owners, one for each implementation of the service.

13.4.2 Responsibility

Irrespective of ownership, the customer is always responsible for how the service is used within the context of his or her organization's business processes. This kind of responsibility is separate from the responsibility of the provider for ensuring the availability of the service and from the responsibility of the supplier for actually running the service. For example, in using a currency translation service the customer is responsible for ensuring, say, that the service is used in the international funds transfer process and not for domestic credit control.

These distinctions are sometimes overlooked. For example, an insurance organization outsources its claims services but fails to ensure that they link up properly with in-house fraud detection procedures, with the result that there is an excess of fraudulent claims. Again it is incumbent on the customer to take responsibility for all services. For example, as explained in chapter 14, Queensland Transport offers a vehicle registration service to motor dealers. This service effectively becomes part of each motor dealer's registration process. The legal onus to ensure that a vehicle is properly registered remains with the motor dealer acting as the service customer.

13.4.3 Service funding models

There are two major types of funding model: tactical and strategic. In reality, a combination of the two is often sensible.

The *tactical* funding model assumes a basket of money provided by a group of business units with like-minded interests to reduce costs through common services. Motivation is often provided in these cases by a need to leverage existing systems or databases. For example, there may be several business units that use different product databases that contain duplicated or fragmented information. This results in an attendant drain on resources to maintain consistency across the different databases. A basket of money is set aside to unify and rationalize them. Designing a set of services to provide unified data services on top of the existing databases might be a good logical first step toward unification.

At the same time, there must also be sufficient funds to cover up-front investments in business process redesign, SOA, education, and planning. Such costs should be factored into the return on investment (ROI) calculations.

The advantage of a tactical approach is that ROI can often be demonstrated reasonably quickly. The disadvantage is that there are risks in overlooking wider requirements and that business unit managers may still balk and argue about their own interests. If one manager is paying most of the money then he or she may influence requirements to the detriment of the common interest.

The *strategic* funding model assumes an enterprise-wide investment that is used to provide a jump-start to business units seeking to reduce costs through reuse of services. The corporate fund must be sufficient to cover up-front investments in business process redesign, SOA, education, and planning There is also the investment in asset inventory management tools and techniques to help sow suitable services that may be harvested by the business units. We can expect a significant up-front rise in costs to cover the up-front expenditure. It may not be possible to recoup these costs directly and some organizations may simply write them off, although they can be factored into ROI calculations.

The advantage of the strategic funding model is that because business units do not pay for service usage, they are more likely to embrace it! The disadvantage is that it may be a long time before any ROI is realized.

13.4.4 Service charging models

Charging of services is a young and emerging science. Developments in technology are opening up possibilities for revenue collection over the Internet that were impossible not so long ago. At the same time, it is very important to have a clear charging model upon which charging is based. The details must be agreed between customer and provider, and documented within the billing details section of the SLA (see 12.3.1).

Broadly, there are two models for charging of services: flat-rate and usage-based, with a variety of combinations. For example, there can be a monthly minimum fee (to cover overheads) plus usage charges above a specified threshold.

In addition, if an aggregated service is offered then a plan is required for settlements – a way to share revenues with the provider's upstream service providers.

13.5 Market pragmatics

By constructing an SOA, an organization is much better able to pinpoint its services with respect to its line of commoditization and to make much more rational choices regarding sourcing and usage of services than would otherwise be possible. Very broadly, an organization may take a customer-driven approach to sourcing some services, a provider-driven approach to sourcing other services, and a collaborative approach to sourcing yet further services. Many organizations will carry out a mix of these different approaches for different types of business.

In this section, we provide some guidelines on these different approaches with respect to the SLAs and roles already described.

As with the role catalog, the guidelines are not intended always to be adopted in their entirety. For example, it may well be that some of the market contexts overlap. The guidelines are intended to stimulate readers to think about the relevant issues with respect to the needs of their organizations and to adapt the guidelines where necessary.

13.5.1 Services versus software services revisited

Before we look at some different approaches to creating SLAs it is worth quickly recapping the distinction between *software* services and services *in general*.

Usage of software services should be distinguished from usage of services in general. On the one hand, a set of services may be used in the manner of a business process, such as insurance claims. On the other hand, it may be a set of software services that is used. For example, an organization whose sourcing strategy is based around a particular package vendor wants to move toward using a new set of modules offered by the vendor that offer an existing software package as a set of Web services.

Similarly, provision of software services should be distinguished from provision of services in general. On the one hand, a set of services may be offered in the manner of a business process, such as insurance claims. On the other hand, it may be a set of software services that is offered. For example, an organization such as a package vendor wants to upgrade and modularize an existing software package as a set of Web Services.

In our present context, we are particularly interested in usage and provision of software services. These services may be offered as software services through a service catalog or library. Unless specifically stated otherwise, in the following sections we use the term "service" to mean software service.

13.5.2 Reality kicks in

The vision of service orientation is of global software services advertised in UDDI directories, in which the "best" service will be selected dynamically at run-time. Providers will advertise their service-levels in a uniform way, so that they can be compared in the run-time selection process. Or off-line agreements may be reached with certain suppliers, covering all their service offerings, and at run-time the dynamic service selection is limited to these suppliers.

In this scenario, there is no dominant role. The customer–provider process is evenly balanced. Communication from customer (or user) to provider starts with searches and requests for quotations and ends with agreements between the customer and provider to purchase chosen services. Communication in the other direction starts with service catalogs and specifications and ends with delivered services.

In real life for the foreseeable future we must be much more pragmatic. There are situations in which there is a more even spread of control, with customer and provider working in collaborative fashion. Indeed, we provide some guidelines for dealing with this type of relationship below. However, the reality is that in most situations either the customer or the provider takes the dominant role. Either the customer drives the process, or the provider drives the process.

It is also important to consider the particular business relationship context. For example, consider the SLAs offered by your domestic electricity or water supplier. As compared to a large company, as a domestic consumer, you pay relatively little for those services, and service outages are broadly speaking inconveniences and not major incidents. For an organization such as a hospital, however, loss of water can be life-threatening. Similarly if a call center loses its telephone service its business is severely disrupted. In situations such as these, it is common sense for the customers to carefully scrutinize, if not negotiate, their SLAs.

Whether or not an SLA is offered (or negotiated) depends on the business relationship between the customer and the provider. For example, a global package transporter is much more likely to offer package tracking services to its major accounts than to its occasional small customers.

The reader should keep these qualifications in mind in considering the approaches described in the following sections. Also keep in mind that the market patterns that follow tend toward the situation in which the customer and provider roles belong to different organizations. However, it is important to realize that these approaches can apply equally where the roles belong to the same company, though generally with less criticality.

13.5.3 Customer-driven approaches

A customer-driven approach is predicated on a compelling business case for services to be used as part of an organization's business improvement plans.

We have recommended iterative and incremental approaches to business improvement and SOA throughout this book. The SLA process pattern described in chapter 12 also follows this approach. Assessing the availability of external services should be made throughout these iterations, and not left until the implementation stage, with

Table 13.14 *Strategic usage approach*

Dominant role	Customer
Description	A strategic usage approach looks to reuse software services in response to a particular class of business problem embodied in a corporate program. The approach involves the evolution of an SOA. The value proposition underlying the business need must be understood by senior management, resulting in commitment to the approach. Examples of such value propositions range from reduced development costs through software reuse or through better control of service providers to reduction in disruption to the business caused by changes to technology. The service-oriented business process redesign pattern (see chapters 5 and 6) is used as a basis on which to evolve the SOA (see chapters 7–10). In particular, the line of commoditization informs strategic decisions on whether to develop services internally, to reuse existing internal services (from legacy systems), or to acquire or subscribe to external services.
SLA	In a strategic customer-driven approach, a customer publishes an SLA in the manner of a formal invitation to tender to suppliers. As these invitations are part of a wider strategy, it is important that they are based on clear SLAs, so that appropriate providers may be chosen. It is important that the services are specified with maximum reusability in mind and in such a way that they are robust enough to survive business changes.
Organization type	This approach is likely to appeal to large organizations keen to use software services for strategic business improvement. Such organizations are likely to have an "early adopter" technology adoption profile and be powerful enough to pre-stipulate their service requirements to the market.

SLAs bolted on as something of an afterthought. A balance must be struck between the functionality of available services and the ideal business requirement. This judgment also involves comparing the time and cost to acquire the services versus that of delivering services internally. In fact, gap analysis is applied at increasing levels of detail through the stages of a project, at decreasing levels of abstraction, not just once at the specification stage as is sometimes mistakenly thought. Service orientation affects the software's entire life cycle, as we emphasized at the beginning of this chapter.

Good business knowledge is an absolute necessity in making trade offs as an organization moves through this life cycle. A set of vendor-supplied software services that supplies 80% of business functionality in a quarter of the time it would take to build from scratch is hugely attractive. But how critical is the missing 20%? Is the business prepared to wait for the required extensions?

Many variants of the customer-driven approach are possible, depending on the company's sourcing strategy. On the one hand, a strategic approach may be taken, as described in table 13.14.

On the other hand, a much lower-key tactical approach may be taken, as described in table 13.15.

Table 13.15 *Tactical usage approach*

Dominant role	Customer
Description	A tactical usage approach should be based on a compelling business case for acquiring or subscribing to services. A service-oriented business case will typically be to minimize costs of market participation and/or to focus energies on exploiting core competencies. A tactical usage approach looks to reuse services to gain quick business pay-back in terms of cost savings or business process efficiency improvements. As with the strategic usage approach, decisions on whether to reuse services are based on an evolving SOA, using the techniques described in chapters 4–10. However, unlike the strategic approach this is essentially a short-term approach: typically, two or three increments of a software solution may be produced over a 3–6-month timeframe. "Just enough" SOA is used. There are many potential routes to achieving these gains. Typically, however, an opportunity exists to streamline an existing business process either by reusing internal services or by acquiring or subscribing to external services.
SLA	If internal services are reused, although service descriptions should be produced, it is unlikely that SLAs are justified. A common example is the integration of two (or more) legacy systems. In contrast, acquiring or subscribing to external services to improve the business process may involve a limited invitation to tender exercise based upon outline SLAs. However, again the essence of a tactical as opposed to a strategic approach to service usage is that specifications should be kept as light as possible in the knowledge that the solution is short term. In a tactical customer-driven approach, an SLA is used in a less stringent sense with the accent on finding a quick business solution.
Organization type	This approach is likely to appeal to large organizations keen to explore services to solve specific business problems, at low risk. Such organizations are likely to have an "early majority" technology adoption profile and have a significant legacy portfolio that is difficult to manage.

13.5.4 Provider-driven approaches

A provider-driven approach is predicated on a compelling business case for developing a service or set of services to take to market. A provider-driven approach seeks to take services to market, often publishing services over the Internet. Many variants of the approach are possible, depending on the company's usage strategy, as exemplified in tables 13.16–13.18.

13.5.5 Collaborative approaches

This approach, which is increasingly popular, may take a great many different forms. Table 13.19 provides a general description.

Table 13.16 *Opportunistic provisioning approach*

Dominant role	Provider
Description	The provider may offer a single very specific service. This is usually a commodity service: for example, a currency conversion Web service offered via a UDDI compliant catalog. In this case, the provider and the supplier are very likely to be the same entity. The services offered in an opportunistic provisioning approach are anonymous: in other words, the services operate without requiring the identity of the consumer at run time. An example of an anonymous service would be the retrieval of a stock price: the authentication, authorization, and metering of the service is delegated to the hosting system, so anonymous services typically need no additional provisioning.
SLA	For the largely commodity services offered by this approach it may be a case of the customer either accepting the terms and conditions stated by the provider in the directory, or not using the service. So the customer will need to check that these terms and conditions cover all the aspects that they deem important.
Organization type	This approach is attractive to organizations keen to use Web services technology to realize immediate market opportunities. Such organizations are likely to have an "innovator" technology adoption profile.

Table 13.17 *Market provisioning approach*

Dominant role	Provider
Description	The provider builds an SOA, along the lines described in this book, with a target market in mind. In this case the provider and the supplier are very likely to be different entities. The SOA reflects a generic business process that the organization is expert in and wants to take to market. It is likely that the SOA would be designed with large-grained sets of services. In contrast to the customer-driven approaches, this approach is likely to involve a much longer period of up-front work. The company does not want to be issuing new releases every few weeks: although this does not preclude an incremental style of development, the approach is naturally focused on larger increments.
SLA	Market provisioning requires fully detailed SLAs. It is vital that the interfaces offering the services are specified with maximum reusability in mind and in such a way that they are going to work effectively in consumer environments. They must also be robust enough to survive business changes. The provider specifies the software units required to realize these interfaces and decides the supplier that will be engaged to develop or supply the software units.
Organization type	The market provisioning approach is probably best exemplified by B2B applications offered as Web services. The Galileo airline reservation application offered to the travel industry is a case in point. This approach is attractive to organizations keen to use Web services technology to realize market opportunities. Such organizations are likely to have an "innovator" technology adoption profile.

Table 13.18 *Strategic provisioning approach*

Dominant role	Provider
Description	An increasingly popular provider-driven approach is to offer services in an evolutionary fashion as part of the business strategy. The techniques described in this book are especially appropriate in this context Often the approach is used as part of a legacy systems migration strategy in an effort to rationalize and then expose existing software functionality as a set of services.
SLA	The services offered do not generally require detailed SLAs. Usually it is a case of the customer either accepting the terms and conditions stated by the provider in the directory, or not using the service. So the customer will need to check that these terms and conditions cover all the aspects that they deem important.
Organization type	This approach is best exemplified by B2C applications. BT's Friends and Family service offered over the Internet is based upon existing legacy systems which have been rationalized and then exposed as Web services for use in new Internet applications. Another example is Queensland Transport's offering of a service to book driver tests on the Internet and through the call centre. They created a generic booking service component and then this service was consumed by both the driver examination booking system and the vehicle inspection booking system (giving reuse of the service). This approach is attractive to organizations keen to use Web services technology to realize market opportunities. Such organizations are likely to have an "innovator" or "early adopter" technology adoption profile.

Table 13.19 *Collaborative approach*

Dominant role	Service-level manager
Description	A collaborative approach is predicated on a compelling business case for companies to work together to use services to improve a shared business process, or to take advantage of a market opportunity.
SLA	The provider and customer work together in partnership to agree an SLA Often the boundaries between the responsibilities of the two are not clear. There is an important element of exploration and negotiation involved that raises the profile of the service-level manager role and that calls for clear SLAs.
Organization type	This approach is likely to appeal to large organizations that have very close links with partners. Examples range from Queensland Transport's offering of vehicle registration services within the context of motor dealership business processes (discussed in detail in chapter 14), to the hosting of partner services by Amazon.com.

13.6 Where to next?

Service orientation is a business phenomenon that involves massive cultural change. Many of these changes concern breaking down the traditional barriers between business and IT, and between software development and operations. In addition, groups that traditionally worked within their own comfort zones, within the organization firewall, are finding those comfort zones eroded as organizations are forced to work in ways that transcend the old corporate boundary lines.

The IT industry has been prolific in its attempts to invent processes to address the long-standing problems of division of business from IT concerns and the gulf between software development and operations. What has been lacking is a clear approach to definition of deliverables that bridges these divides. Throughout this book we have therefore emphasized the importance of *specification* – a clear approach to defining services, SLAs, and other deliverables. The underlying view has been that it does not make sense to tackle any kind of cultural change until people are agreed on a definition of the terms involved in that change.

In this chapter, we have made a start in laying out some of the strategies required to address these cultural changes of service orientation by examining some of the roles that people play in this emerging world and discussing the various market contexts that organizations increasingly find themselves in.

However, we have probably come as far as the scope of the book will allow in providing cultural guidance in the form of templates, process patterns, and the like. The case studies that follow, in chapters 14 and 15, put some flesh on the bones of this book by discussing how to assess your current situation and transition forward in way that is in tune with the organization's needs.

Part 5 Case studies

Part 5 of this book provides two case studies in service orientation.

In chapter 14, Sam Higgins and Paul McRae describe how Queensland Transport has developed a service-oriented approach that reflects service orientation as the business phenomenon described in this book. In particular, they look at how the approach is used to allow third-party software developers to integrate Queensland Transport regulatory processes seamlessly within commercial motor vehicle dealer business processes. The results are impressive. For example, the organization can now execute a service-based transaction at an activity-based cost of less than A$1.5 compared with the more traditional face-to-face approach which costs between A$4 and A$10. The department now collects over A$260 million (or 12%) of its total revenue using an SOA, with no direct customer contact. Services are available 23.5 hours per day, 6.5 days per week, and 80% of all services respond in less than 2 seconds.

In chapter 15, Hermann Schlamann describes how Credit Suisse has evolved from an organization encumbered by a large and complex legacy software portfolio to one which can now use significant parts of its legacy systems for business value. The organization has worked hard to rationalize these legacy systems into sets of reusable services that are aligned with business objectives. Again, the results are impressive. For example, money transfers for a customer are now enabled over different channels: via the internet with "online banking", via a "teller terminal" in a branch, or via "personal contact" at a branch counter. In addition, services are being used in more entrepreneurial ways. For example, various foreign exchange and accounting services are offered, at a charge, to the smaller banks. This brings in revenue and at the same time lowers the total cost of ownership.

Both case studies represent ongoing improvement programs that have successfully applied the core techniques explored within this book. Each shows how architectures and models can be nurtured, with a focus on getting the most out of the existing software assets. In both cases, service orientation has become integral to the organization's business strategy.

14 Queensland Transport: a case study in service orientation

Sam Higgins is now with Forrester Research, Inc. as a senior analyst. Formerly he was a director of Encode Services, an Australian information and communications technology consulting firm, and managed application architecture direction for the Innovation and Planning unit of Queensland Transport's Information Services Branch. He has over twelve years' experience in both traditional and model-driven environments, and is a popular international speaker. Sam was a driving force behind Queensland Transport's adoption of the Java 2 Enterprise Edition (J2EE) platform and subsequent Web services implementations.

Paul McRae is the application architect for the Innovation and Planning unit of Queensland Transport's Information Services Branch. He has almost twenty years' experience in the IT industry, was a key player in the introduction of an SOA within Queensland Transport, and was instrumental in developing Queensland Transport's CBD initiatives, J2EE applications, enterprise architecture, and Web services.

In this chapter Sam and Paul share their experiences in Queensland Transport's transition to service orientation.

14.1 Background

Queensland is Australia's north-eastern state, covering nearly 1.7 million km^2, and home to approximately 3.6 million people. Known as the "Smart State," Queensland is a center of significant leading-edge technology developments. Queensland Transport is a department within the State Government of Queensland.

Queensland Transport has a budget of approximately A$1.3 billion and over 4,700 staff throughout the state, and is responsible specifically for the regulation of land, sea, and aviation transport in Queensland. Its mission is to achieve better transport for Queensland by connecting people, goods, and services to enhance the economic and social well-being of all Queenslanders.

14.1.1 Profiling a service-oriented organization

In 1997, Queensland Transport's strategic plan (Queensland Transport, 1997c) heralded a profound change in the way the department viewed delivery of service to customers:

> Meeting our clients' changing needs and requirements in a diverse regional state means involving clients in identifying needs and in delivering services in a manner that is timely, convenient, easily accessible and cost efficient while still being individualized. The use of new information technology to speed up and improve service delivery, as well as extend the availability and range of services we offer, will enable our clients to be better serviced and informed 24 hours a day.

Queensland Transport's service delivery business began to focus on this new approach with a simple, but powerful motto: "Doing business where it happens."

In response, the department's IT branch offered the business exactly what it requested, a new kind of information technology solution – the "*Service-Oriented Architecture*."

However, this solution and its associated benefits did not merely appear. Instead it is the result of an ongoing rationalization and improvement program that has successfully applied the core techniques explored within this book.

Today, Queensland Transport's investment in service orientation – from business requirements through to implementation – has, achieved impressive results, and the service execution remains aligned to the strategies set down almost eight years ago.

Queensland Transport's service-based approach allows third-party software developers to integrate Queensland Transport regulatory processes seamlessly within commercial motor vehicle dealer business processes and the software that supports these processes.

Queensland Transport can execute a service-based transaction at an activity-based cost of less than A$1.5 compared with the more traditional face-to-face approach which costs between A$4 and A$10. The department now collects over A$260 million (or 12%) of its total revenue using an SOA, with no direct customer contact. Services are available 23.5 hours per day, 6.5 days per week, and 80% of all services respond in less than 2 seconds.

The Queensland government recently published a government-wide approach to IT planning based on Queensland Transport's own service-oriented approach. This, coupled with the inclusion of SOA concepts within the government IT standards and polices, has set the stage for the further expansion of service orientation within the Smart State.

Our case study reveals how Queensland Transport embarked on its journey; it shows the application of key techniques, offers a view of the critical success factors, and provides some tips to follow and traps to avoid for achieving a truly evolving SOA.

14.1.2 The need for a service-oriented approach

In early 1992, Queensland Transport began the strategic replacement of five separate core systems (developed over thirty years) with a single integrated system – the Transport Registration and Integrated Licensing System (TRAILS).

Most of Queensland Transport's business activity occurs within this core operational system. TRAILS processes an estimated 60–70% of all Queensland government face-to-face transactions and covers thirteen major business systems such as accounting, vehicles and vessels, customer and involved parties, plates, licensing, infringements, registrations, and vehicle inspections. TRAILS is considered to be the largest single line of business application in Queensland government today.

TRAILS represented a major leap for Queensland Transport from an uncoordinated information systems environment to a single platform from which the organization could continue to evolve.

TRAILS was developed using the Information Engineering method fathered by Clive Finkelstein and popularized by James Martin in the late 1980s and early 1990s (Finkelstein, 1989; Martin, 1991). The method itself is well documented, and for brevity is not covered in any detail here.

At Queensland Transport, we focused heavily on the core business process re-engineering techniques included within the Information Engineering method: data, process and interaction analysis.

Queensland Transport was four years and almost 300 person-years of effort into the project when it became apparent that we needed to implement an SOA. Up to that point, we were building TRAILS using a "best practice" monolithic mainframe-based methodology.

Several major drivers emerged that caused us to look for a new approach. The large financial investment in TRAILS (the original funding was over A$100 million) meant that reusability was a priority as there was unlikely to be any further large-scale funding provided for future business initiatives. The business also expected TRAILS to have a long useful life, so the application needed to be adaptable and agile.

The ongoing activity (process) analysis in preparation for releases 3 and 4 (due in 1997 and 1999, respectively) revealed that the majority of Queensland Transport registration transactions were triggered by motor vehicle dealers on behalf of their customers. Queensland Transport was just one element of a much larger complex value chain.

Queensland Transport had changed its strategic view of service delivery away from a traditional single-channel model to a multiple-channel, highly available, customer-focused approach. We wrote a technical paper to address our changing business environment (Queensland Transport, 1997a).

The paper assessed the implications of the changed business environment on TRAILS and the TRAILS system architecture, in order to improve functionality and reduce

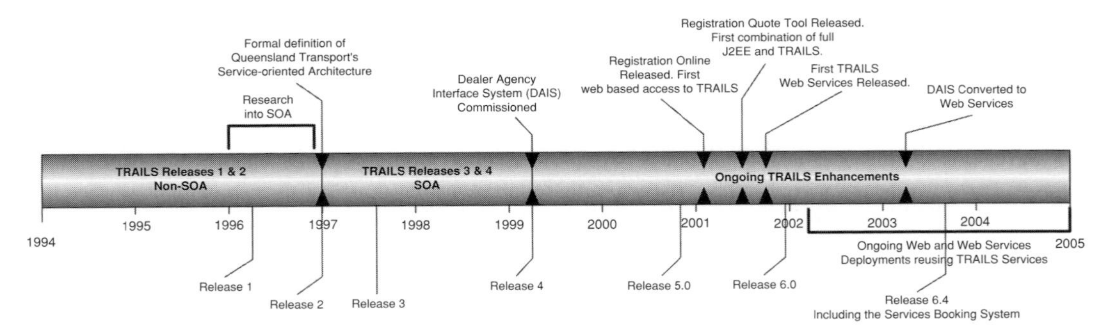

Figure 14.1 Major milestones in Queensland Transport's TRAILS development

the cost of the environment. We argued that service enablement of TRAILS included more than just new interfaces; instead it required re-engineering TRAILS to deliver functionality through several interfaces. This would include Web enablement as a short-term solution, but in the longer term this new architecture would provide sufficient agility to support delivery using other channels such as telephony, wireless mobile devices, and other interfaces as required.

Today, it is easy to see that the paper and the plans that followed provided Queensland Transport with a clear business–IT alignment assessment using some elements of the SOV7 in response to the new strategic direction. It was this clear alignment of the business need for a major shift in Queensland Transport's technology approach that ultimately prompted approval for the TRAILS project to start a detailed investigation and analysis of the style of architecture that could meet these requirements, a search that ultimately led to service orientation and an SOA.

At the conclusion of the implementation phase in 1999, TRAILS was officially treated as a departmental asset and assigned a value of A$60 million with an estimated useful life of thirty-plus years.

Figure 14.1 shows the major milestones of our TRAILS development. This shows the continued evolution of the system post-1999, made possible through the agility provided through service orientation.

After making this scale of investment, imagine explaining to senior management how you plan to remain agile and provide the customer a one-stop shop, all the while remaining channel-neutral, without a service-oriented approach!

14.1.3 Adoption of the SOA

Having gained approval to re-engineer a half-completed multi-million-dollar asset, we started defining a new foundation for the remainder of the development effort.

Our original investigation for the "new" architecture resulted in the discovery of two research notes published by Gartner Group in 1997 (Shulte, 1996; Shulte and Natis,

1996), whose key concept was the creation of a *service tier* within the overall structure of an application. These research notes became the backbone of Queensland Transport's own approach (Queensland Transport, 1997b). This new specification required the development teams to establish software services based upon our existing analysis deliverables – specifically, the existing process, data, and interaction models. These were provided to the development team at the conclusion of the detailed business area analysis stage of Queensland Transport's information engineering-based software development life cycle.

The new approach included design-time guidelines for the behavior expected from all layers within the application architecture, with specific emphasis on the core behaviors of services. This defined the behavioral dimension of the core or standard contract required for all Queensland Transport services.

We defined a standard interface that would be implemented by all software services within TRAILS. This defined the interface dimension of the core or standard contract required for all Queensland Transport services. We also specified the security policies to be applied at each layer of the application architecture, and policies for re-engineering of the early elements of TRAILS to ensure that they become SOA compliant.

The new SOA elements, coupled with existing analysis techniques already embodied within the TRAILS project, gave rise to an overall service-oriented approach based on the need to:

- Continue to produce business process models and the associated data model in accordance with the existing analysis approach, while increasing focus on service orientation
- Use the business process models for identification of supporting services, independent from the access – or "presentation" – mechanism, then define the service interfaces in alignment with the data model and allocate services to components within TRAILS
- Analyze and assign business rules, then assign to services and/or access mechanism
- Develop specifications (in "use case" style) for access mechanisms such as online screens, batch procedures, and system-to-system exchanges that use the defined services
- Realize the services using AllFusion® Gen, and then implement the services within their consuming access mechanisms.

The approach was applied to the development of new software units and to the tactical renewal of pre-1997 elements of the system The overall adoption strategy was to extend TRAILS in the following order (as illustrated in figure 14.2):

1. Implement services internally within TRAILS, within both the online and batch processing environments
2. Expose these services for consumption within the department by new and existing internal applications outside TRAILS
3. Expose services to external trusted partners for consumption

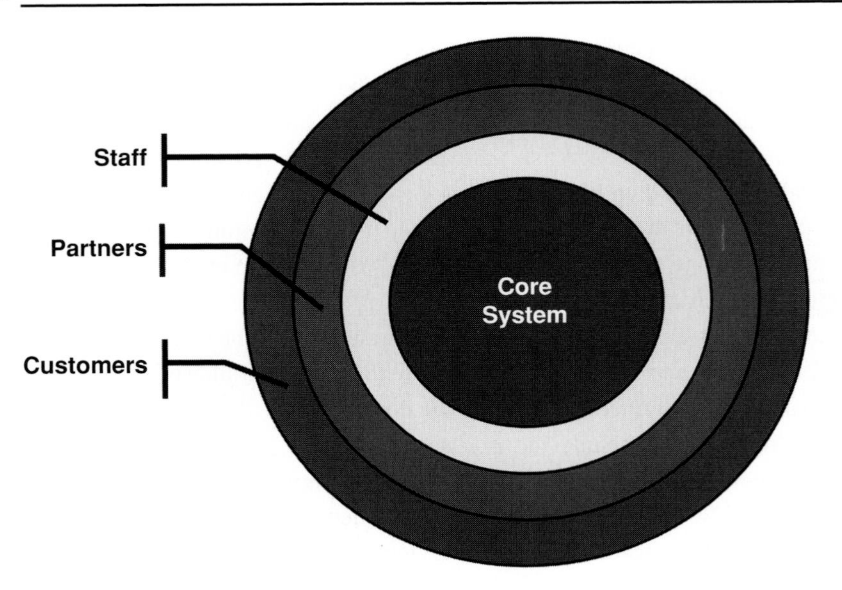

Figure 14.2 Queensland Transport's strategic expansion of access to TRAILS services

4. Expose services directly to the department's customers, through access mechanisms provided either by partners or the department.

Interestingly, Queensland Transport did not question the need to continue to perform extensive and significant activity or business process analysis and modeling during the strategic shift in the business and technologies towards an SOA. Instead, the real change was the simple realization that service orientation was more of a business model than any new or overly complex technological implementation.

Today we have evolved and refined the service-oriented approach within Queensland Transport. We will now explore this evolution and provide examples, lessons learned, and a glimpse into the future for Queensland Transport.

14.2 First steps

Many organizations see the benefits of a service-oriented economic community and realize that at some point they, too, must make the fundamental shift. Yet it is never easy to take the first step. Even if that step is towards joining the growing numbers of the new millennia's most successful, like the almost legendary Amazon.com!

So how did we make the first step? How did we transition to this new goal architecture, this new business model?

Before we discuss how Queensland Transport now tackles each of the key elements within a service-oriented approach, let us explore the initial first steps to see how we obtained early business value.

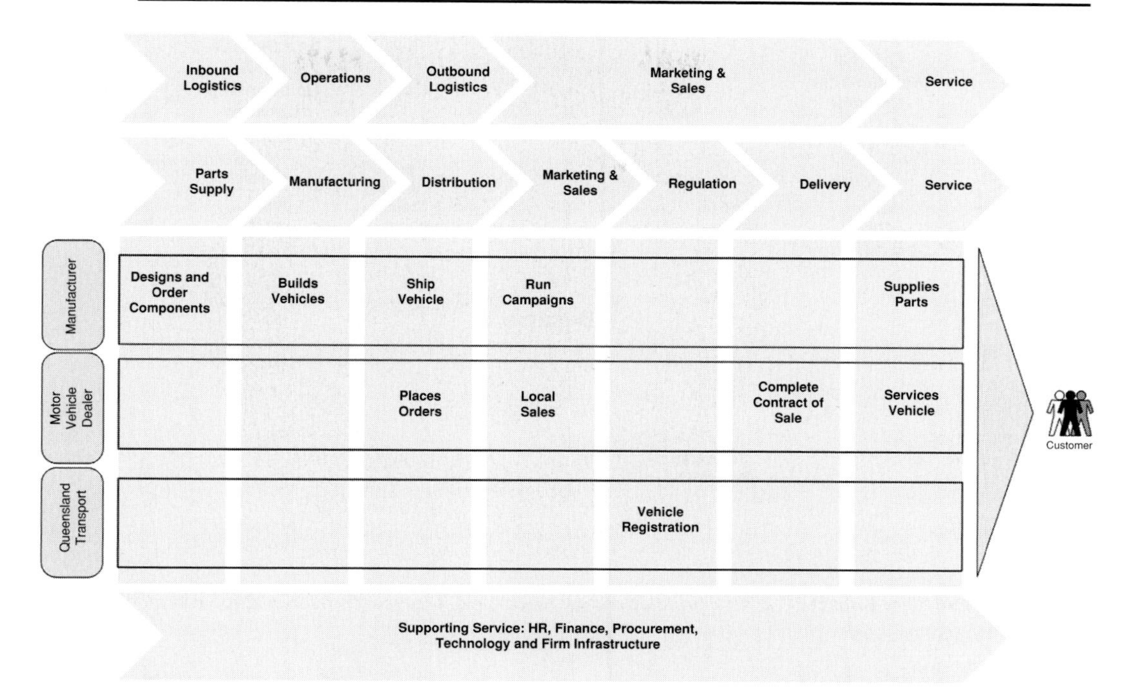

Figure 14.3 The original motor vehicle sales value chain

14.2.1 The Dealer Agency Interface System

After the SOA was defined in 1997 and we had completed the majority of business analysis for TRAILS releases 3 and 4, we saw an opportunity to exploit the newly defined SOA beyond the boundaries of Queensland Transport.

TRAILS release 4 included the replacement of an existing email-based e-commerce application introduced in the early 1990s known as the Dealer Interface System (DIS). The resulting system assessment of DIS and the existing business analysis initiated a detailed investigation of the value chain surrounding registration of new vehicles by motor vehicle dealers.

The investigation showed that, while motor vehicle dealers had an extensive set of business processes which spanned the act of registration, Queensland Transport's registration process was a regulatory imposition that diverted this otherwise straight-through process. Essentially, the value chain included a requirement for the customer (or the motor vehicle dealer on their behalf) to interact with Queensland Transport (as shown in figure 14.3), an activity that added little value to the purchase transaction itself.

We also identified that motor vehicle dealers who dealt in new cars had significant investment in industry-specific IT systems to support their core business (order processing, inventory, integration with their motor vehicle manufacturer). These IT

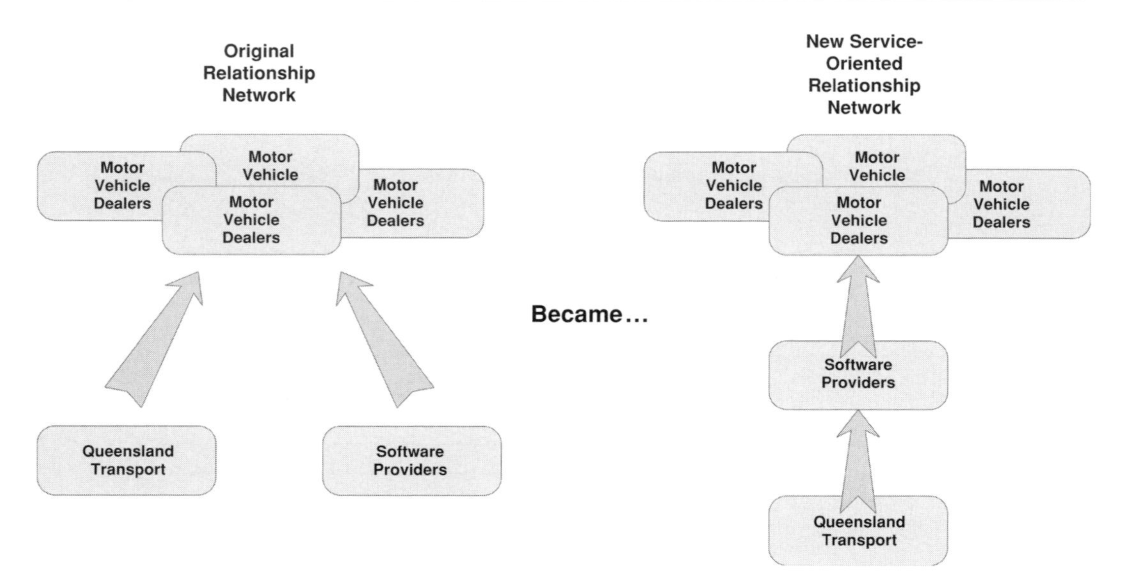

Figure 14.4 Impact of service orientation on Queensland Transport's relationship with motor vehicle dealers

systems were purchased from only four companies. Additionally, the majority of data collected by motor vehicle dealers was identical to that collected by Queensland Transport.

Queensland Transport then proposed a model to grant access to the four companies who supplied the motor vehicle dealer software, rather than supplying each individual motor vehicle dealer with access to TRAILS services. The four software providers could then integrate the Queensland Transport services into their own applications. Queensland Transport was finally positioned for "doing business where it happens."

In addition to seamless integration, the department was no longer supplying services to numerous motor vehicle dealers. Instead we needed to deliver services only to the four software providers. The effect of this model on the relationship between the department, motor vehicle dealers, and their software providers is illustrated in figure 14.4.

This model became a reality in 1999 with the introduction of the Dealer Agency Interface System (DAIS). DAIS did not require any additional services other than those already designed and built to support Queensland Transport's standard face-to-face service delivery model, yet the result was that motor vehicle dealers could now perform all aspects of the "register a new vehicle" process via their application interface. This effectively allowed the entire "sell new vehicle" business process to occur via our system. A new shorter value chain was introduced, which no longer included Queensland Transport as part of the customer interaction, as shown in figure 14.5.

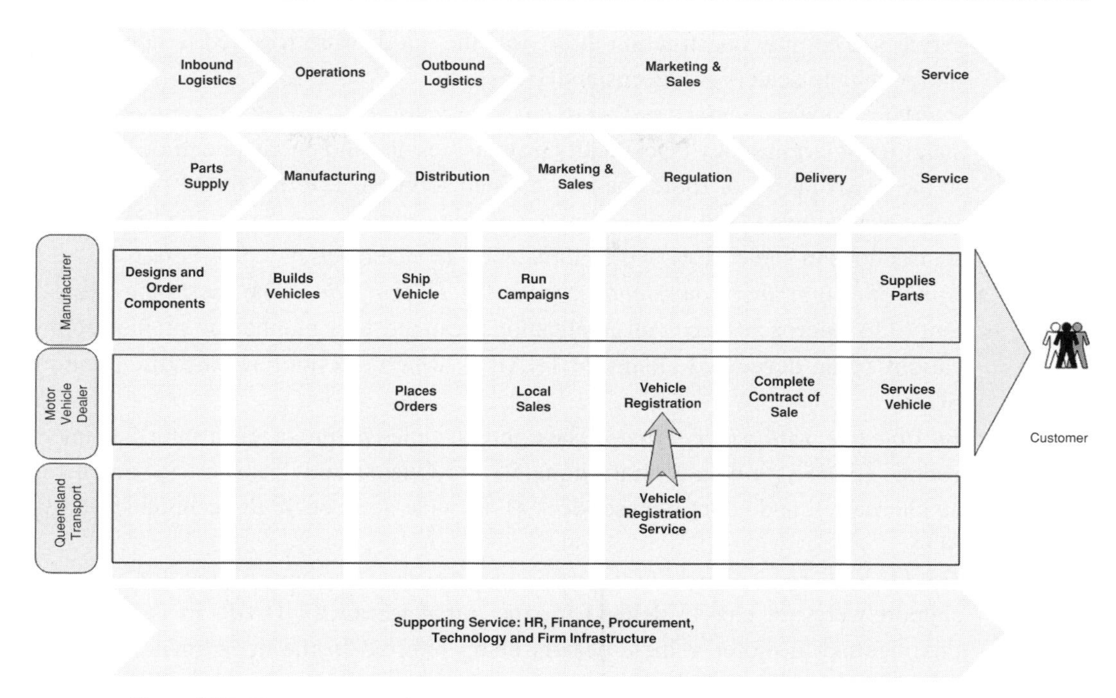

Figure 14.5 The new streamlined motor vehicle sales value chain

14.2.2 Lessons learned

Let's crystallize the nine key lessons that we learned during these first steps:

- *service orientation is a business phenomenon* This understanding meant that the SOA was applied from the beginning of the development life cycle, rather than during implementation.
- *A shift in mind-set is required for effective business process modeling* The opportunity to become a service provider to motor vehicle dealers was identified because we extended the scope of process modeling beyond our own processes.
- *Design for agility* Introducing the service layer and associated models between business analysis and design stages (as outlined in 14.1.3) ensured that the services could be reused. Beginning the design stage without a service layer led us directly to "hardwired" processes, without the necessary service- and component-level abstractions. In other words, the traditional interaction analysis within our methodology led to a fixed or "point-in-time" view of the interaction between process and data, rather than a more agile approach where the interaction occurred through services.
- *Clear policy is required for service design* The mandatory creation of services for all new functionality within the application ensured that DAIS could reuse existing

services. The only construction effort was the middleware required to connect the motor vehicle dealers to Queensland Transport.

- *Establish common understanding of basic terminology* A lack of common understanding of terminology (specifically process, service and public operation) resulted in the development of some coarsely grained services. The reusability of these services was reduced, and their complex and long-running nature resulted in queued transactions in some cases, which threatened service levels.
- *Apply the approach consistently across the IT portfolio* The lack of a consistent SOA approach across all applications resulted in a number of smaller applications being developed alongside TRAILS, which did not have the discipline of SOA.
- *Flexible integration is required* The tightly coupled nature of electronic data interchange (EDI) as a data format ultimately led Queensland Transport to search for an alternative, and adopt Web services as the wide-scale enabling technology for its SOA.
- *Establish clear responsibilities between all parties* Responsibilities for end-user support were not clearly defined for the initial release of DAIS. In some cases, this resulted in motor vehicle dealers being referred to the department's internal TRAILS help desk with software problems, even though Queensland Transport had no exposure to the software being used. Queensland Transport eventually established "schemes" and "scheme management" functions to control and support groups who consumed Queensland Transport services.
- *Accredit software consumers* Originally Queensland Transport took a hands-off approach when it came to the software released by the software providers. However, one of the providers was motivated to capture new compulsory third-party insurance business, rather than register vehicles. This motivation meant that the provider "skimmed" on testing the vehicle registration process, focusing instead on the insurance business processes. The resulting software did not properly comply with the service interface and associated behaviors published by Queensland Transport, and exhibited such a high rate of business errors that it began to impact the overall service levels.

Having distilled the main lessons learned as a result of our early steps toward a service-oriented architecture, we now move on to discuss how we built on these experiences.

14.3 Service-oriented process redesign

Queensland Transport's experience with applying service orientation to TRAILS has evolved. Today, Queensland Transport's approach begins with strategic and tactical IT planning which embodies the service-oriented spirit through the use of focal points, identification and reuse of existing assets, and planned incremental process improvement.

The approach sets the scope and priority for service-oriented process redesign which is then performed in detail as part of Queensland Transport's software development cycle. The more detailed and traditional analysis does not begin until this detailed planning step has been completed.

14.3.1 Business–IT alignment: three useful techniques

The discussion in 4.2.5 and 5.2.2 explains how the scoping of service-oriented process redesign requires an understanding of the business theme that dominates your organization's direction, using techniques such as focal points. This in turn led to the creation of the SOV7 approach to scoping the process redesign and developing measurable goals.

At Queensland Transport, we apply three key ideas to help align IT and business:

- The Value Discipline model
- The Competitive Strategy model
- The McFarlan Grid.

These techniques are used in the same spirit as both focal points and SOV7 to enable a service-oriented approach to business process redesign. They show what is involved in scaling up early successes in alignment with overall business needs.

The Value Discipline model

The Value Discipline model proposed by Treacy and Wiersema (1995) outlines three areas in which an organization can choose to excel. These areas are:

- *Product leadership*, which focuses on the organization's ability to develop new products and services that are valued by customers. In government terms this is the ability of the enterprise to develop new policies and services.
- *Operational excellence*, which focuses on efficiency, cost control, accuracy, quality, and speed.
- *Customer intimacy*, which focuses on how close an enterprise can get to its customers:
  - Does the enterprise understand individual customer requirements?
  - Can the customers serve themselves?
  - Can the customers easily communicate with the enterprise, from any distance and at any time?
  - Do the services of the enterprise have synergy with the needs of the customers?

To compete – or, in the case of a government enterprise like Queensland Transport, to perform well – organizations *must be competent in all three of these disciplines.*

To be seen as exceptional, an enterprise must perform very well in just *one* of these disciplines *while being competent in the other two*. Enterprises that attempt to excel in

Table 14.1 *System requirements for product leadership*

How well do the systems contribute to product leadership?	
Capability maturity	Developing and sustaining capabilities in service design and delivery
Intellectual asset leverage	Developing and leveraging intellectual assets for improved service design and delivery.
Responsiveness	Improving response times and reducing cycle times in service design and delivery.

Table 14.2 *System requirements for operational excellence*

How well do the systems contribute to operational excellence?	
Organizational performance	Making operations more time efficient so that existing resources and inputs can produce more. Reducing reliance on human effort.
Quality	Uncovering, correcting and eliminating any problems in established practices that lead to wasted execution time, pointless costs, inferior service quality, or undesirable outcomes.
Cost containment	Making operations more cost efficient by eliminating unnecessary expenses. Maximizing the use of available resources. Moving the creation or delivery of services to third parties who specialize (are more cost-effective) in that particular field.

Table 14.3 *System requirements for customer intimacy*

How well do the systems contribute to customer intimacy?	
Reach and range	An increase in service delivery channel freedom. The provision of self-service. An increase in the reach of services. An increase in the range of services.
Improved cycle times	A reduction in waiting times. An improvement in quality.
Product development	The ability to identify the services required by clients.

all three areas will likely run into organizational and process conflicts. In other words, trying to excel in all three disciplines does not add any value.

From a service orientation perspective it is critical to be clear on a value discipline specialization, since each discipline carries a different set of requirements for the services and their associated technology implementation. This is similar to the questions posed when using SOV7, as outlined in section 5.3.

Strategic	R&D
Systems that will change the way we do business	Systems that might be valuable in the future
Systems that support core business functions	Systems responsible for house keeping functions
Key Operational	**Operational**

Figure 14.6 The McFarlan Grid applied to Queensland Transport systems

Tables 14.1, 14.2, and 14.3 show how the requirements differ depending on the discipline specialization chosen. Each value discipline specialization requires a corresponding specialization in the business's supporting services and IT.

The Competitive Strategy model

The second technique applied by Queensland Transport is the Competitive Strategy model proposed by Michael Porter (1980). Porter describes three fundamental competitive strategies – differentiation, cost leadership, and niche. We use this model to supplement the Value Discipline Model described above.

The McFarlan Grid

To complete the picture, Queensland Transport also determines the line of commoditization using a technique similar to that described in 4.5. In Queensland Transport's case, each initiative is classified according to the McFarlan, Grid (Cash, McFarlan, and McKenney, 1992), shown in figure 14.6. This yields a coarse view of whether the business perceives the processes and associated services as being operational (commodity), key operational (territory), or strategic (value-add). McFarlan also includes a fourth classification: research and development (R & D). This area includes current initiatives that are considered to be "ventures" and may not involve a full service-oriented implementation in the short term. As outlined in 4.4.1 "innovation that comes from such functionality is best exploited quickly, without incurring the overheads inherent in servicization."

Determining this line of commoditization provides invaluable information for the process of service sourcing – in other words, a clear guide to the approach that should

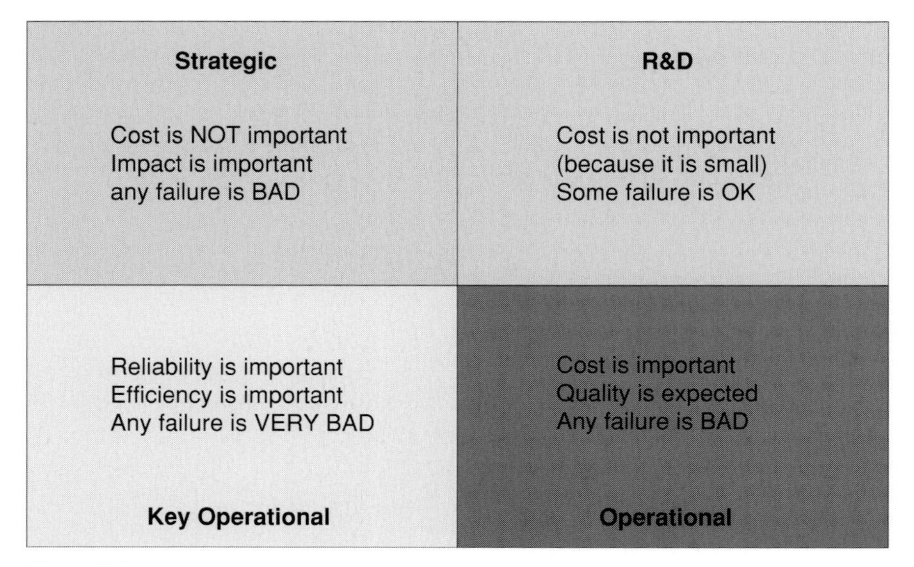

Figure 14.7 Queensland Transport's use of the McFarlan Grid relating to service sourcing

be taken when sourcing services in line with a "reuse before buy, buy before build" mentality. We are again able to use the McFarlan Grid to provide some high-level SOA policies regarding reliability in relation to service sourcing, as shown in figure 14.7.

14.3.2 Scoping process redesign

Queensland Transport's IT to business alignment method requires each business initiative to be identified and scoped. As part of this scoping, three further techniques are applied which assist in supporting service orientation:

- Process classification
- Domain analysis
- Identification of technology enablers/drivers.

Process classification

Each business initiative is initially scoped using processes selected from a fixed list, which in Queensland Transport's case is the American Productivity and Quality Centre (APQC) process classification framework (1996). This allows Queensland Transport to perform two important service-oriented checks:

- Identify *potential collaboration between business initiatives*. That is, identify opportunities to provision services once and reuse them immediately to achieve two different outcomes.

- Identify *potential legacy*[1] *services* from the asset inventory that can be used to reap early business value. The application of process classification to existing asset inventory to facilitate this process is discussed further in 14.3.3.

Queensland Transport's use of the APQC is a supplement to the recommended use of patterns in 4.2 to classify processes and provide process scope. Queensland Transport augments the use of APQC with domain analysis to provide early identification of services.

Domain analysis

This technique uses domain analysis concepts and relies on the understanding that a service represents a point of interaction between a process and the data required to support its execution. These interaction points manifest themselves in software. Software can be defined and classified, which provides insight into services and data.

Queensland Transport used an approach to domain analysis described in Biscotti and Fulton (2002) as the basis for classification of new and existing applications (both internally developed and off-the-shelf) to identify existing assets required to support new business initiatives.

Each domain shown in figure 14.8 is defined in terms of the services or functions and data it offers. This approach demonstrates the power of business domains, as previously explored in 6.3, to identify and scope services.

Identification of key technology enablers/drivers

Queensland Transport also assesses *key technology enablers* (in the spirit of SOV7) such as self service, any time anywhere, personalization, etc. which we perceive as being critical to the success of our business initiatives.

Having compiled these strategic, but tactically valuable assessments (known simply as business system assessments), Queensland Transport applies a *prioritization process* to determine the incremental changes for both strategic (four-year) and tactical (current financial year-) windows.

14.3.3 The importance of an asset inventory

As we have seen, service-oriented process redesign relies on the identification of existing legacy services through the analysis of the asset inventory. Queensland Transport's baseline asset inventory was initially compiled through a series of interviews and is now maintained through a series of governance processes. In order to identify the potential for legacy service reuse, each legacy asset is subject to classification using

[1] Queensland Transport prefers the word *heritage*, which includes the idea of preservation and ongoing maintenance of something worth retaining, rather than implying something left behind.

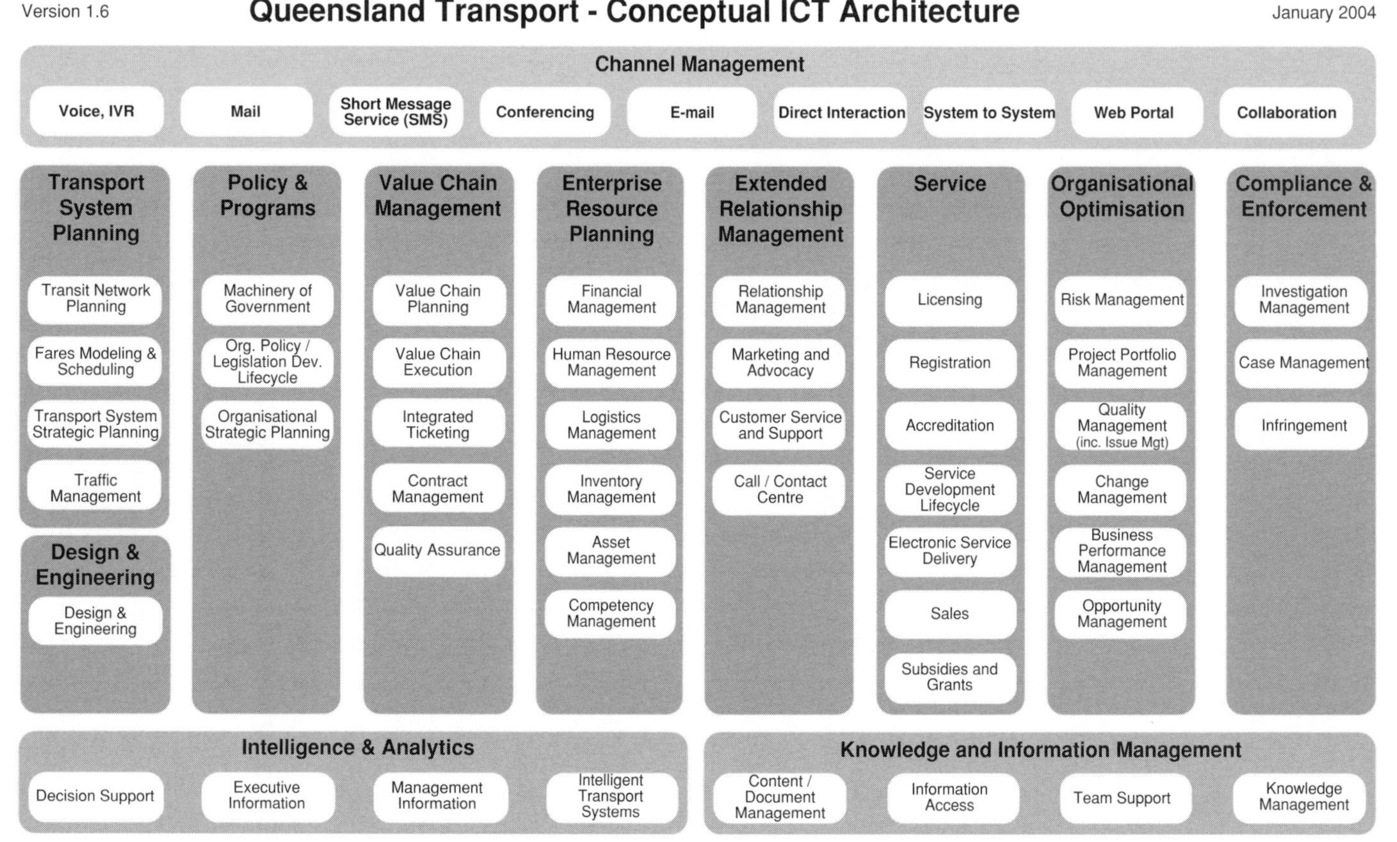

Figure 14.8 Queensland Transport's application domains

Classification	Generic Action
Nurture	Focus on keeping the existing process appropriately resourced and managed this may require the use of a Streamline or Leverage approach from time to time to maintain position in the Nurture quadrant
Streamline	Focus on improving the efficiency of the existing process by: Lowering the cost of the system – simplify Identifying and adopt lower cost means of supporting/providing the system
Leverage	Focus on improving the effectiveness of the existing process by: Improving the process – modernise or renew the process Increasing the depth of value – use the existing system to address higher value (more important) business requirements Increasing breadth of value – use the existing system to address a greater range of business requirements Increasing adoption – increase the enterprise-wide use of the existing system
Reposition	Focus on moving the system into another value/cost quadrant by: Adopting either a streamline approach or a leverage approach (simultaneous adoption of both approaches is possible but difficult)
Terminate	Terminate the application Removing the system from the portfolio altogether

Figure 14.9 Technical condition assessment of existing assets

the same schemes used for new business initiatives, namely the APQC and application domains.

It is this consistent classification that allows Queensland Transport to identify existing assets as the basis for new business initiatives. Classifying all assets using the same classification schemes that are used for scoping new business initiatives means that we can bring together the "new" scope and the "existing" scope (in the form of existing assets), thereby identifying overlaps. These overlaps then allow us to indicate potential reuse of existing assets.

In addition to the standard classification, Queensland Transport also performs a *technical condition assessment* using a technique originally published by MetaGroup (Robertson and Sribar, 2002). This simple alignment technique compares the business value of an application to its current technical state. This results in one of five generic actions that apply to assets, namely: nurture, streamline, leverage, reposition, terminate. This is shown in figure 14.9.

This additional information is invaluable for identifying opportunities to either improve or simply commoditize elements of the department's asset portfolio. For example, there is little value in identifying an existing system as a potential source of services if the condition of the asset requires it to be "streamlined." It may be more appropriate in this circumstance to recommend the sourcing of the supporting services from outside the department.

14.3.4 Service sourcing and usage strategy

As outlined in 4.6, an organization must understand both its *sourcing* of services (sourcing strategy) and its *provision* of services (usage strategy). A critical element of this process is understanding an organization's line of commoditization.

Our primary strategic focus at first appeared to be one of service sourcing (i.e. outsourcing of the functions). For example, we have outsourced the collection of registration renewal payments to Australia Post. However, further analysis revealed this is not necessarily a true case of service sourcing. Let us elaborate.

Queensland Transport is a government regulator so the services that underpin its business processes are typically created as additions to existing value chains which, for various reasons, require government intervention. After all, government regulation itself is typically a layer of process above existing processes that would otherwise operate essentially *laissez faire*.

As a result, the focus for Queensland Transport is on service usage. In this context, Queensland Transport's primary stance is one of a provider to trusted partners.

Although Queensland Transport's relationship with Australia Post initially appears to be one of service sourcing, in reality it is more subtle. The business process is definitely occurring externally to Queensland Transport, in that payments are made at Australia Post. However, at a service execution level, Australia Post is responsible only for the services associated with cash management elements of the business process. Australia Post must still invoke Queensland Transport services to achieve the actual payment of the registration renewal itself.

It is critical that an organization does not confuse the concepts of service usage and sourcing, as the roles and responsibilities with regard to SOM are very different. It is also important, as shown by the Queensland Transport example, that the strategic decisions regarding sourcing and usage are separate from the operational decisions impacting SOM.

14.3.5 Model-based development and business rules

As discussed in 6.2.4 the business rule dimension of any analysis is critical to the success of service orientation.

For Queensland Transport, the use of the Information Engineering method forced this separation though process logic analysis within the interaction analysis stage, which required the creation of action blocks to capture business rules. In addition, the tooling support for this method, the AllFusion® Gen product from Computer Associates, allowed the creation of these action blocks in a reusable and technology platform independent modeling notation known as Action Diagramming.

This early exposure to model-based development, and particularly the concept of separating business rules from process while reliant on the data being processed, was another factor that helped Queensland Transport achieve the balanced approach between process, domains, and policy required for effective service orientation.

Queensland Transport continues to use both this approach and the AllFusion® Gen product after almost ten years. However, it is only now that the critical need to integrate process, domain, and rule is beginning to emerge.

For example, with the advent of the Model Driven Architecture (MDA), the Object Management Group (OMG)™ have begun serious efforts to define more formal and consistent notations for business rules to compliment the highly effective domain modeling of UML. These emerging standards, if coupled with BPMN and traditional UML, have the potential to further assist in the shift away from the technology towards the business dimension of service orientation.

14.4 Service-oriented architecture

As outlined in chapter 7 an SOA comprises various elements, at both a software and an infrastructure level. Ultimately, it is the objective of any SOA to provide a reality for the promised benefits of service orientation using the various business models which have been developed and now require implementation.

For Queensland Transport, successful delivery of service-oriented benefits is based on a solid approach to service design that began within TRAILS and has since been expanded to new developments within the department.

We now provide some insights into our service design approach using the types of techniques described in this book. To do this, we shall explore one of our more recent developments, the Queensland Transport Services Booking System (SBS).

14.4.1 Creating an agile asset

As outlined in 14.1.2 TRAILS began in the early 1990s, almost ten years before many of the modern service design techniques discussed in 9.2 were available. However, even in those early days the design principles of TRAILS embodied elements such as "generic," "modular," and "reusable." These principles led to an object-based (rather than an object-oriented) design approach that has stood the test of time in helping us to create an agile set of TRAILS services.

This object-based nature of TRAILS related primarily to its treatment of data and the user interface. Experience in the late 1980s and early 1990s showed that having the system focus on business objects resulted in a far more user-friendly system from a business view point.

Under this approach, TRAILS recognized the concept of an "object" in business terms – e.g. Customer (Individual), Customer (Organization), Vehicle, Registration, etc. These object concepts were then present both within the application (e.g. Maintain Customer Form/Screen) and the data model (e.g. Customer and Individual entities and their associated supporting entities).

The technique also extended to the use of "actions" against "objects." That is, we maintain an object via actions against the object (e.g. the Add Vehicle Registration function from the Maintain Vehicle Registration form/screen). In some cases, the actions themselves were enumerated within the data model (e.g. Registration Action and Plate Number Action entities).

Queensland Transport's SOA, when introduced in 1997, provided a clearly defined interface layer between the "object-based" functions within the system and the business process implemented within the forms/screens used to access the functions. As a result, the business objects and associated functions effectively became the logical services of TRAILS.

Beneath these logical services, TRAILS was divided into a series of actual implementation areas known as *business systems*. These business systems represented the various functional components through which the TRAILS services were realized. Taking the object-based approach, combined with a clearly defined interface layer, allowed simple identification of potential reuse of functionality within TRAILS.

14.4.2 Realizing business agility

In 2002 the State Government of Queensland initiated a program called aligning services and priorities (ASAP)[2]. This program aimed to review existing operations to ensure that services provided by agencies were aligned with the Queensland Government's priorities for the community.

In Queensland Transport's case, this meant reviewing and then improving the effectiveness and efficiency of existing operations. During this review, two situations arose which prompted the department to reconsider the way it dealt with inspections of heavy vehicles and the practical testing of drivers.

The first situation arose because, while TRAILS was responsible for recording and charging for heavy vehicle inspections, each regional area within Queensland Transport had its own independent scheduling and booking system to manage the workload of inspectors. As a result, vehicle owners who were unable or unwilling to predict where their vehicle would be on a given date started booking the same vehicle into multiple centers. They would then turn up at the closest center on the preferred date. Multiple inspectors would experience a no-show, while one inspector performed the work. At the time, Queensland Transport charged only for successful inspections. You can no doubt see the issue!

The second was a similar situation in which TRAILS was responsible for recording the pass or fail for a practical driving test, but the scheduling of examiners was undertaken on a regional basis. In this case, the high level of demand for examiners meant that they were often double-booked, leaving customers to reschedule. This was not the type of service delivery we were aiming for!

In both cases, the booking required our customers to perform the transaction face-to-face at a Customer Service Center. In most cases, this meant two separate visits, one for booking and the second for performing the actual inspection or test.

[2] Further information on the ASAP program is available from the Queensland Treasury website, http://www.asapreviews.qld.gov.au/.

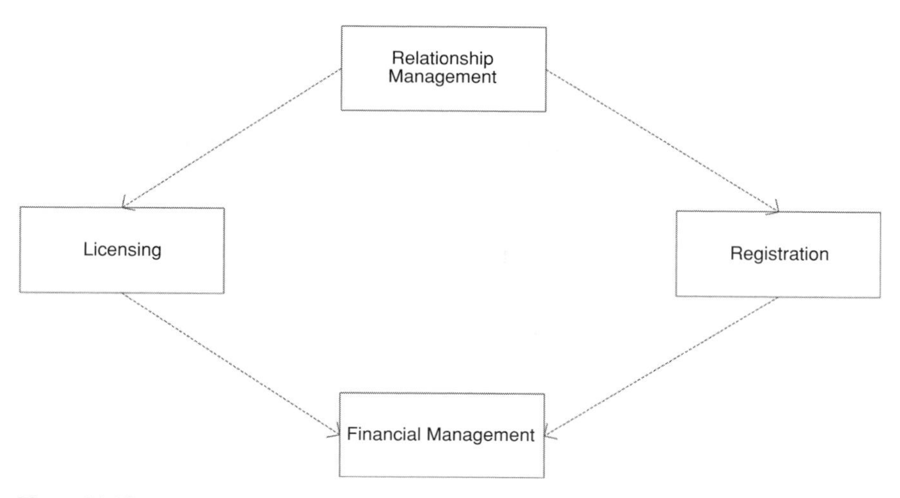

Figure 14.10 Original domain dependency diagram

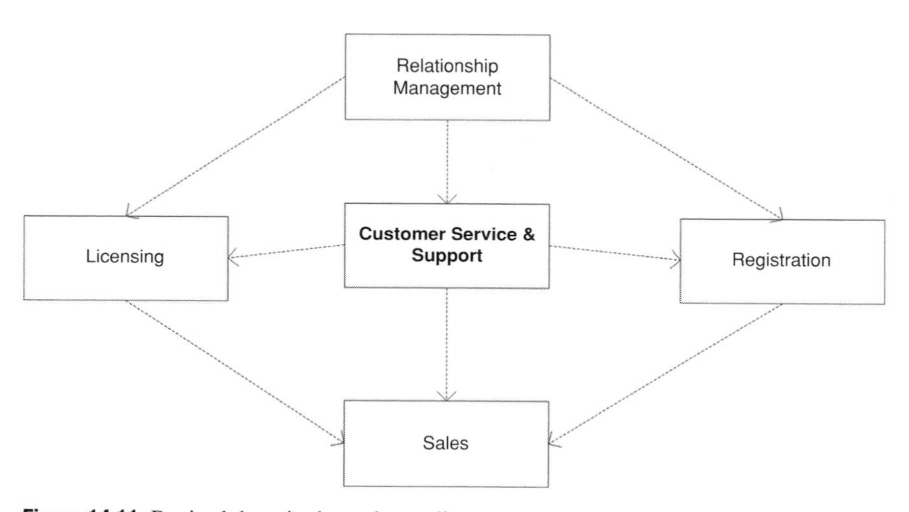

Figure 14.11 Revised domain dependency diagram

Using TRAILS as the basis, the department undertook a project to correct these problems with a single business model based on a centralized booking approach which could be accessed via the Web or call center. This eliminated the need for customers to use the face-to-face channel to perform a booking and also avoided the no-show and double-booking situations by centralizing data.

Having completed the early analysis it was clear that the current domains were insufficient to meet the stated business goals. This is shown in figure 14.10.

Instead, the solution required the addition of a customer service and support element, as shown in figure 14.11.

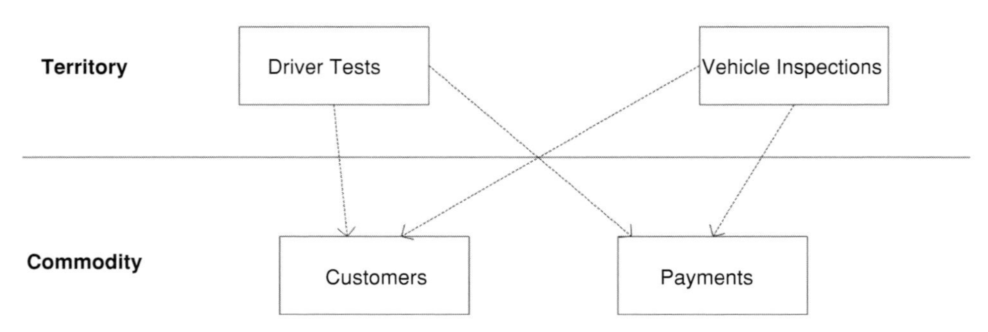

Figure 14.12 Original service/sub-domain dependencies

Further analysis showed the need to establish a single booking function within Queensland Transport to handle not only situations associated with licensing, but also registration and any other similar scenarios where Queensland Transport staff or resources could be booked. Effectively, Queensland Transport had identified the need to provide the booking services as value-add functions on the traditional "territory services" of licensing and registration via face-to-face. The analysis of the dependencies between these services also highlighted the need for Queensland Transport to change its approach to collection of payment. Instead of taking payment when completing the transaction, it became clear that Queensland Transport could collect payment for the booking. This change in business meant that the likelihood of no-shows would be greatly reduced, thereby maximizing Queensland Transport's resource usage, effectiveness, and efficiency. The contrast between these two models is shown in figure 14.12 and figure 14.13.

Having identified the high-level services, Queensland Transport's approach was to realize the services using a software unit architecture comprising:

- TRAILS vehicle inspection and customer services
- A combination of TRAILS and an existing J2EE-based application known as the Generic Payment System for payments
- TRAILS and a new component for driver testing
- A new J2EE component to provide a generic booking system.

Since the original TRAILS approach Queensland Transport has adopted more modern techniques for development of software unit specifications, specifically using the Computer Associates Component Standard Version 3.1 (CS/3.1) as the basis for specification of all major software units. These are realized using a component-based development approach facilitated using AllFusion® Gen to target both the existing TRAILS COBOL platform and the department's newer J2EE environment.

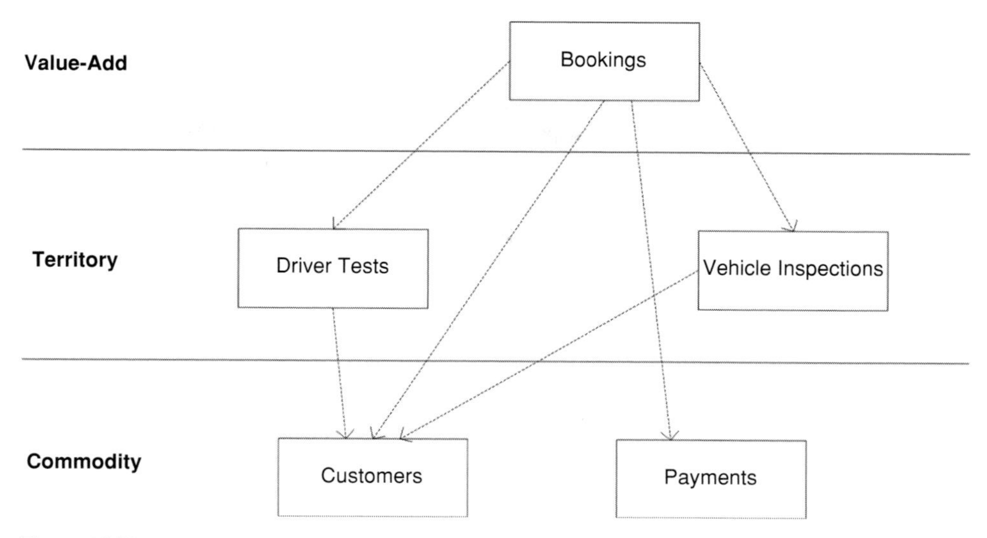

Figure 14.13 Revised service/sub-domain dependencies

Using these newer approaches combined with the existing TRAILS services, the entire SBS project took just over a year (thirteen months) to deliver the core components, with only four person-years of effort.

Today, SBS is positioned to be expanded within Queensland Transport from practical driving tests to written tests. In addition, the core components of SBS, such as the Generic Booking System, have been identified as a potential source of services for whole-of-government initiatives, such as booking mobile public dental health clinics. It is this ability to evolve and expand using an SOA that represents our ability to realize business agility.

14.5 Service-oriented management

When we initially invested in SOA we did so in response to business needs. We were focused on moving into this new paradigm; SOM (and its components SLAs, SLM, and SEM) were not high on our agenda. The absence of these elements within our initial ventures into providing services for consumption by trusted partners caused us pain on a number of separate occasions.

What started off as a more convenient service had become a victim of its own success. A number of separate incidents convinced us to monitor services usage more closely (Longworth, 2004). These are discussed in more detail in 14.5.3.

The five steps outlined in this book as the basis for SEM (deploy, run, monitor, analyze, and adjust) are important for any software module, whether it is a component, an application, or a service. With an SOA, these steps are *critical*, especially the monitoring. A service can be a part of multiple applications, inside and outside an organization, and therefore the ability to monitor, analyze, and adjust are very important.

Three key factors now help the department manage service usage:

- Understanding and agreeing on "service support"
- Accrediting of consumers before authorizing access
- Understanding service demands – you can't monitor what you don't measure.

14.5.1 Understanding and agreeing on "service support"

The discussion in 13.4.2 describes the responsibility for services and outlines how the sourcing or usage of a service does not change the responsibility for services.

One area not normally considered part of SOM is end-user support. However, by now it should be clear that service orientation is not just about software. As outlined in 14.1.3, Queensland Transport was expanding services from internal users to partners, and ultimately to our customers. This expansion of the business function from internal users to end-users required a shift in business process support.

When an internal user conducts a transaction and encounters a problem they call system support or the business process support area – they call a help desk. When an end-user of a third-party application encounters a problem and the problem relates to a function which the third party consumes from you, who does the user call? The answer is not that clear.

For Queensland Transport, the lack of a clear answer to this simple question (at least in our early years of service orientation) led to a serious underestimation of the impact of business support.

Before DAIS was introduced, Queensland Transport's motor vehicle dealers would perform their internal transactions using their own software, and then take the required vehicle registration paperwork to a local customer service center for processing. If any problems arose with the processing of the registration transactions a customer service operator would contact our internal business process support area, known simply as "Help 4000."

With the introduction of DAIS, motor vehicle dealers no longer needed to deal directly with Queensland Transport, since they could access a TRAILS service from within their own software. However, business process problems do not vaporize just because you connect two pieces of software via Web services. In fact, in the case of Queensland Transport, the inexperience of the staff processing the transactions meant that their need for support was unusually high.

The result was the emergence of informal lines of support, as shown in figure 14.14. Motor vehicle staff would contact customer service centers to ask questions. Unfortunately, the customer service officers were not familiar with the software being used by the motor dealers and, in many cases, the terminology and functionality was completely different to that of the native TRAILS environment.

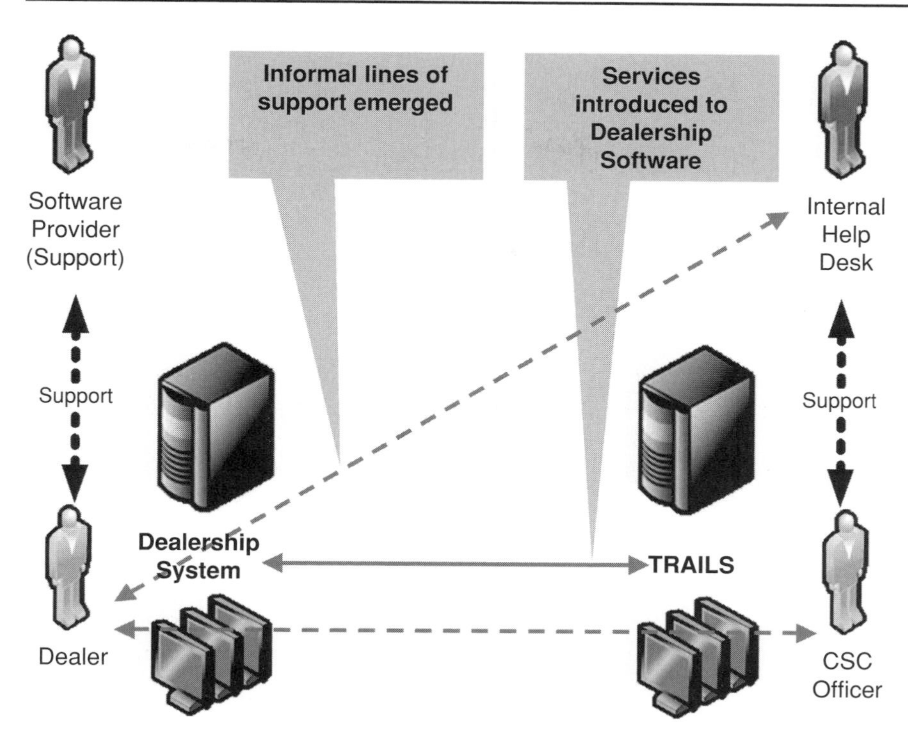

Figure 14.14 Advent of services created informal and inappropriate lines of service support

So these customer service officers called Help 4000. Again, the staff within Help 4000 were not trained to support the motor vehicle dealers' software – that was the job of the software providers. In some cases, the help desk staff were asked questions that didn't even relate to Queensland Transport services, but were about non-Queensland Transport functions within the software. Ultimately the internal number for our help desk was released to the dealers, and the support network flooded. Not a happy day!

It was clear from this experience that we had an obligation to support our business process regardless of who did the work, but that this support could be offered only in the context of the software being used. The result was a tiered support model that was built into our SLAs. All end-users were directed to contact their software provider as first-level support. The software provider's help desk could then contact Queensland Transport. This model is shown in figure 14.15

This "funneling" of support has since allowed Queensland Transport to provide software providers with access to certain standard operating procedures and even internal TRAILS transactions. This has improved the quality of support they provide, and in turn reduced the load on Queensland Transport.

This experience shows not only the need to increase service support alongside the actual services, but also the importance of agreeing how services are to be supported. It is critical when defining your organization's SOA policies (such as agility, capacity,

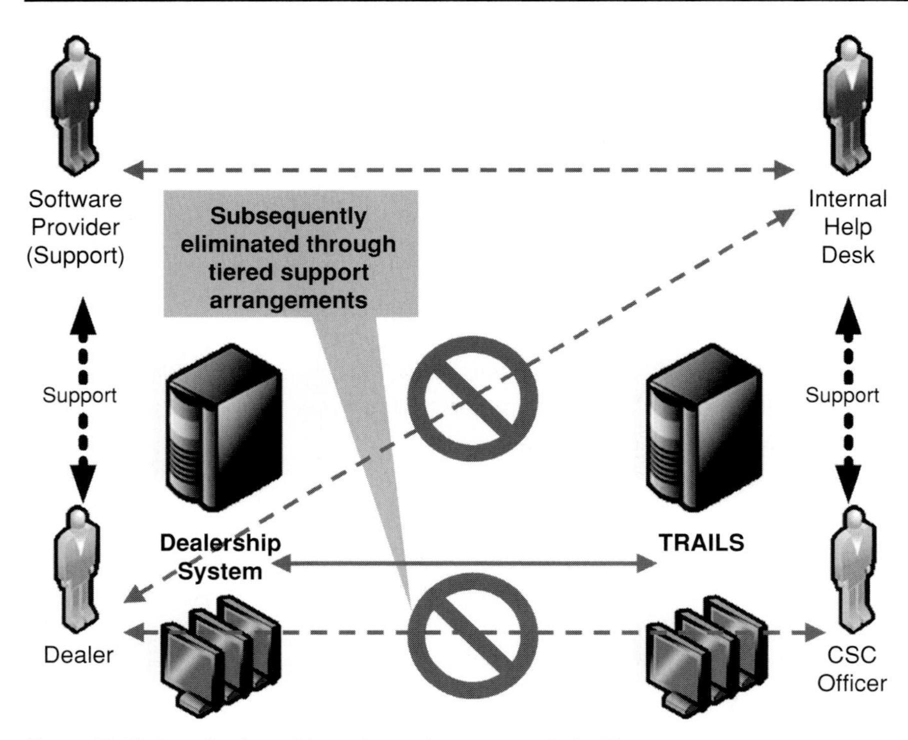

Figure 14.15 Introduction of formal two-tier support within SLAs

availability, and security) that you do not merely deploy a service that will meet technology-oriented dimensions. Policies such as "minimizing the number of user errors through effective use of error messaging" and other service support-oriented policies are also critical.

14.5.2 Accreditation of consumers

As outlined in 14.2.2 one of Queensland Transport's early lessons was the need to ensure both the quality of our own services, and also the quality of the use of consumed services.

As previously discussed, one of the original software providers was motivated to capture new compulsory third-party insurance business, rather than register vehicles. This motivation meant that the provider "skimmed" on testing the vehicle registration process, focusing instead on the insurance business processes. The result was software that did not properly comply with the service interface and associated behaviors published by Queensland Transport, and had such a high rate of business errors that it began to impact the overall service levels.

Queensland Transport understood it had to improve its control over participants in any extended service network. Today, Queensland Transport controls access to what it calls a "scheme" for end-users through a process of accreditation.

Under these accreditation schemes, Queensland Transport provides information on the practices that the participant must comply with when it performs or consumes services on behalf of the department. Queensland Transport then follows up this accreditation with compliance audits and checks. Typically, these checks occur during the initial development and rollout of services, culminating in approval for participation in the scheme. The checks continue on either a regular or random basis.

Today, DAIS represents exactly such a scheme, with Queensland Transport staff involved in application testing, reviewing, and providing standard operating procedures to motor vehicle dealers in conjunction with their software providers.

In a service-oriented environment, accreditation schemes along the lines of those used by Queensland Transport provide an extremely active method to ensure that all elements of an SLA are being followed.

14.5.3 Understanding service demand

In the first iteration of service orientation, Queensland Transport's service offerings used EDI as the basis for integration. EDI formats, while demonstrably secure and reliable, were burdensome for everyday transactions. We could see that Web services offered an opportunity to move to a lighter, more flexible alternative.

Queensland Transport began using Web services in 2001 and by the beginning of 2005 had more than fifty Web services in production. The majority of these services relate to DAIS, allowing motor vehicle dealers to perform vehicle registration transactions. The other major area involves the Theft Reduction Initiative, which includes services to allow wrecking yards and insurance companies to declare a vehicle as "written-off," which ensures that it cannot be illegally brought back on the streets. Other services include the ability to suspend a driver's license, check the availability of personalized plates, and other things.

However, two uncomfortable experiences with service demand made us realize that "you can't monitor what you don't measure."

The first example involved another Queensland government department. After opening up our services beyond DAIS, we had granted this department access to a particular service as part of a step in their online processes. They then conducted an overhaul of their systems, which included introducing batch processing in addition to online processing. This dramatically increased their use of our service, and caused an unexpected spike in demand for which we were not prepared. We were suddenly flooded with requests for the service which our system at the time could not handle. We wasted a lot of time trying to identify the cause of the sudden increase in service demand.

While an SLA might not have prevented this department from changing the way it consumed the service (see the warranty analogy in 12.1.1), monitoring the service would almost certainly have highlighted the anomaly, and allowed us to proactively manage the threat to other users.

In a separate example, another consumer of a service, this time an external partner, had predicted its expected demand, but actual usage in the first twelve months turned out to be ten times greater. In the case of this partner, monitoring could have identified the change in their consumption level and either refused requests or limited the number of requests.

These examples also show how the consumer of a service can impact the quality of a particular service and potentially all services dependent on the underlying infrastructure.

More recently, our lack of service monitoring, while not proving costly, did highlight the deficiencies in our SOA. A partner attempted to use an unauthorized service and we became aware of it only when a member of our production support staff reviewed the system logs. The partner was performing system testing, but was unfortunately directing service requests to our production environment. Monitoring of services would have immediately alerted someone that the unauthorized request had occurred, which would have been beneficial to both ourselves and our partner.

The solution is to anticipate such "anomalies" using monitoring software. Our experiences have led us to explore this type of software and in 2004 Queensland Transport implemented a beta version of the CA's Uncenter® Web Services Distribution Management (WSDM) product, Unicenter WSDM. We now consider this type of monitoring an essential part of any service-oriented infrastructure.

14.5.4 Service orientation and ITIL

Chapter 11 described how important problem management, incident management, change management, and financial management are to a service-oriented approach and how the ITIL guidelines can be enhanced and augmented to assist in this area.

ITIL has been adopted by Queensland Transport as the process framework for IT Service Management. We also follow the SEM life cycle and have integrated a number of ITIL guidelines. While it is still early days, we are confident that the ITIL process framework will prove invaluable in our overall IT service management, especially working in harmony with SEM. As previously discussed, the initial lack of quality service monitoring tools inhibited our service monitoring, and equally our ability to ensure that SLA conditions for both consumer and supplier were being met. SEM has provided us with the opportunity to improve the quality of both our services and our SLAs.

Queensland Transport has recently placed greater emphasis on SLAs. We now have a service development and measurement manager, which is equivalent to the service-level manager, discussed in 13.3, and are currently in the process of creating and refining SLAs for all existing services, while any new service must have an SLA before it is made available for consumption.

14.6 Cultural factors

As highlighted throughout the book, a successful shift to service orientation is not merely the application of a series of techniques or technology practices. Rather, it is the combination of these elements with a clear understanding of what it means to be service-oriented. This understanding of service orientation must be widespread; it must become part of the beliefs, values, and knowledge which constitute the shared basis for taking action. Queensland Transport, like any organization, has its fair share of inherited software development ideas. We shall now briefly explore some of the cultural factors encountered as we made the shift to service orientation.

14.6.1 Evolving cultural awareness

Queensland Transport's experience is interesting not only from a technical perspective, but also from a cultural aspect. As outlined in chapter 13, good professionals know their roles and the relationship their role has with the business outcome. Queensland Transport has many good professionals, yet between 2001 and 2003 Queensland Transport's move from the EDI environment to a J2EE-based integration platform meant a changing of the technical guard. Java had arrived and with it came practitioners whose exposure to model-based approaches was limited. The result was a period of cultural strain.

By then, most staff were intimately familiar with the SOA, and it was eventually assumed that everyone was familiar with it. This of course was not the case. Queensland Transport found itself drifting away from the best practices it had adopted. Why? Many of the new staff were skilled in new technologies (such as Java) and with these skills came new approaches to analysis that were more technology-driven. Our failure to recognize the need to embed our existing integrated service-orientated practices within these new techniques began to result in applications which resembled pre-SOA TRAILS.

This resulted in some situations where newly created applications contained duplication of functions already present in existing TRAILS services, and hardwired business processes with insufficient service exposure. Maintenance efforts for these new applications skyrocketed.

Queensland Transport's response to this situation has been to ensure that our approach to service orientation embraces the latest technology approaches. We have re-introduced technical governance in the form of a Technical Advisory Committee (TAC) that had previously existed during the TRAILS project. The TAC included a mixture of newer staff and those from the original TRAILS days, and provided a forum to discuss issues or concerns regarding architectural directions, standards, and policies.

Table 14.4 *Queensland Transport's service-oriented role support*

Roles of service orientation (see 13.2)	Queensland Transport's equivalent role
Business process architect	Enterprise architect, Business process modeler, Application architect, Planning analyst
Service architect	Application architect, Planning analyst
Service infrastructure architect	Systems architect, Infrastructure architect, Security architect, Network architect
Legacy advisor	Application architect, Solutions architect (Principal advisor – Applications)
Sourcing and usage manager	Strategic partnership manager
Service funding and charging manager	Finance management advisor, Principal manager (Customer relations)
Service funding and charging manager	Finance management advisor, Principal manager (Customer relations)
Customer	Customer
User	User
Service level manager	Manager (Service development and measurement), Release manager, Principal manager (Host systems)
Provider	Manager (Service development and measurement)
Supplier	Not a centralized role.
Asset inventory manager	Release manager, Manager (Service development and measurement)
Service execution manager	E-Biz manager

We have also improved our education programs through various mechanisms including presentations, intranet-based discussion forums, and working groups. This response has allowed Queensland Transport to maintain the service-oriented path that had been so carefully plotted.

14.6.2 Role support

We have nurtured a set of roles that have been critical in maintaining a service-oriented culture. These roles map to the ones described in chapter 13, as shown in table 14.4. Although the roles names differ, the concepts are very similar.

14.6.3 A business phenomenon

It is interesting to observe that Queensland Transport's culture now reflects service orientation as the business phenomenon described in this book. Within Queensland Transport's IT branch, we no longer use the word "product." Instead, our management practices are based solely on the concept of providing services. Even our business unit's name, the Information Services Branch, embodies the theme of service orientation, rather than the more traditional or commonly used name "IT Branch."

14.7 Where to next? A service-oriented future

The early success of DAIS and other subsequent service-oriented business solutions prompted the whole nature of Queensland Transport to change. This shifting mind-set has led to some amazing strategic planning discussions. One such discussion painted a scenario which, for Queensland Transport, would represent the ultimate service-oriented fulfilment of a regulatory outcome.

First let's set the context. If you drive a motor vehicle in Queensland, you must be licensed. Why? Licensed drivers have been trained and are therefore statistically safer drivers. One of Queensland Transport's regulatory aims is to make sure transport is safe, so licensing is a safety outcome. The license document (a card) is a physical representation of your authority to operate a motor vehicle. This authority is granted once you have been positively identified by Queensland Transport and attained a set of competencies through a combination of theory and practical driving experience and testing. The presence or absence of a license is used by police to ensure that only authorized drivers are permitted to operate motor vehicles. This is a pretty standard driver regulation framework worldwide.

However, imagine a purely service-oriented approach where Queensland Transport is informed (via the invocation of a service) that your daughter has just completed a theory test for a motor vehicle at a registered testing center, such as her local high school or college. In response, Queensland Transport verifies her identity by cross-checking the details provided with the Registry of Births, Deaths, and Marriages and the Passport Office. Sometime later, Queensland Transport again receives a service invocation, this time from an accredited motor vehicle driving school. They inform Queensland Transport that your daughter has passed a practical test. Queensland Transport checks its records and issues her with a license to drive. She receives her license by post and uses it in combination with her key to start her motor vehicle. Immediately the motor vehicle invokes a Queensland Transport service to verify her authority to drive. The response is positive and the engine roars to life . . . and not once did she meet anyone from Queensland Transport as they were all out managing the network, authorizing her college as a testing center, accrediting her driving instructor, and developing wireless connectivity with her car company. What about the police? Well, they were out doing their job – preventing real crime.

It could happen. And one day it will.

14.8 Acknowledgments

Queensland Transport, Paul McRae, and Sam Higgins would like to acknowledge the early work of Greg Smith who was the TRAILS system architect at the time Queensland Transport adopted the SOA. Without his initial analysis Queensland Transport's path to service orientation would undoubtedly have been a longer one. We would also like to thank our director, Paul Summergreene, and all the staff (past and present) of the TRAILS project and Information Services Branch who made our contribution to this case study possible.

The views expressed in this case study are those of the authors and do not necessarily reflect the opinions or policies of the Queensland Department of Main Roads, Queensland Transport, TransLink or Maritime Safety Queensland, or the Queensland Government.

15 Credit Suisse: a case study in service orientation

Hermann Schlamann is a senior architect in the architecture group of Credit Suisse (CS), having worked as a consultant in the banking industry in Switzerland since the mid-1990s. Since starting his career in 1971, as a programmer of applications in technical engineering, Hermann has built a deep and extensive understanding of the practicalities of using methodologies to produce successful software.

In this chapter, Hermann shares his experiences in CS's transition to service orientation.

15.1 Introduction

Like many large financial organizations, CS has grown significantly, from relatively small beginnings since the 1970s. This period has witnessed huge changes in business and technology, which continue unabated. Service orientation has emerged as an important part of our strategy for managing these changes. However, it is an ongoing journey that has required careful nurturing. It has not happened "overnight." In this chapter, we examine some of the lessons that we have learned along the road to service orientation and at how CS is responding successfully to the continuous demand for change.

15.1.1 Historical IT background

In 1974, the IT system of Schweizer Kreditanstalt consisted of one IMS mainframe system. In 1975, for reasons of capacity, the system was split into two IMS mainframe systems to support regional areas. In 1980, a new application, Wertschriften System Asset Management (WS80), was introduced. Three IMS systems were necessary for reasons of independence and the growing demand. In 1985, greater demands for availability and capacity, as well as 7/23 hours operation, made a fourth IMS system necessary.

Subsequently, in 1993, the acquisition of Schweizer Volksbank doubled the number of IT systems. By 1996, the focus had moved to a horizontal business unit split, though not to physical or logical restructuring of the system portfolio.

In parallel with these developments, the restructuring of the Swiss banking market alongside a climate of great business and technological change led to many tactical IT projects, some of which were left incomplete. The technology portfolio continued to grow throughout the 1990s. The result was a jungle of different overlapping systems and mixed technologies that were not responsive to change.

Various attempts were made to turn this situation around with the goal of achieving agile software solutions that were much better aligned to the changing requirements of the business. Greenfield approaches were tried, but failed under the weight of the challenge.

The strategic options were reappraised in 1998. One option was to move to software packages, but it was decided that the packages on offer were too rigid and not sufficiently aligned to the CS business. Using the IT capability of another bank was another option that was discarded on the grounds that this would severely restrict CS' competitive differentiation. Leaving things as they were – approaching projects in a highly reactive and opportunistic fashion – was a risk that could not be taken, as this was seen as likely to further exacerbate the situation.

It was against this background that CS decided, in 1998, to embark on a strategy of service orientation and to introduce SOA as the first step along this road. This led subsequently, in 2002, to a restructuring program, the main goal of which was to manage system complexity using one logical system. This involved applying what we had learnt about SOA in a business context: decoupling the implementation of physical systems from the interfaces offered by those systems. It was a strategy that involved a major restructuring and rationalization of the legacy system portfolio.

15.1.2 Market context

The major business motivation behind the CS transition to service orientation is to minimize the costs of market participation and to focus the company's energies on exploiting core competencies. Services must provide quick business pay-back in terms of cost savings or business process efficiency improvements.

Decisions on whether to reuse services are based on an evolving SOA. However the need for business results means that "just enough" SOA is used. With regard to market context, the CS approach is therefore perhaps closer to the tactical usage approach (described in table 13.15) than the strategic usage approach (described in table 13.14).

CS is perhaps typical of many large organizations that have significant legacy system portfolios that are difficult to manage but at the same time recognize the considerable investment that has been made in these systems. We are keen to explore services to solve specific business problems, at low risk. At the same time, as we see during the course of this case study, CS are equally conscious of the need to use commodity services from third parties, where appropriate. Conversely CS has also made significant inroads into offering their own services to other companies – for example, offering our established territory accounting services to smaller banks.

15.1.3 The overall approach

As we have stated, the scale and complexity of our legacy system portfolio rendered a greenfield approach to SOA impossible. Not all systems could be changed at once and set online in a "big bang." We needed to migrate to an SOA in an incremental and evolutionary way so that business processes were improved at manageable risk.

Our overall approach was to reduce complexity and inter-dependency of the legacy system portfolio by decoupling applications, restructuring them where necessary, and making their functionality and information available to other applications over well-defined interfaces.

In addition, the number of different operating systems, hardware systems, and different kinds of middleware posed a further challenge. Maintenance of all these system components was difficult and costly. Reducing the number of system platforms, operating systems, and middleware promised remarkable cost savings and increases in system quality. Moreover, if we could standardize on the technical infrastructure we could take much greater advantage of commodity applications such as human resources (HR) software.

At the same time, business processes had to be improved, new business goals addressed, and changes in business practices supported quickly and accurately. Our approach to SOA had to be pragmatic enough to meet the requirements of software projects tasked with meeting these business needs. Moreover the business needs themselves often changed. Increasingly, we found that systems were never "complete" and that "nothing is more stable than change".

The great question was how to rationalize both our legacy system portfolio and the technical infrastructure, while at the same time satisfying business requirements in a world of continuous change. An IT Architecture group was founded to address these challenges. This group developed principles, techniques, and guidelines to support the transition to service orientation.

15.1.4 Managed evolution

A major requirement from senior business management was that IT should be more flexible to support new banking products. Investigation showed that great parts of these new products could be supported by already existing IT functionality. However, currently implemented applications were not flexible enough to meet the new requirements.

As we have seen throughout this book, where common functionality is reused, access to this functionality has to be regulated more strictly. All consumers of the functionality

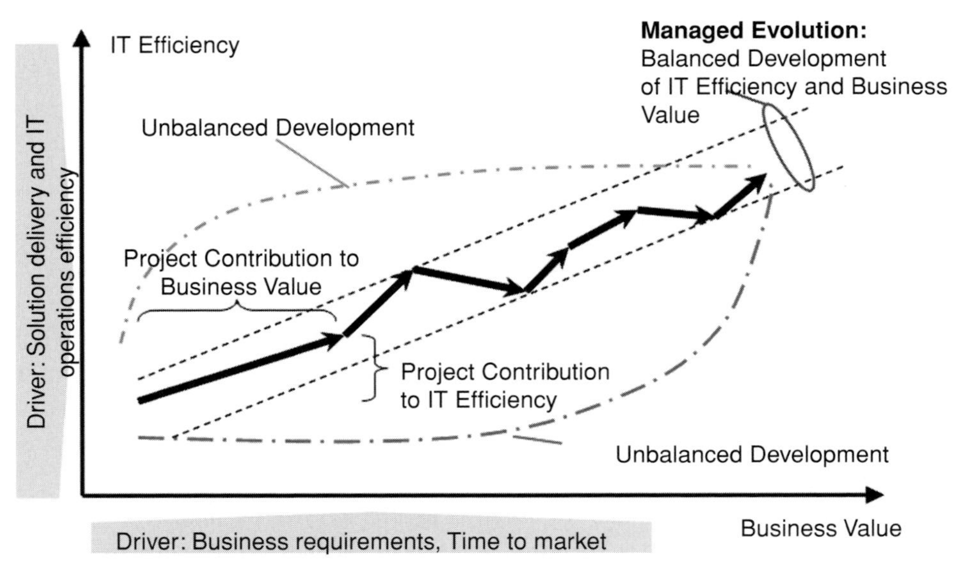

Figure 15.1 Managed evolution to balance business value and IT efficiency

should access the service via a well-defined interface and adhere to the rules and condition of that interface.

One axis of our strategy is to *restructure our legacy system portfolio* into services – applications with well-defined interfaces – as part of an SOA. However, while this strategy is designed to bring more IT flexibility and efficiency it is much more difficult to justify in terms of immediate business impact.

The second axis of our strategy is to *provide demonstrable business value* by improving our business processes through use of services. Business value must be provided in timely fashion, ideally by combining these services (from both existing and new applications).

In running software projects we have to take care that both dimensions are in balance. We termed this strategy "managed evolution": it is designed to balance the need for quick business value with the need for a good SOA (Murer, 2004).

15.1.5 Introducing the case study

In the pages that follow, we distill some of the main themes of our transition to service orientation. It has been a long road, involving hard-won experience. Inevitably, we have had to simplify some of our experiences in order to express the lessons we have learned.

We begin in 15.2 by recalling some of our first steps in servicizing our legacy systems and building a technical infrastructure to support the services, while in parallel achieving some useful business results. This echoes the two-tier approach described in the book.

In 15.3, we describe how we consolidated these early developments. In particular, we focus on two key tasks: defining and building the asset inventory and developing ISBs. We then move on in 15.4 to consider how we leveraged some business process redesign initiatives based upon what we had learned, before summarizing some of the related developments in integrating SOM into our managed evolution approach in 15.5.

A need to improve our approach emerged as a result of our evolving work. As more business success was achieved, so the case for service orientation grew stronger, helping to enable these changes in approach, some of which we describe 15.6.

15.2 First steps

Our first steps along the road to managed evolution centered on exposure of existing functionality across different channels: the SOV7 multi-channel capability view (see 3.4.3). This strategy helped to achieve useful business results in small chunks at low risk. At the same time, we had to rationalize both our existing software portfolio in the form of an SOA and our technical infrastructure to support the SOA.

15.2.1 Understanding domains and applications

The legacy system portfolio comprised groups of tightly coupled applications with a great many relationships between the applications. Our long-term strategy was to gradually decouple these applications, starting with the creation of central customer information services that would provide functionality such as consistent address information and quick and accurate checking of creditworthiness; we discuss this further in 15.4.1.

We applied the concept of "domains," as described in 6.3.1, in a top-down fashion. An example of this technique, for the customer information requirements, is shown in figure 15.2. Each domain provides a service offered through a single interface (though it is possible for a domain to offer more than one interface). Figure 15.2 shows the groups of operations offered by each interface.

The Customers and Accounts domains provide commodity services. Financial Instruments provides territory services. It contains all those applications which define and maintain information about shares, warrants, derivates, and the like. The basic data model of "financial instruments" was a common development of several Swiss banks in the mid-1990s, known as the Financial Instruments Object Model (FIOM). Contracts and Customer Credit are considered to provide value-add services. Both are domains which represent core competencies of the enterprise.

At the same time, we worked bottom-up in analyzing and grouping the existing applications into these domains as part of our overall "application landscape architecture." The domains are in effect clusters of applications which have a strong inter-relationship

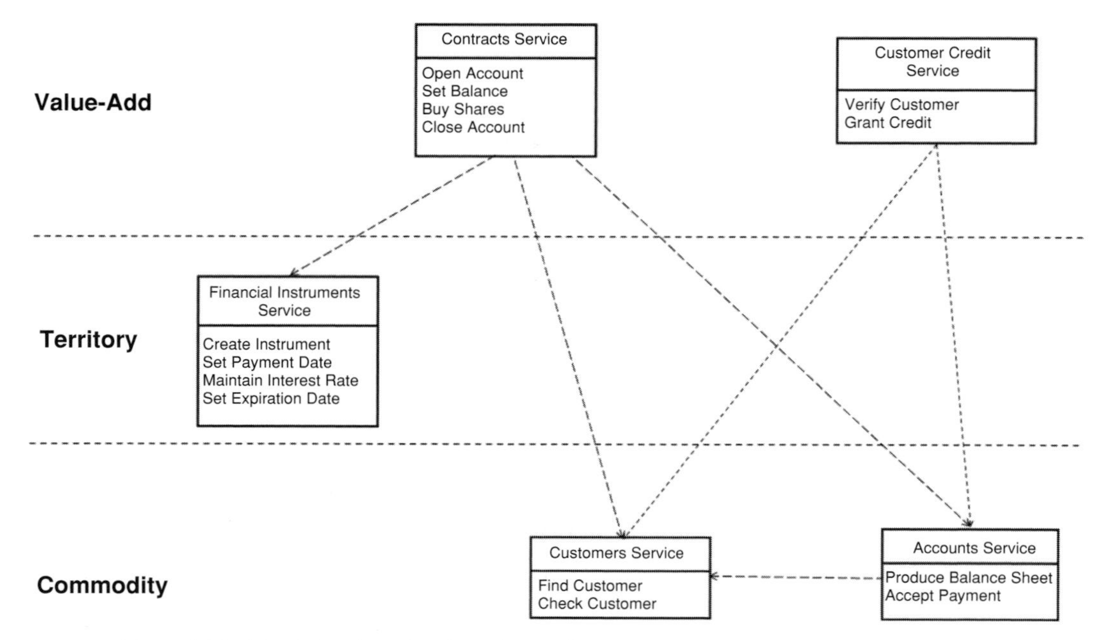

Figure 15.2 A high-level service dependency diagram

to each other. The applications are grouped around core entities like Customer or Account.

Our approach was to use domains pragmatically, to restructure our existing assets with respect to the wider application landscape architecture shown in figure 15.3. For example, the Customer application would need to be restructured to offer the Customers Service depicted in figure 15.2.

15.2.2 Rationalizing the legacy portfolio

Having structured the problem space in terms of domains and interfaces, we used the idea of a logical "integration bus" to allow applications to exchange data and communicate with one another; this idea is akin to the concept of BSB described in 9.4.3. Our approach would be to decouple the applications first and provide access between applications only over well-defined interfaces, using the "integration bus."

Information exchange between applications of different domains could happen only via a defined public interface over the integration bus. Functionality offered by a provider application could be used by consumers in other domains over those interfaces.

To keep things manageable, we minimized the number of interfaces offered by each domain via the integration bus, as shown in figure 15.4. Our strategy was to gradually

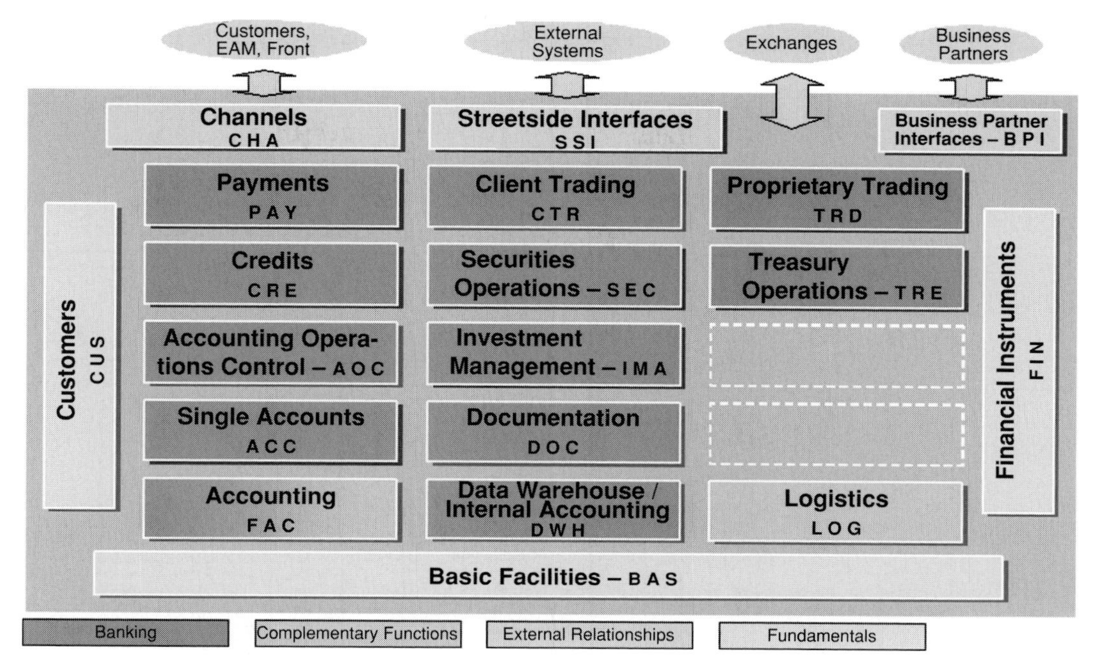

Figure 15.3 Application landscape architecture

reduce the number of provider systems, in cases where there were multiple providers for the same interfaces. This would lower maintenance costs, increase quality, and encourage governance by clarifying ownership.

15.2.3 Rationalizing the technical infrastructure

While defining domains and using an integration bus helped to rationalize our application landscape, we also needed to rationalize our technology portfolio. This meant focusing on the technical infrastructure in terms of both software (for example, operation software, or middleware) and hardware. Reducing the number of operating systems and the necessary support for the systems promised large cost savings. We needed to define and implement a technical infrastructure architecture (TIA), which specified the platform for the applications, the operating system software and middleware to be used for new or changed applications.

We structured the available platforms and defined the TIA as shown in figure 15.5. Let's look at the figure from left to right:

- *Host Transaction (TX) Processing*: A mainframe environment for the core banking applications. In most cases, these are legacy software, such as Book Keeping (BS70) and WS80.

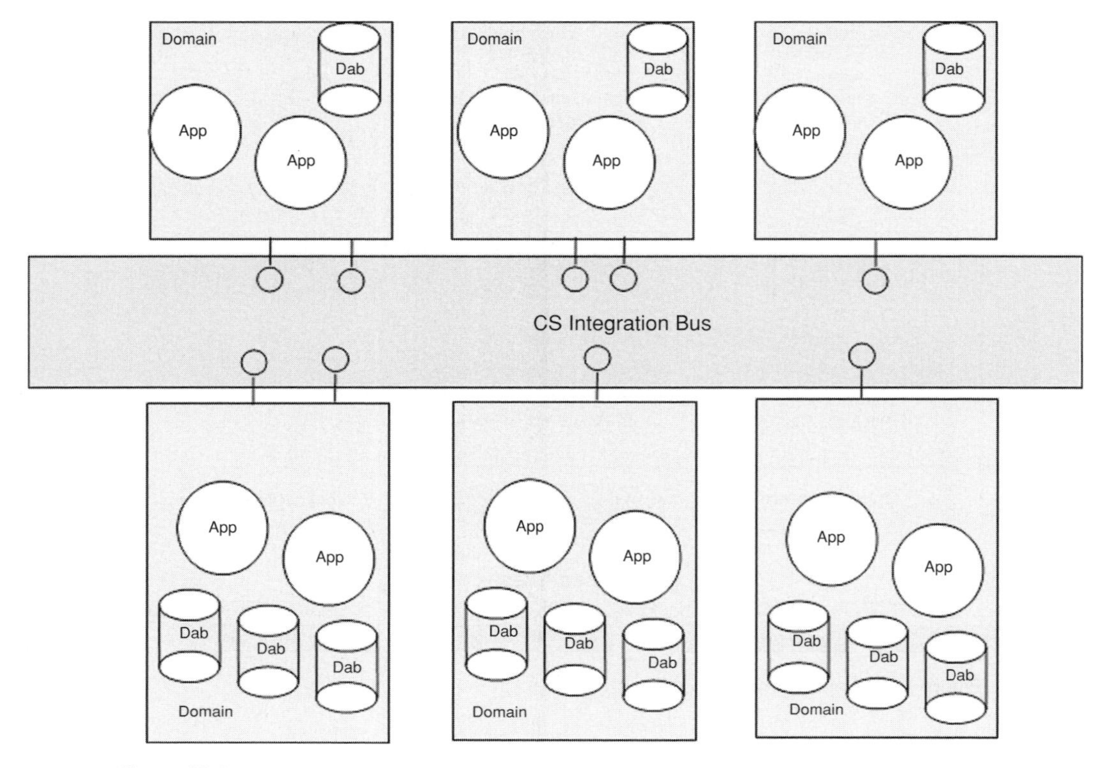

Figure 15.4 Access to applications via domain interfaces

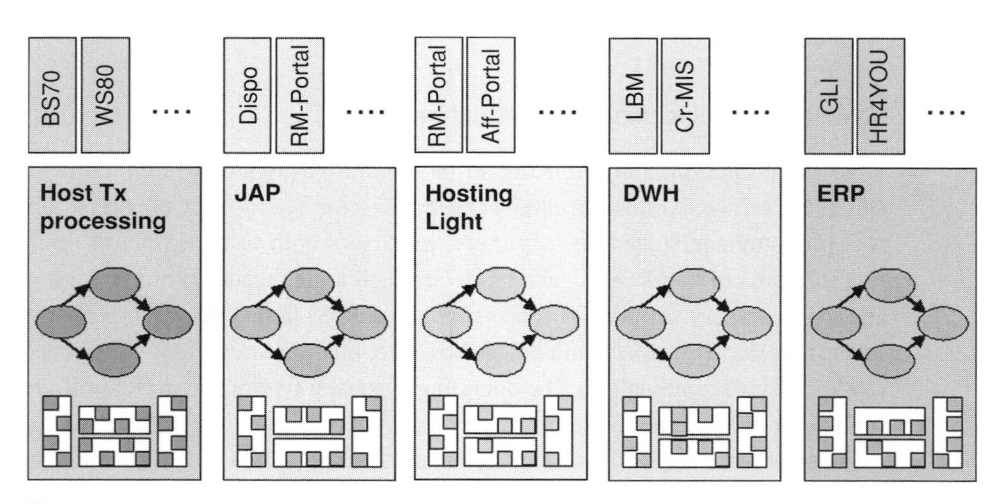

Figure 15.5 Technical infrastructure architecture

- *Java Application Platform (JAP)*: A platform based on Java and Unix to satisfy the demand for decentralized processing. Most applications on this platform are already renewed legacy applications and conform to the rules we set up for the SOA. These applications include money transfer (Dispo) and relationship management portal (RM-Portal).
- *Hosting Light*: A platform to "host" applications which are special to certain bank business and provide short-term value-add support for small applications. It is mainly based on standard Windows, which does not need special support like the affiliate portal (AFF.-Portal)
- *Data Warehousing (DWH)*: A mixture of centralized and decentralized systems like loyalty based management (LBM) and the credit management information system (Cr-MIS) that needs special support and is therefore defined as a separate platform. This platform is still incomplete and needs further refinement.
- *Enterprise Resource Planning (ERP)*: mainframe-based software packages that provide support of commodity applications like general ledger (GLI) and human resources (HR4YOU).

As we developed, the TIA we encountered a number of technical challenges. One of these concerned operation invocation. As discussed earlier in this book (8.4.4), there are three basic types of operation invocation: synchronous, asynchronous, and bulk transfer. We devised various mechanisms to provide these different types of operation invocation, the details of which are covered in 15.3.4.

15.2.4 Achieving early business results

Most of our early business results were based around the idea of exposing existing functionality as services for use in different channels: the SOV7 multi-channel capability viewpoint. For example, the Stock Exchange Rate Service exposes existing functionality that customers can use on a subscription basis. This service has been developed for reuse over different distribution channels, such as mobile phone, internet, and digital television.

One of the operations that comprise this service is Report Stock Alert, which can be offered via a Wireless Application Protocol (WAP)-enabled cellphone. This operation reports if a stock rate for a certain security reaches a specific value and may be individually based on a personal portfolio or publicly on general stock exchange rates.

Another is the Access Stock Rate operation, which is used where customers choose to run their individual banking business over the Internet on the basis of a special business contract. Access Stock Rate is offered over an individually secured Internet connection between the bank and the customer.

A third operation, Report Rates, is used to publish stock exchange rates for interested parties and is embedded in a quotes portal where general information on securities is shown.

15.3 Consolidation

The above examples all provided "quick wins" that demonstrated business value, while at the same time allowing the requisite technical knowledge and capabilities to be developed. A more complex example, introduced in 15.2.1, is the functionality of "customer information," spread across several legacy systems and used by various banking applications. The benefits of offering this functionality as a service were plain to see, especially in terms of greater reuse and easier maintenance.

At the same time, the challenges were not to be underestimated. In order to address these challenges and to scale up our early successes in line with the principle of managed evolution, we needed first to consolidate what we had learned so far. Four of the most important tasks, that we discuss in this section, were as follows:

- Encourage reuse of commodity services by software projects
- Clarify our sourcing and usage strategy
- Define and build an asset inventory
- Develop ISBs to cater to different types of operation invocation.

15.3.1 The line of commoditization

One of our strategic goals in consolidating our early experiences was to allow CS to concentrate its own resources on its core competencies, and to encourage reuse of commodity services that were found to represent cost of market participation. We were very much aware of the need to keep track of our line of commoditization (see 4.5).

In 15.2.1 we distinguished services as commodity, territory, or value-add. The question was how to implement these ideas pragmatically at project level. We now introduced the idea of the line of commoditization as part of the cost benefit analysis performed for the business in the first phase of each project. Even if the short-term cost benefit ratio is negative, where the long-term cost benefit ratio can be well justified to the project review board, the project gets started to secure the long-term goal of implementing service orientation. In this way, we began to encourage projects to reuse commodity services, and also be aware of how territory and value-add services could best be implemented.

15.3.2 Sourcing and usage considerations

Because of the scale of the challenge to rationalize our legacy system portfolio, we found that initially most of our services were offered and used only internally. However, as the SOA developed so areas became apparent where we could offer our banking services to the "outside" world. These considerations actually led to the idea that

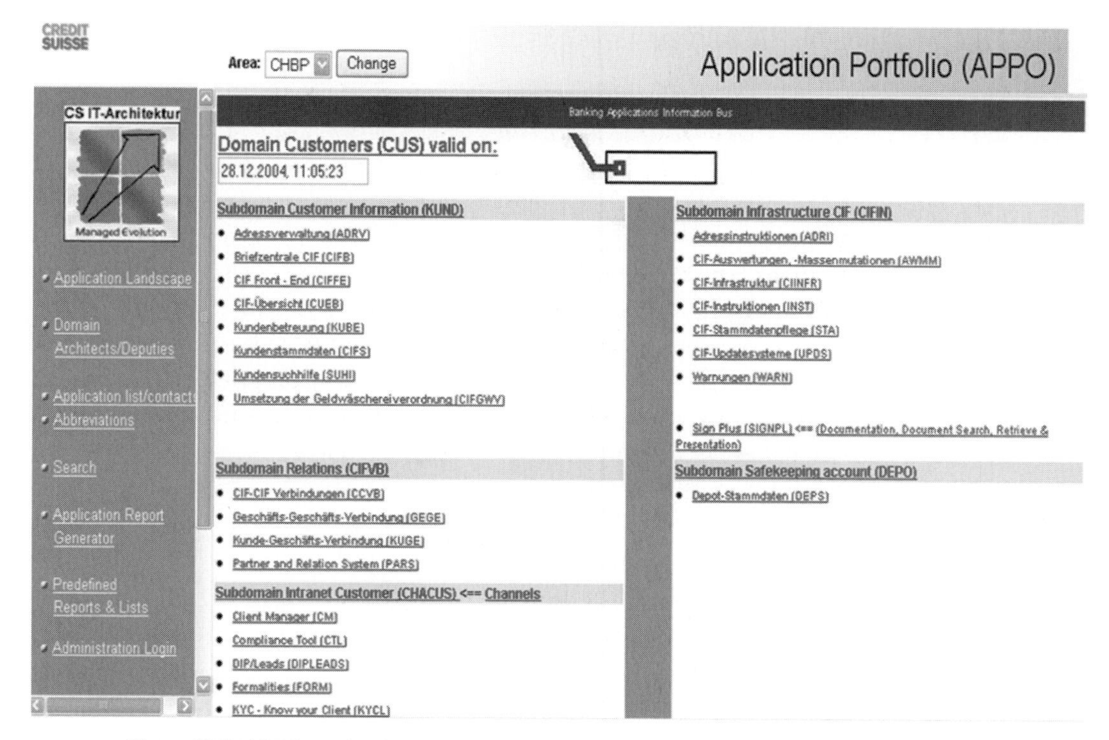

Figure 15.6 APPO entries for the Customer domain

if territory services could be made "multi-entity capable" (i.e. accounting, customer bookkeeping) then we could offer the applications as business services, at a charge, to other banks. This approach would also lower the total cost of ownership (TCO) of our IT operation.

15.3.3 The asset inventory

If we were to build on our early successes and extend the use of services across different projects within the enterprise, then an asset inventory was a practical necessity.

Two important inventories were introduced:

- The *Application Portfolio* (APPO), where all applications and their domain membership are recorded and maintained
- The *Service Database* (SDB), where the specification of all services and their interfaces and operations are recorded and maintained.

The APPO is organized in terms of the domain structure and is decomposed further into sub-domains, as shown in figure 15.6. Each domain has a business definition and a list of

Service Datenbank
Listen
Services
Alle Services
Services in P0
Services in IT
Services in ET
IDL Dateien
Basisdaten
Bearbeiten
Suche
Reports

CORBA Service Datenbank - Version 2.2

Service-ID = + − Filtern

1 - 20 von 2282

ID	Servicename	Interface	Operation	Status	Typ
ACC_1001	Liefere Bewegungskomponenten	ACCT_Bookings_2_0	getBookComponents	P0 / new	Public
ACC_1002	Liefere Bewegungsdetails	ACCT_Bookings_2_0	getBookCompDetails	P0 / new	Public
ACC_1010	Liefere Limitendaten aus OTIS/Asx zu Maestro-Karte	DCA_AuthSystemLimits_1_0	getLimitDataFromOtis	P0 / new	Private
ACC_1020	Liefere Limiten zu Konto	AccountLimits_1_0	getAccountLimits	P0 / new	Public
ACC_1021	Liefere Limiten zu Konti	AccountLimits_1_1	getAccountLimits_1	P0 / new	Public
ACC_1030	Liefere Vormerkungen zu Konto	VR_ProvBookings_1_0	getProvBookings	P0 / active	Public
ACC_1031	Liefere Vormerkungen zu Konto	VR_ProvBookings_2_0	getProvBookings	P0 / new	Public
ACC_1040	Liefere ESTV-Steuerprodukte	VAT_FTAProducts_1_0	searchFTAProducts	P0 / new	Public
ACC_1110	Eröffne Konto	Acct_AccountCreate_1_0	createCurrentAcct1	P0 / active	Public
ACC_1111	Eröffne Konto	Acct_AccountCreate_1_0	createCurrentAcct2	P0 / active	Public
ACC_1112	Eröffne Konto	Acct_AccountCreate_1_0	createCurrentAcct3	P0 / active	Public
ACC_1113	Eröffne Konto	Acct_AccountCreate_1_0	createCurrentAcct4	P0 / active	Public
ACC_1114	Eröffne Konto	Acct_AccountCreate_1_0	createPrivateAcct11	P0 / active	Public
ACC_1115	Eröffne Konto	Acct_AccountCreate_1_0	createPrivateAcct12	P0 / active	Public
ACC_1116	Eröffne Konto	Acct_AccountCreate_1_0	createSavingsAcct21	P0 / active	Public
ACC_1117	Eröffne Konto	Acct_AccountCreate_1_0	createSavingsAcct22	P0 / active	Public
ACC_1118	Eröffne Konto	Acct_AccountCreate_1_0	createProvisionAcct91	P0 / active	Public
ACC_1119	Eröffne Konto	Acct_AccountCreate_1_0	createGuaranteesAcct32	P0 / active	Public
ACC_1120	Eröffne Konto	Acct_AccountCreate_1_0	createGuaranteesAcct31	P0 / active	Public
ACC_1121	Eröffne Konto	Acct_AccountCreate_1_0	createSavingsAcct25	P0 / active	Public

Figure 15.7 SDB entries for CORBA services

core entities belonging to the domain. There is no application in production which is not categorized and does not have a link into a domain. Each application has an additional list of service operations which provide "public" services to other applications from other domains. Application ownership is clearly regulated, and responsible organizations can be identified from the entries.

Constructing this inventory has been a challenging task. The requirements for the inventory stated that the schema of the inventory should be enhanced dynamically without unloading and loading the data when schema changes are necessary. We developed a generic database schema, where both new items (entities) and new associations (relationships) are instances. This allowed us to expand the APPO database dynamically. We used an associative model for inventory data to assist this process (Williams, 2002).

The SDB (see figure 15.7) is organized using a service request template which describes the service operation to be developed and contains a full specification of the service operations and their associated interfaces. Each operation will have an identifier, a description, the involved core entities, pre- and post-conditions, and signature (input and output structures). This information is reviewed and checked against our architectural rules for service constructions. The results are verified and translated into CORBA Interface Definition Language (IDL) and stored with the information from the service request in the SDB.

From the CORBA IDL out of the SDB, the interface part for the provider and the consumer application will be generated. Because the provider and consumer of an interface have to use the same signature, we have most of our interfaces versioned to support the demand for change.

We support and maintain, if necessary, three versions of the same interface in the production environment. This is implemented by the major and minor versioning mechanism in CORBA IDL.

15.3.4 Developing ISBs

Earlier in this chapter we mentioned the use of a logical integration bus to assist in the process of rationalizing our legacy systems portfolio. We also noted that three different types of operation invocation – synchronous, asynchronous, and bulk transfer – were required to cater for different requirements.

As part of our managed evolution approach, we have developed three further different types of buses (these are ISBs, as described in 10.5) to deal with each service interaction type as follows:

CS Service Bus (synchronous)

We manage synchronous service calls using the CS Service Bus. The consumer of these services waits until the results are provided. Implementation of this type of interaction is mostly done via CORBA.

CS Event Bus (asynchronous)

The CS Event Bus supports asynchronous services. These services are not triggered by a call from the consumer; instead they are triggered by events. These events are of two types: *temporal* events that signify that a certain point in time has arrived and *status change* events that signify a change in status (for example, a stored attribute has reached a threshold value). Asynchronous services present many more challenges than synchronous services.

We distinguish two basic types of asynchronous services: peer-to peer and publish–subscribe. Peer-to peer services involve only two partners, one provider and one consumer. With publish–subscribe services the message is sent to several consumers with the same interest, who have subscribed to the service. Publish–subscribe services may be implemented directly point-to-point, which leads to many connections that must be maintained.

Alternatively, these manifold connections may be implemented by a message broker, which organizes the routing (and, eventually, the translation) centrally. The message may be forwarded to a message broker system once and then be distributed by the message broker to several consumers. So routing is an issue that must be considered when building asynchronous messaging services.

Another feature of an asynchronous service is that the message body may be translated into another code or enriched with certain fixed information. All this will be handled by the middleware chosen to implement asynchronous services. Message transformation is possible at both ends of the connection, and is best done when no business logic is involved. If additional business logic is needed, translation can be done in the middle tier at the integration broker, which routes the message. Performance considerations will determine where it is best to do the transformation. If several messages are combined into a chain of messages in which each message is triggered by the completion event of the previous one, this is called *multi-step processing*. There will be no central control of the processing. This can be enhanced to a workflow system when central control of processing is needed and the right middleware is chosen.

At CS, a great number of messages are exchanged either through one-to-one or one-to-many connections. In most cases, partners in these types of connections are applications that run on application servers representing business processes. In rare cases, end-users are directly involved.

Bulk Integration Bus (bulk transfer)

Bulk transfer of large amounts of data requires special treatment. Implementation of a bulk transfer service is achieved using a File Transfer Protocol (FTP) or the Secured File Transfer Protocol (SFTP). Bulk transfer services may be either "push" or "pull" services, depending on which side of the connection issues the request. With a pull service, the caller waits until he or she gets the full file. With a push service, the transfer begins when the pre-defined event happens.

The practical benefit of a bulk transfer service stems from implementations of B2B concepts. Especially for the finance industry, a secured bulk transfer interface over the Internet is of high importance. However, we see the greatest benefits internally when we use bulk transfer interfaces for batch processing and data warehouse feeding.

By way of summary, figure 15.8 shows the various bus types in context with some examples of the technologies used to support them.

In the case of synchronous services we have three types of implementation: CORBA, Remote Method Invocation (RMI), and direct module calls on the mainframe. Only CORBA is relevant from the view of SOA. The other two implementations matter only when tight coupling of applications is meaningful.

The implementation of asynchronous services is based on MQ-series middleware for the inter-domain information exchange. Again, other forms of tight coupling may be used; for example, the Java Message Service (JMS).

For the transfer of mass information (that is, from some megabytes to several gigabytes) FTP or SFTP is implemented. For a more complex file transfer, the translation product "Connect: Direct" (from the company "Sterling Commerce") is used.

Monitoring in production will cover each type of operation invocation. The logging information from production monitoring (SOM, see 15.5) is correlated to the specification information at development time (SOA) and is the basis for performance improvements.

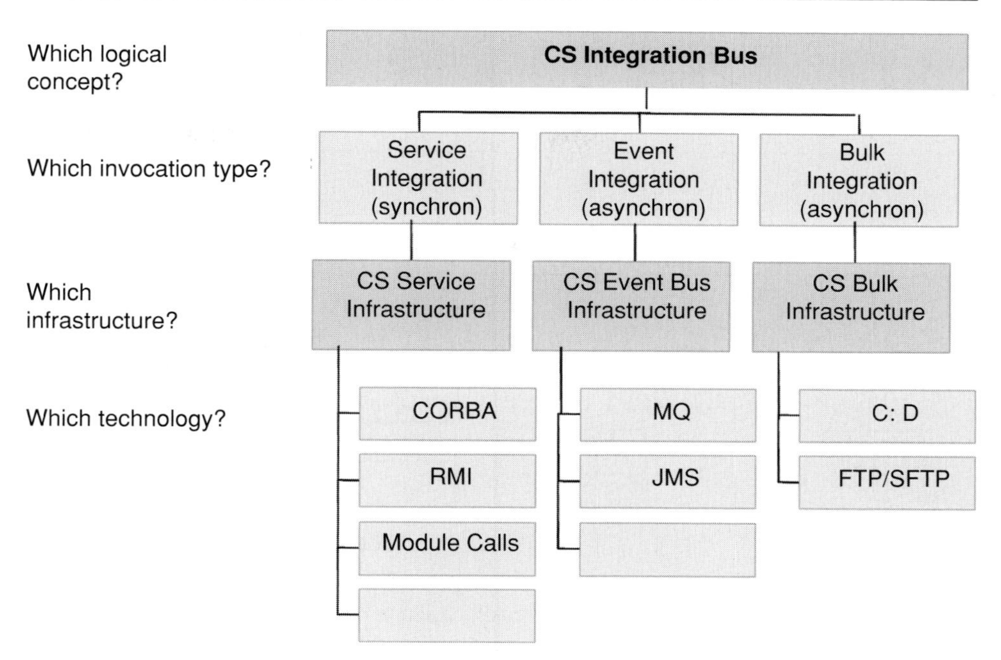

Figure 15.8 CS bus types

15.4 Delivering business value

For each requirement to improve a business process, a *cost benefit analysis* is performed for different target solutions. Costs for development and for run-time are evaluated for each variant solution.

We found that sometimes the steps of a complete business process could be implemented using a workflow, running already implemented services. In these cases, with only a few changes and additions, a message-based service (via the CS Event Bus) can trigger the next process step.

In this section, we look at some examples of how we delivered business value using the services that we had developed thus far and at how we began to expand these services to cover wider business requirements.

15.4.1 Developing customer information services

We have stressed that some of our legacy applications were monolithic giants and offered identical functionality in several places. We had to clean up that functionality and encapsulate these applications by defining and developing clean interfaces, which could be offered to consumer applications. The customer information services, introduced in 15.2.1, were a case in point. Before making progress in this area we had to perform much work in renovating and rationalizing the relevant existing systems.

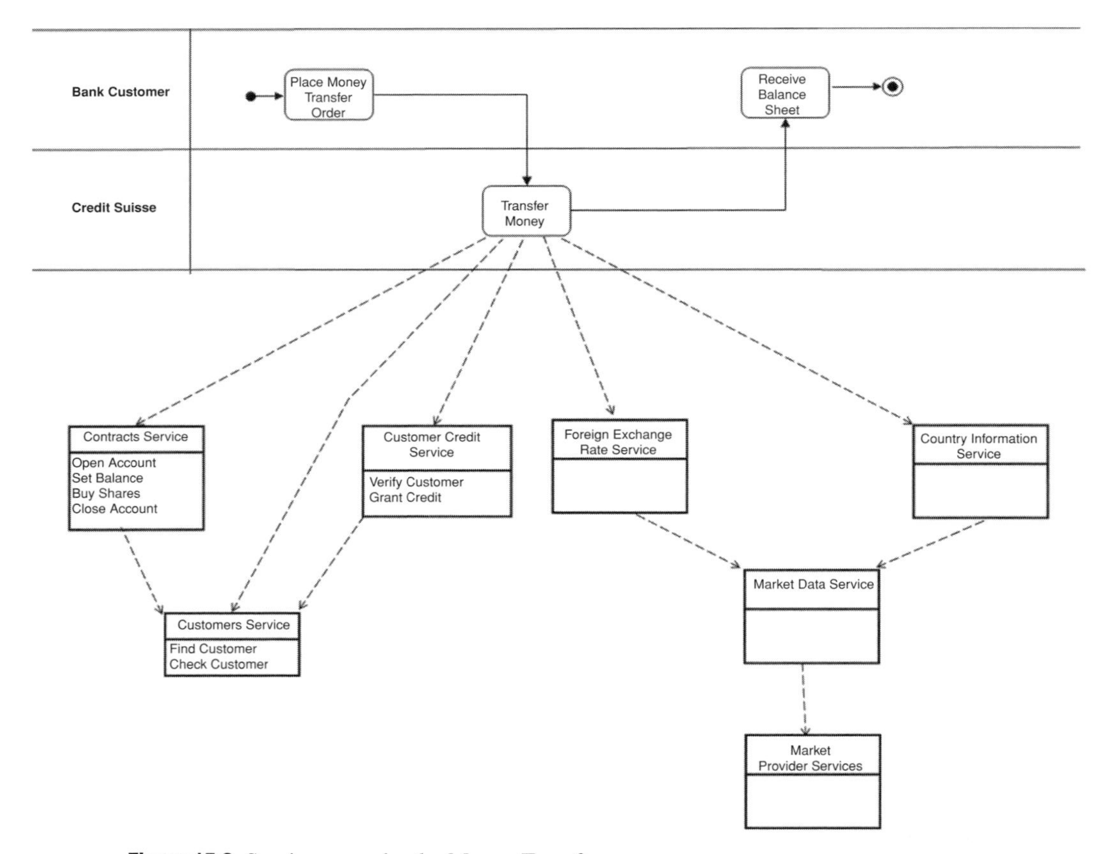

Figure 15.9 Service usage by the Money Transfer process

As a result of this work we were able to ensure that the contractual information between a bank customer and CS was offered via a single interface to the Customer domain.

15.4.2 Money transfer example

In this example, we recall how services were used to enable money transfer for a customer over different channels – for example, via the Internet with "online banking," via a "teller terminal" in a branch, or via "personal contact" at a branch counter. No matter how the access for the Money Transfer process is invoked the process executed is the same, as shown in figure 15.9.

Notice that the Contracts Service, Customer Credit Service, and Customers Service (from our earlier example in 15.2.1) are reused in the context of the Money Transfer process.

In executing the activity Transfer Money it may be necessary, in the case of an international transaction, for the official exchange rate to be calculated and for country

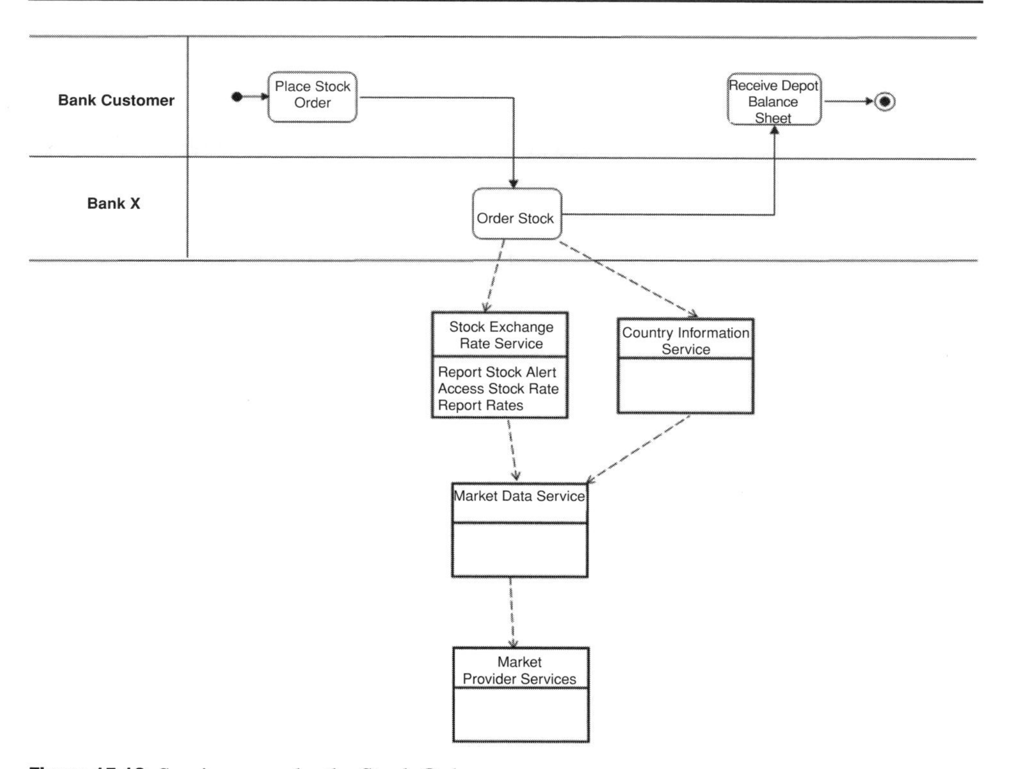

Figure 15.10 Service usage by the Stock Order process

details to be verified. We provide these capabilities using the Foreign Exchange Rate Service and Country Information Service offered on the IT-platform of CS, which we call the "Swiss Banking IT-platform."

Both these services are aggregated services with enhancements provided by our Market Data Service which collects market data from various financial information providers such as Reuters or Telekurs. The Market Data Service includes further central data-oriented services which access and maintain information about worldwide bank identifiers, as well as market identifiers (such as the codes for the stock exchanges) and worldwide country codes.

15.4.3 Stock order example

Our next example shows how we used service orientation to provide territory services to other banks. The Stock Order process, shown in figure 15.10, involves customers ordering stock at other banks ("Bank X"). In executing the activity Order Stock it is necessary for stock exchange rates to be calculated based on official market information and for country details to be verified. The Stock Exchange Rate Service and the Country Information Service are necessary territory services in banking operations, regardless of the size of bank.

Notice that we are reusing services we had already developed. The Country Information Service was developed in 15.4.2, and the Stock Exchange Rate service was developed in our earlier example in 15.2.4.

Our intention is to offer these territory services, and the Foreign Exchange Rate service (from 15.4.2) to other (smaller) banks. This is cheaper for the smaller banks and lowers the TCO for CS. In addition, the reuse factor of services – one of our strategic goals – is clearly increased.

15.5 Service-oriented management

Our approach to SOM evolved alongside our SOA efforts as part of the managed evolution approach. Monitoring of services in production involved coverage all types of operation invocation (described in 15.3.4). We had to deal with a number of heterogeneous systems, underlying the services, each playing its role in a wider context. As a result of these efforts SOM is now integrated with SOA. In this section, we briefly discuss a couple of examples in terms of SLM and SEM.

15.5.1 SLM

Our approach to SLM follows the ITIL guidelines (discussed in chapter 11) in clearly distinguishing between SLAs and OLAs. SLAs define the class of service in business terms and from an end-user perspective.

If IT commits to an SLA, such as delivering 99.9% service availability for a particular application or to a particular business unit, there are potentially many supporting OLAs that have to be in place in order to meet this commitment. Operational-level objectives are gained by systematically breaking down service-level objectives along layers of technology, processes, and functions.

There are several QoS levels (see 15.6.6) that must be monitored and controlled as part of our approach. *Business continuity planning* is a particularly important aspect for CS as the organization is extremely dependent on a seamless functioning set of services. Each service is therefore categorized in terms of a business continuity planning level that defines for an emergency case what time frame is at maximum allowed until the service is available again.

15.5.2 SEM

We have been particularly concerned with the run-time monitoring and control of services. For example, logging information from production service monitoring is now correlated to the specification information developed as part of the SOA in terms of SLAs, including QoS requirements.

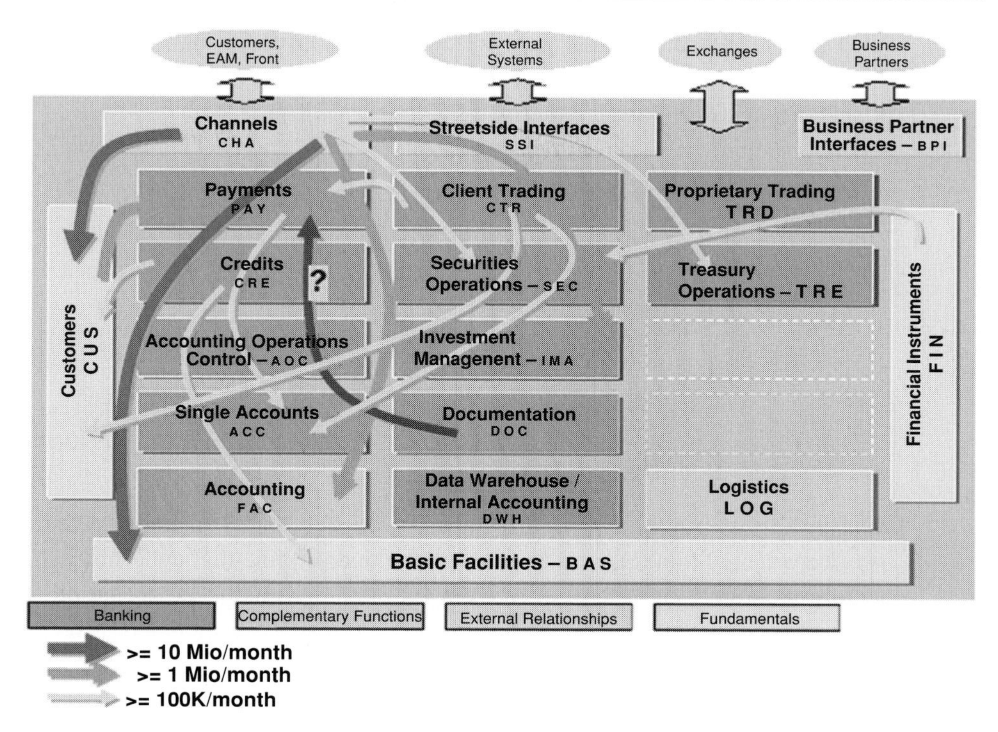

Figure 15.11 Number of service calls per domain

For the CORBA services we have a monitoring system in place which measures the number of client calls for a certain interface. The total number of clients per interface is counted, as well as the number of calls per client in a certain time frame. From the SOA point of view, only cross-domain calls are of interest, as indicated in figure 15.11.

We keep a close eye on these calls across domains. For example, the calls indicated by the arrow with the question mark in figure 15.11 look like a good candidate for further investigation. This is because the domain Documentation normally functions in the manner of an information source and custodian. Therefore we would want to investigate why so many service calls are originated in that specific domain.

15.6 Improving the approach

Our move toward service orientation has not happened "in one go" but is a continuous process that we expect to take several more years. Managed evolution is an *incremental* and *iterative* process.

We needed to make various improvements to our approach in parallel with delivering business value. In this section, we discuss eight of the most important improvements:

- Introducing SOA design patterns
- Understanding resistance to change
- Introducing interface specifications
- Dealing with poorly structured legacy systems
- Improving the software process
- Defining QoS levels
- Developing awareness
- Improving organization structure.

15.6.1 Introducing SOA design patterns

The managed evolution approach has allowed developers to become familiar with service-oriented thinking and to get used to the discipline that underlies it. The architecture team collected experience from the field and published their findings in the form of *design patterns for service specifications*; these included "exception handling," "naming standards," "type of service operations," and "usage patterns for update services."

15.6.2 Understanding resistance to change

Our managed evolution approach had proved to be an effective way to handle the demand for continuous change in our business systems. We defined a special term for an application's capability to respond to continuous change: "resistance to change." An application that is well structured along service-oriented lines is likely to have a low resistance to change.

We wanted to use resistance to change as an indicator to measure SOA compliance. Our approach was to view resistance to change as a compound indicator influenced by several parameters. One parameter which influences the resistance to change is the *project cost*. We tracked the costs of similar projects in the domains between different years and recognized a correlation between lower costs and lower resistance to change. Reuse of services was found to be an associated factor that influences the resistance to change. The higher the reuse factor, the better the service is suited to the business needs.

The emerging importance of reuse led us to gather reuse data in order to strengthen the business case for service orientation. We found that with a reuse factor of 1.7 (average for all services in production) we could save an average of two developer days on the development costs of each service operation. The current reuse factor we measured is approximately 2.5.

15.6.3 Introducing interface specifications

The interface(s) of each domain are the doors that provide access to the domain. The interface is the "contract" to which the provider and consumer must adhere. To give access to the applications of a domain, an interface must be specified very clearly. The *interface specification* proved to be the key deliverable in our software process.

We documented the interface specifications in our asset inventory. Potential consumers of a service could then check the asset inventory to see whether a proposed interface corresponded to their business requirements.

15.6.4 Dealing with poorly structured legacy systems

In analyzing our legacy system portfolio, we inevitably encountered some legacy applications that did not have well-structured interfaces. Their data access was often direct and not encapsulated. Sometimes applications in separate domains shared data by using one common database.

To handle this situation, we needed to evaluate and prioritize requirements with reference to the capabilities of the current application landscape architecture. This involved various changes to the SOA, such as the splitting or merging of applications, as well as specifying and building new domain interfaces.

In other situations where the existing applications could not be changed (for example, purchased third-party software), *"wrapper" programs* were introduced to provide an interface to those applications. In these cases, we allowed a limited degree of data redundancy by replicating data in one domain from another domain, provided it was clearly defined that the updating of original data was to occur only where the master data was located.

15.6.5 Improving the software process

We needed to apply quality assurance to new and changed services. An important aspect of consolidation was to introduce *architectural checkpoints* into our software development process. We established three quality checks, all of which must be passed before a service goes into production, as shown in figure 15.12.

Let's look at the checkpoints in a little more detail:

- *Quality Check 1 (QC1)*

  The requester of a new service describes the business requirement. This minimal description, set up quite early in the process, enables the architects to determine whether there is reuse potential or whether similar services exist whose interface(s) could be enhanced. We also check if the request complies with the overall domain architecture. The findings of the review contain hints, remarks, enhancements, and a final decision as to the next step.

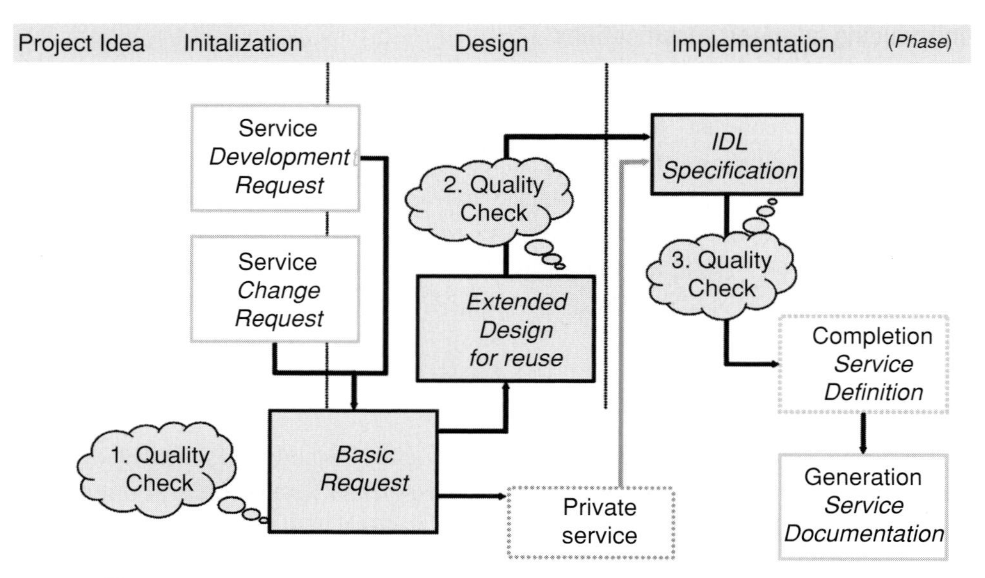

Figure 15.12 Service review quality checks

- *Quality Check 2 (QC2)*
  For a public service, QC2 is mandatory. For internal interfaces, a precise description of the interface is left to the developer, so QC2 is skipped.
  We complete the interface specification with the operation signatures (input and output structures) and the pre- and post-conditions. Reviewers check compliance of the interface specification with our standards and decide whether to proceed to QC3 or to apply corrections.
- *Quality Check 3 (QC3)*
  Reviewers check the translation of the interface specification from QC2 into IDL, as well as conformance to the naming conventions and design rules. Eventually, we add new data formats to the dictionary of base structures.

15.6.6 Defining QoS levels

We defined the QoS levels for capacity and availability and introduced monitoring of these QoS levels in production. For example, regarding capacity, we now measure the response time of each operation of a service in the CORBA environment as well as the number of calls by the consumers. Load balancing may trigger a further operation queue if a single queue gets overloaded. We found that even the transfer of a few megabytes via a CORBA operation does not influence the response time too much; it stays in the tolerance time of 3 sec for the end-user.

Regarding availability, it was important to achieve a high degree of fault tolerance of the services. Because the operations of services are stateless, a chain of operations may run into a problem when the first operation in the chain waits until the last operation completes. If the last operation for any reason fails, the whole chain of operations is blocked. As a consequence of load balancing new queues will build up, until the total numbers of available queues are exhausted. This may block a system completely.

For a Swiss bank security is a key QoS type. Each service going into production is designed with a built-in authorization check for a specific role, which is controlled by a central security administration service. Consumers must pass this check before the operations of the service can execute.

15.6.7 Developing awareness

Our achievements have not been for free. We have had to make considerable investments in developing awareness of the need for the managed evolution approach, and in education and communication programs, to enable the transition to service orientation.

We found that, with the exception of very special types of training in specific topics, a consistent education program that fitted the needs of CS could not be provided externally. CS now therefore runs its own education department, which was recently enlarged and reorganized to form an award-winning business school. This has helped CS to take a very agile approach to education in which sessions can be quickly tailored to changing requirements and delivered on an as needed "just-in-time" (JIT) basis.

Education is provided at both management and technical levels. Our own IT architects often take the role as coach in education sessions and project-oriented lessons. This helps promote the SOA and take feedback from the projects in parallel. Sessions are regularly updated to reflect changes in technology, business, and our own overall approach.

15.6.8 Improving organization structure

The need for clear definition and communication of the managed evolution approach has also required some organizational changes. For example, an IT-Architecture department in conjunction with various review boards now takes care of the maintenance and governance of the SOA. The role of a "domain architect" has been established to promote the ideas of SOA throughout the organization and to work with architects to develop and enhance architectural standards.

In addition we have introduced a matrix organizational structure within the IT-Architecture department. The horizontal axis of this matrix distinguishes and defines the technical and application work aspects of SOA. The vertical axis consists of three main streams of work:

- *Integration* focuses on consistency of interfaces and software units, across the application and technical levels, and includes procedures to measure the static and dynamic behavior of the interfaces in production.
- *Security* includes authentication, authorization, data confidentiality, auditing, and monitoring of services across the application and technical levels.
- *System management* is responsible for the run-time management of services in line with SOM principles (as described in 15.5) across the application and technical levels.

A separate team for the "architecture process" now takes care of publishing architectural standards and guidelines and organizes information events about architectural principles.

15.7 Summary

Today, in CS, we have 700 applications running with 20 million lines of source code. They contain data structures in databases, internal functions and relationships, software developed in-house, software purchased from external vendors, and a host of relationships with external sources. Transitioning this legacy system portfolio to service orientation is a long-term and ongoing task. However, whereas before we were unable to see the wood for the trees we are now in a much better position where we now actually can see the wood. We have progressed to a situation where we can start to exploit technology as an asset instead of having to tolerate it as a liability.

It is interesting to note that this need for agility reflects the overall market trend in the banking business to move from several banks to "one bank." The number of mergers and acquisitions throughout the worldwide banking and finance industries are testimony to this trend. The likelihood is that this trend will continue for the next five–seven years: the long-term requirement is for an *increasingly agile business in the face of change*.

Our managed evolution approach is geared to the goal of an agile business. It has provided an effective approach for dealing with transitioning monolithic software systems to smaller building blocks with standardized interfaces that can be used in different combinations to solve business problems. The building blocks may even be replaced with other building blocks, as long as the interfaces are kept stable.

A major benefit of this approach is that we have been able to replace certain in-house implementations of services with commodity software or to outsource implementation to an external provider at lower costs. Adopting this strategy allows us to reduce costs by raising the line of commoditization, while maintaining control and responsibility for services through the SOA. It is important to note that this has been possible only because our applications and services are accurately managed – specified, recorded, and maintained – as assets of CS, within the asset inventory. To underline a key theme of this book: "You cannot manage what you do not measure, and you cannot measure what you do not specify."

15.8 Acknowledgments

Many of my colleagues from the architecture department in CS have contributed to the information presented in this chapter. My thanks to all of them, especially to our chief architect Stephan Murer.

References

Ahern, D. M., Clouse, A., and Turner, R. (2003), *CMMI Distilled: A Practical Introduction to Integrated Process Improvement*, 2nd edn., Addison-Wesley

Allen, P. (2000), *Realizing e-Business with Components*, Addison-Wesley Longman

(2001a), Light Methodologies and CBD, *Component Development Strategies*, **11**(4)

(2001b), Heavy Methodologies and CBD, *Component Development Strategies*, **11**(5)

(2002), Component Quality: The Great Debate, *Component Development Strategies*, **12**(3)

Allen, P. and Frost, S. (1998), *Component-Based Development for Enterprise Systems: Applying the SELECT Perspective*, Cambridge University Press/SIGS Publications

American Productivity and Quality Centre (1996), *Process Classification Framework*, APQC International Clearing House

Austin, R. and Devin, L. (2003), *Artful Making: What Managers Need to Know About How Artists Work*, Financial Times Prentice Hall

Bass, L., Clements, P., and Kazman, K. (2003), *Software Architecture in Practice*, 2nd edn., Pearson Education

Beck, K. (2000), *Extreme Programming Explained*, Addison-Wesley Longman

Best, J. (2003), Half of All Outsourcing Deals Failing, *Silicon.Com*, November 21, http://management.silicon.com/itdirector/0,39024673,39117009,00.htm

Cambray, D. and Hodgkiss, G. (2003), *IT Security Management*, Version 1.0, *it*SMF Ltd

Carr, N. (2003), IT Doesn't Matter, *Harvard Business Review*, May, 41–9

Cash, J., McFarlan, W., and McKenney, J. (1992), *Corporate Information Systems Management: The Issues Facing Senior Executives*, Richard D Irwin

Cheesman, J. and Daniels, J. (2000), *Component Modeling with UML*, Addison-Wesley

Chisholm, M. (2003), *How to Build a Business Rules Engine: Extending Application Functionality through Metadata Engineering*, Morgan Kaufmann

Chrissis, M. B., Konrad, M., and Shrum, S. (2003), *CMMI: Guidelines for Process Integration and Product Improvement*, Addison-Wesley

Coplien, J. (2004), Back to the Source: Putting the Business Back in OOD, *Cutter IT Journal*, **17**(1), 5–11

DeMarco, T. (1982), *Controlling Software Projects*, Prentice Hall (Yourdon Press)

DeMarco, T. and Lister, T. (2003), *Waltzing with Bears: Managing Risk on Software Projects*, Dorset House

Dodd, J. (2004), Web Services: Time Off for Good Behavior, *Cutter IT Journal*, **17**(5), 22–30

Erl, T. (2004), *Service-Oriented Architecture: A Field Guide to Integrating XML and Web Services*, Prentice Hall

Finkelstern, C. (1989), *Information Engineering: Basic Principles*, Addison-Wesley Longman

Foreman, J. (1996), Product Line Based Software Development – Significant Results, Future Challenges, *Proceedings of the Fifth International Conference on Software Quality*, Salt Lake City, April 23

Gamma, E., Helm, R., Johnson, R., and Vlissides, J. (1995), *Design Patterns: Elements of Reusable Object-Oriented Software*, Addison-Wesley

Gilb, T. (1988), *Principles of Software Engineering Management*, Addison-Wesley

Haeckel, S. H. (1999), *Adaptive Enterprise: Creating and Leading Sense-and-Respond Organizations*, Harvard Business School Press

Hammer, M. and Champy, J. (1993), *Reengineering the Corporation: A Manifesto for Business Revolution*, Allen & Unwin

Harmon, P. (2003), *Business Process Change: A Manager's Guide*, Morgan Kaufmann

Hughes, J., Bader, J., and Corrigan, P. (2004), Characteristics of Service-Based Infrastructure and the Role of Utility Computing, *Cutter IT Journal*, **17**(5), 37–43

JISC infoNet, RAEW Analysis (2004), Information Kit Notes, Northumbria University; http://www.jiscinfonet.ac.uk/InfoKits/process-review/process-review-9.7

Kalakota, R. and Robinson, M. (2003), *Services Blueprint*, Addison-Wesley Longman

Kaplan, J. (2005), Backsourcing: Why, When, and How to Do It, *Cutter Sourcing and Vendor Relationships Advisory Service*, **6**(1)

Kaplan, R. S. and Norton, D. P. (1996), Using the Balanced Scorecard as a Strategic Management System, *Harvard Business Review*, January–February, 75–85

Kaye, D. (2003), *Loosely Coupled: The Missing Pieces of Web Services*, RDS Press

Kiepuszewski, B., Paluskiewicz, M., and Stokalski, B. (2004), Service-Oriented Architecture Applied, *Cutter IT Journal*, **17**(5), 14–21

Longworth, D. (2004), Web Services Quicken Pace of Queensland's SOA, *Loosely Coupled Monthly Digest*, September, 8–9, http://www.looselycoupled.com/pubs/digest

Luckham, D. (2002), *The Power of Events: An Introduction to Complex Event Processing in Distributed Enterprise Systems*, Addison-Wesley Longman

Lycett, M. (2001), Understanding Variation in CBD: Case Findings from Practice, *Information and Software Technology*, **43**(3), 203–13

Macfarlane, I. and Rudd, C. (2003), *IT Service Management*, Version 2.1b, *it*SMF Ltd

Martin, J. Gaall (1951), *Information Engineering: Basic Principles Book 1*, Prentice Hall PTR

Meyer, B. (1997), *Object-Oriented Software Construction*, 2nd edn., Prentice Hall

Morgan, T. (2002), *Business Rules and Information Systems: Aligning IT with Business Goals*, Addison-Wesley

Murer, S. (2004), *Credit Suisse IT Architecture Swiss Banking IT Platform – An Overview IT Architecture*, Concept CO_0058, CS internal paper, Zurich, August

Office of Government Commerce (OGC) (2000), *ITIL Service Support* (CD-ROM), version 2.2 (0-11-330867-1)

(2001), *ITIL Service Delivery* (CD-ROM), version 2.1 (0-11-330893-0)

(2002), *ITIL Application Management* (CD-ROM), version (2.1 0-11-330904-X)

OMG (2004), *UML 2.0 Superstructure FTF Convenience Document*, OMG; http://www.omg.org/cgi-bin/doc?ptc/2004-10-02, October

O'Rourke, C., Fishman, N., and Selkow, W. (2003), *Enterprise Architecture Using the Zachman Framework*, Course Technology

Overton, C. (2001), *On the Theory and Practice of Internet SLAs*, Keynote Systems; http://www.keynote.com/downloads/SLA_Theory-and-Practice-060802.pdf

Penker, M. and Eriksson, H. (2001), *Business Modeling with UML: Business Patterns at Work*, Wiley

Polan, M. (2002), Web Services Provisioning: Understanding and Using Web Services Hosting Technology, in *IBM developerWorks*, January, http://www-128.ibm.com/developerworks/webservices/library/ws-wsht-index.html

Porter, M. (1980), *Competitive Strategy*, Free Press
Prieto-Diaz, R. (1990), Domain Analysis: An Introduction, *ACM SIGSoft Software Engineering Notes* **15**(2), 47–54
Queensland Transport (QT) (1997a), *Technical Implications of a Changing Business Environment*, report prepared by G. Smith
(1997b), *Specification for Service-Oriented Architecture*, report prepared by J. Gallagher and G. Smith
(1997c), Better Community Involvement and Client Service Delivery, *Strategic Plan 1997–2001*, section 3.7
Robertson, B. and Sribar, V. (2002), *The Adaptive Enterprise: IT Infrastructure Strategies to Manage Change and Enable Growth*, Pearson Education
Rosenberg, J. and Remy, D. (2004), *Securing Web Services with WS-Security*, Sams Publishing
Ross, R. (2003), *Principles of the Business Rule Approach*, Addison-Wesley
Rummler, G. A. and Brache, A. P. (1995), *Improving Performance: How to Manage the White Space on the Organisation Chart*, 2nd edn. Jossey-Bass
Schlamann, H. (2004), Service Orientation: An Evolutionary Approach, *Cutter IT Journal*, **17**(5), 5–13
Schulte, R. (1996), *Service Oriented Architectures*, **2**, Gartner Group RAS Services, April
Schulte, R. and Natis, Y. (1996), *Service Oriented Architectures*, **1**, Gartner Group RAS Services, April
Schwaber, K. (2004), *Agile Project Management with Scrum*, Microsoft Press
Sedighi, A. (2004), Quality-Oriented Services, *Cutter IT Journal*, **17**(5), 31–6
Shaw, M. and Garlan, D. (1996), *Software Architecture*, Prentice Hall
Smith, H. and Fingar, P. (2003a), *IT Doesn't Matter Business Processes Do*, Meghan-Kiffer
(2003b), The Third Wave of Business Process Management: Digital Six Sigma, *Business Process Trends*, December; http://www.bptrends.com/publicationfiles/12%2D03%20ART%20Digital%20Six%20Sigma%2DSmith%2DFingar1%2Epdf
SOSA Team (2004), *CA's Service-Oriented Solutions Approach (White Papers and Guidance in the form of Outline, Patterns, Deliverables and Techniques)*, CA Internal Documentation
Sprott, D. (2004), Grand-Central and the Value Added Service Network, *CBDI Journal*, February, 5–11
Stapleton, J. (1997), *Dynamic Systems Development Method*, Addison-Wesley Longman
Taylor, D. (1995), *Business Engineering with Object Technology*, Wiley
Treacy, M. and Wiersema, F. (1995), *The Discipline of Market Leaders*, HarperCollins
Veryard, R. (2001), *The Component-Based Business: Plug and Play*, Springer Verlag
(2004a), Sarbanes-Oxley Drives Web Services Adoption, *CBDI Journal*, April, 4–13
(2004b), The SOA Lifecycle, *CBDI Journal*, July–August, 16–23
Welsh, T. (2004), Market Trends, *Web Services Strategies*, **3**(12)
Wilkes, L. (2004a), Establishing a Lifecycle, *CBDI Journal*, January, 16–22
(2004b), Time to Board the Enterprise Service Bus?, *CBDI Journal*, July–August, 4–14
(2004c), The Web Services Protocol Stack, *CBDiForum Reports*, http://roadmap.cbdiforum.com/reports/protocols/
Wilkinson, N. M. (1995), *Using CRC Cards: An Informal Approach to Object Oriented Development*, SIGS Books
Williams, S. (2002), *The Associative Model of Data*, 2nd edn., Lazy Software Ltd

Useful sources of information

Industry bodies and consortia

www.bpmg.org: The Business Process Management Group (founded in 1992) is a global business club exchanging ideas and best practice in business process and change management.

www.bpmi.com: BPMI.org (the Business Process Management Initiative) develops standards for BPM, such as BPMN, as well as working with complementary standards bodies such as the OMG(TM) and OASIS. Note that on June 29, 2005 the BPM/ org merged with the OMG(TM) (see http: //www.bpmi.org/downloads/BPMI-OMG Merger.pdf).

www.dmtf.org: the Distributed Management Task Force, Inc. (DMTF) is the industry organization leading the development of management standards that provide common management infrastructure components for instrumentation, control, and communication in a technology-neutral way.

www.gridforum.org: The Global Grid Forum (GGF) is a community-initiated forum of thousands of individuals from industry and research leading the global standardization effort for grid computing, including OGSA.

www.oasis-open.org: The Organization for the Advancement of Structured Information Standards is a not-for-profit, international consortium that drives the development, convergence, and adoption of e-business standards. These range from Web Services Security to WSBPEL.

www.omg.org: OMG(TM) is a not-for-profit consortium that produces and maintains computer industry specifications for interoperable enterprise applications, which include MDA and UML.

www.projectliberty.org: The Liberty Alliance is an open body working to address the technical, business, and policy challenges surrounding identity and Web services.

www.w3c.org: The World Wide Web Consortium develops interoperable technologies to lead the Web to its full potential, including its Web Services Activity, which covers a range of specifications from WSDL to SOAP.

www.ws-i.org: Web Services Interoperability Organization is an open industry organization chartered to promote Web services interoperability and focusing in particular on profiles to grow best practices in this area.

General references

www.cbdiforum.com: CBDi Forum provides a rich source of information on evolving standards, processes, and architectures in the areas of CBD and Web services; the *CBDI Journal* and special interest reports provide up-to-date information, for example, Wilkes (2004c)

www.bptrends.com: Business Process Trends provides independent advisory services in the BPM space with timely advisors from Paul Harmon and other experts. There is a particular emphasis on applying Web services at the business process level.

www.cutter.com: The Cutter Consortium provides independent advisory services on a broad range of IT areas, including Web services. Of particular interest is the monthly *Cutter IT Journal*, which provides insightful collections of articles and lively debate on critical IT management issues.

www.zapthink.com: Zapthink provides a rich source of information on service oriented architecture and associated standards.

Domain analysis and business rules

www.brcommunity.com: a rich source of information on business rules.

www.businessrulesgroup.org: an independent standards group for business rules.

http://www.sei.cmu.edu/str/descriptions/deda.html: a rich source of information on domain engineering and domain analysis provided by the SEI.

Agile development

www.controlchaos.com: The site for Scrum: an iterative, incremental process for developing any product or managing any work (Schwaber, 2004).

www.extremeprogramming.org: a rich source of information on xP.

www.dsdm.org: the official DSDM Consortium web site.

Index